Solid Edge 2026 Black Book

By
Gaurav Verma
Matt Weber
(CADCAMCAE Works)

Edited by
Kristen

ISBN # 978-1-77459-188-8

NOTICE TO THE READER

DEDICATION

To teachers, who make it possible to disseminate knowledge
to enlighten the young and curious minds
of our future generations

To students, who are the future of the world

THANKS

To my friends and colleagues

To my family for their love and support

Training and Consultant Services

At CADCAMCAE WORKS, we provide effective and affordable one to one online training on various software packages in Computer Aided Design(CAD), Computer Aided Manufacturing(CAM), Computer Aided Engineering (CAE), Computer programming languages(C/C++, Java, .NET, Android, Javascript, HTML and so on). The training is delivered through remote access to your system and voice chat via Internet at any time, any place, and at any pace to individuals, groups, students of colleges/universities, and CAD/CAM/CAE training centers. The main features of this program are:

Training as per your need

Highly experienced Engineers and Technician conduct the classes on the software applications used in the industries. The methodology adopted to teach the software is totally practical based, so that the learner can adapt to the design and development industries in almost no time. The efforts are to make the training process cost effective and time saving while you have the comfort of your time and place, thereby relieving you from the hassles of traveling to training centers or rearranging your time table.

Courses Offered

CAD/CAM/CAE: CATIA, Creo Parametric, Creo Direct, SolidWorks, Autodesk Inventor, Solid Edge, UG NX, AutoCAD, AutoCAD LT, EdgeCAM, MasterCAM, SolidCAM, DelCAM, BOBCAM, UG NX Manufacturing, UG Mold Wizard, UG Progressive Die, UG Die Design, SolidWorks Mold, Creo Manufacturing, Creo Expert Machinist, NX Nastran, Hypermesh, SolidWorks Simulation, Autodesk Simulation Mechanical, Creo Simulate, Gambit, ANSYS and many others.

Computer Programming Languages: C++, VB.NET, HTML, Android, Javascript and so on.

Game Designing: Unity.

Civil Engineering: AutoCAD MEP, Revit Structure, Revit Architecture, AutoCAD Map 3D, ETABS, and so on.

We also provide consultant services for Design and development on the above mentioned software packages

For more information you can mail us at:
cadcamcaeworks@gmail.com

Table of Contents

Chapter 4 : Advanced Solid Modeling

Chapter 5 : Selection Tools, 3D Printing, and Views

Chapter 6 : Surfacing

Chapter 7 : Product Manufacturing Information

Chapter 8 : Assembly-I

Chapter 9 : Assembly Design - II

Chapter 10 : Assembly - III

Chapter 11 : Drawing

Chapter 12 : Sheetmetal Design

Chapter 13 : Simulation Study

Preface

Solid Edge 2026 is a parametric (history based) and synchronous technology solid modeling tool that not only unites the three-dimensional (3D) parametric features with two-dimensional (2D) tools, but also addresses every design-through-manufacturing process. Thanks to its long history of development and refinement, Solid Edge 2026 is exceptionally user-friendly. The Mechanical Designers functionality of the software makes it faster to design mechanical components for manufacturing. The software is capable of performing analysis with an ease. Due to a long journey of development in software, this solid modeling tool is remarkably user-friendly and it allows you to be productive from day one.

The **Solid Edge 2026 Black Book** is the 7th edition of our series on Solid Edge. This book is written to help beginners in creating some of the most complex solid models. The book follows a step-by-step methodology. In this book, we have tried to give real-world examples with real challenges in designing. We have tried to reduce the gap between academic use of Solid Edge and industrial use of Solid Edge. The book covers almost all the information required by a learner to master the Solid Edge. The book starts with sketching and ends at advanced topics like Sheetmetal, Rendering, and Simulation Studies. Some of the salient features of this book are:

In-Depth explanation of concepts

Every new topic of this book starts with the explanation of the basic concepts. In this way, the user becomes capable of relating the things with real world.

Topics Covered

Every chapter starts with a list of topics being covered in that chapter. In this way, the user can easily find the topics of his/her interest easily.

Instruction through illustration

The instructions to perform any action are provided by maximum number of illustrations so that the user can perform the actions discussed in the book easily and effectively. There are about 1400 illustrations that make the learning process effective.

Tutorial point of view

At the conclusion of each concept's explanation, the tutorials ensure the user gains a clear and enduring understanding. Nearly every chapter of the book includes tutorials based on real-world projects. Additionally, most of the tools covered in the book are introduced and explained through hands-on tutorials.

Project

Projects and exercises are provided to students for practicing.

For Faculty

If you are a faculty member, then you can ask for video tutorials on any of the topic, exercise, tutorial, or concept. As faculty, you can register on our website to get electronic desk copies of our latest books, self-assessment, and solution of practical. The answers to MCQs in this book are also available for faculty members. Faculty resources are available in the **Faculty Member** page of our website (**www.cadcamcaeworks.com**) once you login. Note that faculty registration approval is manual and it may take two days for approval before you can access the faculty website.

Formatting Conventions Used in the Text

All the key terms like name of button, tool, drop-down etc. are kept bold.

Free Resources

Link to the resources used in this book are provided to the users via email. To get the resources, mail us at ***cadcamcaeworks@gmail.com*** with your contact information. With your contact record with us, you will be provided latest updates and informations regarding various technologies. The format to write us mail for resources is as follows:

Subject of E-mail as ***Application for resources of .... book***.
Also, give your information like
Name:
Course pursuing/Profession:
College/University:
E-mail ID:

Note: We respect your privacy and value it. If you do not want to give your personal information then you can ask for resources without giving the information. Be assured that we do not share any of the information with any third party. We use the information to improve our books.

About Authors

The author of this book, Matt Weber, has written more than 18 books on CAD/CAM/CAE available in market. He has coauthored SolidWorks Simulation, SolidWorks Electrical, SolidWorks Flow Simulation, and SolidWorks CAM Black Books. The author has hands on experience on almost all the CAD/CAM/CAE packages. If you have any query/doubt in any CAD/CAM/CAE package, then you can contact the author by writing at cadcamcaeworks@gmail.com

The author of this book, Gaurav Verma, has written and assisted in more than 17 titles in CAD/CAM/CAE which are already available in market. He has authored Autodesk Fusion 360 Black Book, AutoCAD Electrical Black Book, Autodesk Revit Black Books, and so on. He has provided consultant services to many industries in US, Greece, Canada, and UK. He has assisted in preparing many Government aided skill development programs. He has been speaker for Autodesk University. He has assisted in preparing AutoCAD Electrical course for Autodesk Design Academy. He has worked on Sheetmetal, Forging, Machining, and Casting designs in Design and Development departments of various manufacturing firms. If you have any query/doubt in any CAD/CAM/CAE package, then you can contact the author by writing at cadcamcaeworks@gmail.com

For Any query or suggestion

If you have any query or suggestion, please let us know by mailing us on *cadcamcaeworks@gmail.com*. Your valuable constructive suggestions will be incorporated in our books and your name will be addressed in special thanks area of our books on your confirmation.

Chapter 1

Starting with Solid Edge

Topics Covered

The major topics covered in this chapter are:

- *Starting Solid Edge*
- *Starting a New Document*
- *User Interface Overview*
- *Setting User Interface*
- *Application Menu Tools*
- *3D Builder*
- *Settings*
- *Mouse Button Functions and Workflow*
- *Assessment*

INTRODUCTION

Solid Edge is a 3D CAD software which works in both Parametric (History based) and Synchronous (direct) modeling style. You can create part, assembly, and 2D views using the software. You can integrate Team Viewer and Sharepoint with Solid Edge for collaborating with others. The software also provides support for FEA and CAM.

DOWNLOADING AND INSTALLING SOLID EDGE STUDENT VERSION

Siemens has started to offer educational version of Solid Edge for free to students in related fields. You can also get your free copy of software if you are currently enrolled as a student in related field. The procedure to download and install the software is given next.

*   Visit the following link in your web browser- ***https://resources.sw.siemens.com/ en-US/download-solid-edge-student-edition***. The webpage will be displayed as shown in Figure-1.

Figure-1. Webpage for downloading software

*   Scroll down in the browser and enter your information in the form. After specifying your details in the form, click on the **Access download** button. The downloading of software will start.
*   Go to default Download folder and double-click on downloaded exe file. Follow the prompts to install the software.

STARTING SOLID EDGE

*   To start **Solid Edge** in **Window 11** from **Start** menu, click on the **Start** button in the Taskbar. The **Start** menu will be displayed.
*   Type **Solid Edge** at the top in the menu. The **Solid Edge** icon will be displayed in the **Start** menu; refer to Figure-2.

Figure-2. Solid Edge icon from Start menu

- Before finishing the installation of this software, if you have selected the check box to create a desktop icon of this software then an icon of Solid Edge will be displayed on desktop.
- If you have not selected the check box to create the desktop icon but want to create the icon on desktop later, then drag and drop the **Solid Edge** icon from **Start** menu on the desktop.

After you have performed the above steps, the **Solid Edge** application window will be displayed with learn page; refer to Figure-3. Select desired link button from this page to access beginner's resources.

Figure-3. Solid Edge application first view

STARTING A NEW PART DOCUMENT

Application Menu: New cascading menu->New

You can create a new document in **Solid Edge** by two ways: by using **New** menu or by **CTRL+N** shortcut key:

1. Click on **New** option from **File menu**. The options to create new documents will be displayed in **New** cascading menu; refer to Figure-4. Click on the **New** button to create a new document. The **New** dialog box will be displayed; refer to Figure-5.

Figure-4. Options to create new documents

2. Press **Ctrl + N** from the keyboard. The **New** dialog box will be displayed; refer to Figure-5.

Figure-5. New dialog box

The template whose name ends with **assembly.asm** is used for starting new assembly files.

The template whose name ends with **weldment.asm** is used for starting new assembly files with weldment options.

The template whose name ends with **part.par** is for starting new part files. 3D models are created using this template.

The templates whose name ends with **sheet metal.psm** are for starting new sheet metal design files.

The template whose name ends with **draft.dft** is for starting new drawing (drafting) file.

* Select desired template from the left area of the dialog box. ANSI Inch template uses inch as unit for length. ANSI Metric uses mm as unit for length. In the same way, other templates represent their standard unit system.
* On selecting the template category from left area of the dialog box, the templates to create different type of files are displayed on the right in the dialog box. Double-click on the **ansi metric part.par** template file to start a part file with ANSI Metric standard. The interface will be displayed as shown in Figure-6.

Note that sometimes Solid Edge fails to start a new document because of plug-ins earlier installed in your system. For example Mastercam interferes with Solid Edge and when you start new document then software crashes. Use this link to solve the issue https://support.sw.siemens.com/en-US/knowledge-base/PL8806753 .

Figure-6. Solid Edge Interface for Part Modeling

Note that after selecting the template, a dialog box will be displayed asking you to set the environment preference; refer to Figure-7. The Ordered environment is chosen automatically, but you can switch to Synchronous if you prefer. Choose **Continue with Ordered** button if you want to set dimensions and constraints for your model, which lets you control its exact shape. Or, select the **Continue with Synchronous** button if you want to create freeform shapes without fixed dimensions or history tracking.

Ordered as Default

Solid Edge starts in the ordered environment.

Learn more about the ordered and synchronous environments.

The environment preference can be set or edited by going to File tab -> Settings -> Options -> Helpers.

☐ Do not show this dialog again.

Continue with Synchronous   **Continue with Ordered**

Figure-7. Dialog box for environment preference

USER INTERFACE OVERVIEW

Various common elements of user interface of Solid Edge are shown in Figure-8. These elements are discussed next.

1. **File button** : Displays the **File** menu which provides access to all document level functions, templates, and standards.
2. **Tabbed Document** : It shows all the document in the form of tiles which you have opened.
3. **Path Finder** : Provides a quick way to identify and select model element.

4. **Quick Access toolbar** : The options in Quick Access toolbar are used to access common tools like Save, Undo, Redo, etc. Use the down arrow in **Quick Access toolbar** to check additional options. The procedure to customize QAT is given next.

Figure-8. User Interface Overview

Customizing Ribbon

*   Click on the **Arrow** button from **Quick Access Toolbar**. A drop-down will be displayed; refer to Figure-9.

Figure-9. Quick Access Toolbar drop down

*   Select **Customize** option from the drop-down. The **Customize** dialog box will be displayed; refer to Figure-10.

Figure-10. Customize dialog box

- Now, select **Ribbon** tab from **Customize** dialog box. After that select tools from left list box and location in right list box where you want to add the tool, and then click on the **Add>>** button. The added tools will be shown in the right list box; refer to Figure-11.

Figure-11. Tool added in Ribbon by customizing

- After setting desired parameters, click on the **Close** button.

5. **Ribbon** : It is the group of tools from where you can use tools to create different features. It is classified into **Tabs** and **Groups**. It can be customized according to user. The procedure to add commands in **Ribbon** is discussed next.

6. **Predict Commands Toolbar** : The tools in toolbar change as per the history of operations performed by the user. This toolbar is a part of Adaptive UI feature.

Recording Solid Edge Screen

- Click on the red colored **Record** button at the bottom right corner of Application window to record video. The **Record Video** dialog box will be displayed; refer to Figure-14.

Figure-14. Record Video dialog box

- Click on **Record** button or press hot keys **Shift+F9** from keyboard to start recording the Solid Edge screen.
- After starting the recording, you can pause the recording anytime by pressing **Shift+F9** from keyboard.
- Click on **Stop** button or press hot keys **Shift+F10** to stop the video.
- User can also record his/her voice. To do so, select the **Record audio** check box from the dialog box. You can select the device for recording audio from the drop-down below this check box.
- You can select the codec for video recording from the **Compression Codec** drop-down.
- After recording the video, click on **Save** button to save the video. The **Save Video** dialog box will be displayed; refer to Figure-15.

Figure-15. Save Video dialog box

- Specify desired name of file in the **File name** edit box and click on the **Save** button to save the recording.

Uploading Recording on YouTube

- If you want to upload the recorded file on YouTube then click on the **Upload** button from the dialog box. You can also use the **Upload to YouTube** button from the bottom right corner of the **Application** window to upload video on YouTube. The **Upload to YouTube** dialog box will displayed; refer to Figure-16.
- Login into your YouTube account by clicking on the **Sign In** button. Set the parameters as desired in the **Upload to YouTube** dialog box and click on the **Upload** button.

Figure-16. Upload to YouTube dialog box

FILE MENU TOOLS

The tools in the **File** menu are used to manage application and file related functions. These tools are discussed next.

Learn Options in File menu

The options in **Discover** cascading menu are used to access various resources to learn about **Solid Edge**. The procedure to access different options in this menu are discussed next.

- Click on the **Learn** option from **Discover** cascading menu in the **File** menu. The options will be displayed as shown in Figure-17.

Figure-17. Learn option in Discover cascading menu

- Click on the **Tour Solid Edge** option in the **Learn** page to check beginners videos on **Solid Edge**.
- Click on the **Learn Solid Edge in a Few Steps** option to know how to create basic objects.
- Click on the **What's New** to find the latest feature added to this version of software.
- Click on **Solid Edge Help** to learn Solid Edge.
- Click on **Add Links** button to add more links in the **My Links** area of the **Learn** page. The **Add or Remove Links** dialog box will be displayed; refer to Figure-18. Add desired link in the **Link** edit box, specify the other parameters, and click on the **OK** button. The new link will appear in the list.

Figure-18. Add or Remove Links dialog box

Community Options

The options in the **Community** cascading menu of **File** menu are used to access community of Solid Edge at various social media platforms and share your designs using team center and E-mail options. Various options of this cascading menu are discussed next.

*   Click on the **Solid Edge Facebook** option from the cascading menu to access Facebook page of Solid Edge in default web browser.
*   Click on the **Solid Edge Home Page** option from the cascading menu to access default page of Solid Edge on Siemens website in your default web browser.
*   Click on the **Solid Edge User Community** option from the cascading menu to access community discussion page of Solid Edge in your default web browser.
*   Click on the **Solid Edge Community Blog** option from the cascading menu to explore the articles written by and for community members.
*   Similarly, you can use the **Teamcenter Home Page** and **Teamcenter User Community** options to access respective webpages in the default web browser.
*   Click on the **Mail Recipient** option from the cascading menu to send current open document as attachment in an e-mail using your default E-mailing program/ software.
*   Click on the **Send Command Log** option from the cascading menu to create log of current commands performed in Solid Edge and send it to Siemens for improving software. Note that command log will be send via e-mail service.
*   Click on the **Pack and Go** option from the **Tools** cascading menu to save the current document and other related documents in a ZIP folder for sharing with others. On selecting this option, you will be asked to save the current document if it is not saved yet and then **Pack and Go** dialog box will be displayed; refer to Figure-19. Select desired check boxes from the **Include** area in the dialog box to include respective files in the zip folder. Select desired radio button from the **Folder Structure** area of dialog box to define how files will be structured in the zip folder. Select the **BOM view** radio button to view components in the view style that represents assembly structure of model. Select the **Save to folder** radio button if you want to save copy of the files in a folder at specified location. Select the **Save to zip** file radio button if you want to save copy of files in a zip file. Set desired location for the zip file and click on the **Save** button to generate the zip folder.

Figure-19. Pack and Go dialog box

New Options in File menu

The options in **New** cascading menu are used to create new documents which are generally part model, sheet metal model, assembly, and drawing. The procedure to create a new document is discussed next.

- Click on **New** cascading menu in the **File menu**. The options to start new documents will be displayed as shown in Figure-20.

Figure-20. New cascading menu

- Click on **New** tool from **New** cascading menu. The **New** dialog box will be displayed with list of templates; refer to Figure-21.

Figure-21. New dialog box

- Select **ansi metric assembly.asm** template to create a new assembly using the default template.
- Select **ansi metric weldment.asm** template to create a new welding design document using default template.

- Select **ansi metric part.par** template to create a new part document using default template.
- Select **ansi metric sheet metal.psm** template to create a new Sheet Metal document using the default template.
- Select **ansi metric draft.dft** template to create a new draft (Drawing) document using default template.
- After selecting desired template, click on the **OK** button from the dialog box.

Tip : You can also start a new document by selecting desired template from the **New** cascading menu.

Note : You can change the list of templates by using **Edit List** button from the **New** cascading menu of **File** menu. The **Template List Creation** dialog box will be displayed. Select desired standard templates to be shown in **New** cascading menu from the left area in dialog box and click on the **OK** button. You will learn about **Template List Creation** dialog box later in this chapter.

Opening a File

The options in **Open** cascading menu are used to open files from local drive; refer to Figure-22.

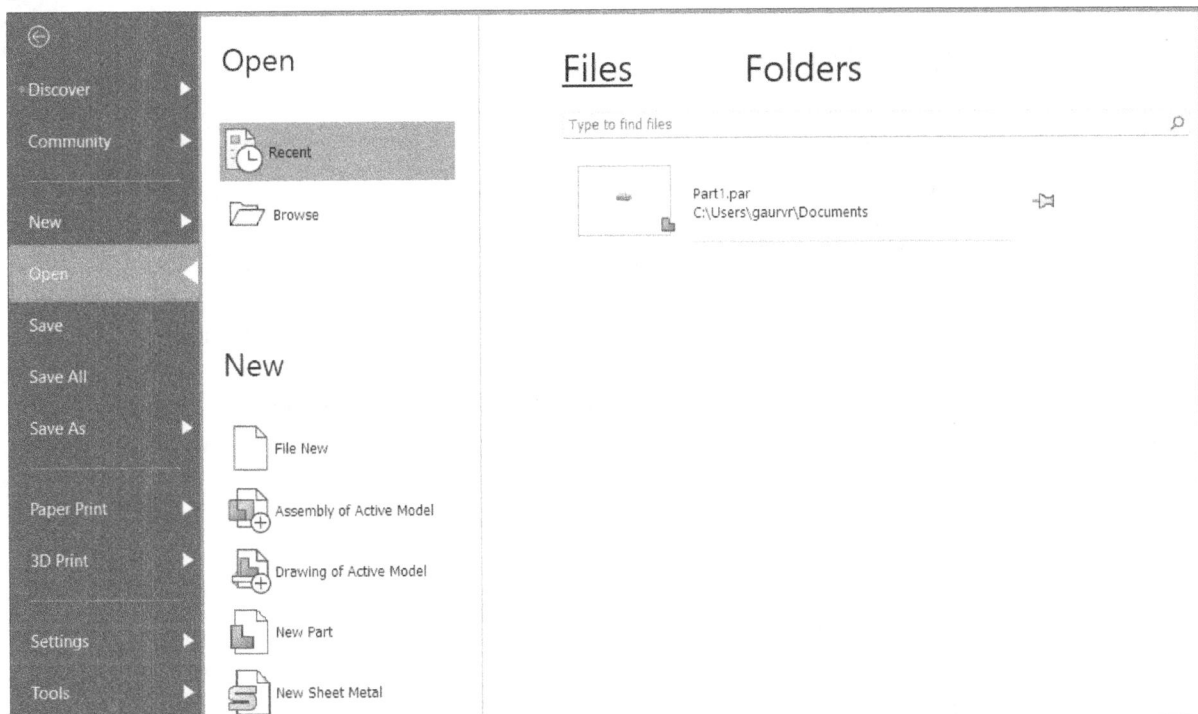

Figure-22. Open cascading menu

- Select the **Recent** option from the cascading menu to check list of recently opened documents. You can find desired document from list of recent documents by typing name in the search box.
- Click on **Browse** button ⌂ from the **Open** cascading menu to browse the local drives. The **Open File** dialog box will displayed; refer to Figure-23.

Figure-23. Open File dialog box

- Click on the **Home** button to check default folder where Solid Edge will save files which is generally **Documents** folder in Microsoft Windows.
- Select the format of file which you want to open in Solid Edge from the **File Type** drop-down.
- Browse to desired location and select the file you want to open in Solid Edge.
- Select the **Open as read-only** check box if you want to open current file as read-only so that you do not change anything in model unintentionally. Select the **First level only** radio button if you want to open current file as read-only while other linked models/files can be edited. Select the **All levels** radio button if you want to open current file and all related documents as read-only.
- Similarly, you can define other options in the dialog box related to assemblies, drawings, and so on.
- You can use the **Search** button in the dialog box to find desired model file using the search function of the software. On selecting this button, the **Search** dialog box will be displayed as shown in Figure-24. Set desired folder/directory location where you want to search the file in the **Look in** edit box or use **Browse** button next to the edit box. Note that using the **Browse** button, you can set multiple locations for searching file. Select the **Search** for results in subfolders check box to look for files in sub-folders of specified directories. Click in the field of **Property** column in the **Criteria** area in the dialog box to define the property based on which search will be performed to find out file and set desired criteria for the property in **Criteria** column. After setting desired criteria, click on the **Search** button from the dialog box. List of documents that match the criteria will be displayed. Select desired document and click on the **Open** button. Click on the **Close** button to exit the dialog box.

Figure-24. Search dialog box

- After selecting file, click on the **Open** button from the dialog box to open the file. The file will be displayed in Solid Edge.

Saving File

The **Save** tool in **File menu** is used to save the model file. If you are saving the file for first time then **Save As** dialog box will be displayed and once you have saved the file then no dialog box will be displayed later after selecting this tool. The procedure to use this tool is given next.

- Click on the **Save** tool from the **File menu**. The **Save As** dialog box will be displayed; refer to Figure-25.
- Specify desired name of file in the **File name** edit box. If you want to save the file in format different from native format of Solid Edge then select desired option from the **Save as type** drop-down.
- Click on the **Save** button from the dialog box to save the file. After saving the file once, modify the model as needed and press **CTRL+S** to save the file.

Note that you should save the file as soon as you have performed modifications in model to make sure you do not loose the work.

If you have multiple files open in Solid Edge and have not saved the files yet then click on the **Save All** tool from the **File** menu.

Figure-25. Save As dialog box

Save As Tools

The tools in the **Save As** cascading menu are used to save the model in different formats with different names; refer to Figure-26. These tools are discussed next.

Figure-26. Save As cascading menu

The procedure to use **Save As** tool of **Save As** cascading menu is same as discussed for **Save** tool earlier.

Saving Model in Image Format

The **Save As Image** tool is used to save a file in image format like JPEG, TIFF, and so on. The procedure to use this tool is given next.

• After creating model, set it in center of viewport using View tools. You can use zoom and pan tools of **View** toolbar to orient model.

- After orienting the model, click on the **Save As Image** tool from the **Save As** cascading menu of the **File menu**. The **Save As Image/VRML** dialog box will be displayed; refer to Figure-27.
- Specify desired name of file in **File name** edit box and select desired format from the **Save as type** drop-down.
- After setting desired parameters, click on the **Save** button. The file will be saved in location selected in the dialog box.

Figure-27. Save As Image VRML dialog box

Similarly, the **Save Copy As** tool in the **Save As** cascading menu is used to store copy of the current active model by a new name in local drive and the **Save Model As** tool is used to save simplified model with no design history of current file.

Save As New Revision

The **Save As New Revision** tool is used to save a new version for the active document. This tool is displayed when you have selected the **Use Solid Edge data management** check box from the **Manage** page in **Solid Edge Options** dialog box The procedure to use this tool is discussed next.

- Click on the **Save As New Revision** tool from the **Save As** cascading menu of the **File** menu. The **Revise** dialog box will be displayed; refer to Figure-28.

Figure-28. Revise dialog box

- Specify desired revision id, document number, and title for the document and then click on the **Save** button. The file will be saved in location selected in the dialog box.

Exporting Solid Edge Model

The **Save As Translated** tool is used to export Solid Edge model in non-native CAD formats. The procedure to use this tool is given next.

- Click on the **Save As Translated** tool from the **Save As** cascading menu of the **File** menu. The **Save As Translated** dialog box will be displayed; refer to Figure-29.

Figure-29. Save As Translated dialog box

- Specify the name and format of file in the dialog box.

- Click on the **Options** button to modify parameters related to translation in selected format. In our case, we have selected **IGES Document** option from the **Save as type** drop-down so the **IGES Export Wizard** dialog box will be displayed; refer to Figure-30. (IGES is one of the most common translation format for CAD models)

IGES Export Options

Figure-30. IGES Export Wizard dialog box

- Select the **Enable logging** check box if you want to generate a log file of translation.
- Click on the **Browse** button and select desired configuration file for translation if you want to change the translation definition.
- Specify desired user information in other edit boxes of the dialog box. Click on the **Next** button from the dialog box. The **Types** page of **IGES Export Wizard** will be displayed; refer to Figure-31.

Figure-31. Type page of IGES Export Wizard dialog box

- Select desired radio button from the **Solid Bodies Export As** area to define how solid bodies will be exported in iges format. Select the **Brep solid as analytics** radio button to use basic analytic surfaces (planar and cylindrical) wherever possible in exported file. Select the **Brep solid as NURBS (Type 128)** radio button to export the model as B-spline surface wherever possible. Select the **Trim surfaces as analytics** radio button to convert solid faces into analytic surfaces in exported file. Select the **Trim surfaces as NURBS (Type 128)** radio button to convert solid faces into B-spline surfaces in the exported file. Select the **Wire frame** radio button to export solid as skeleton of model comprised of lines, arcs, and spline curves.

- Select desired radio button for unit from the **Export** area of the dialog box.

- Select the **Export solids (constructions)** check box to export constructions geometries in solid form to exported file. Similarly, you can use the **Export sheets (construction)** check box to export surfaces and **Export wires and points** check box to export curves used in construction.

- Select the **Japan Automobile Manufacturers Association (JAMA)** check box to export a trimmed version of IGES file as per Japanese manufacturer standards.

- Click on the **Next** button from the dialog box. The **Save** page of **IGES Export Wizard** dialog box will be displayed; refer to Figure-32.

Figure-32. Save page of IGES Export Wizard dialog box

- If you want to save changes in original configuration file then select the **Save the remappings to the original configuration file** radio button. If you want to create a new configuration file based on changes then select the **Create a new configuration file** radio button and click on the **Copy To** button. The **Save As** dialog box will be displayed. Specify desired name of file and click on the **Save** button.

- Click on the **Next** button. The **Finish** page will be displayed.

- Click on the **Finish** button to apply the configuration. The **Save As Translated** dialog box will be displayed again.

- Click on the **Save** button from the dialog box to save the exported file.

Saving Model for Tablet Devices

The **Save for Tablet** tool is used to save model for use in Tablet devices. The procedure to use this tool is given next.

- Click on the **Save for Tablet** tool from the **Save As** cascading menu of the **File menu**. The **Save As** dialog box will be displayed as shown in Figure-33.

Figure-33. Save As dialog box

- Specify desired name of file and click on the **Save** button. You can transfer this file to your tablet for viewing.

Print File

The options in **Paper Print** cascading menu are used to print the model on paper; refer to Figure-34. The procedure to print model is given next.

Figure-34. Paper Print cascading menu

- Select desired printer to be used for printing from the **Printer** drop-down. If you cannot find your printer in the list then install the drivers of your printer.
- Using the **Settings** button, you can adjust the print range and scale of model for printing. Using the **Properties** button, you can define printer properties like orientation of paper, type of paper, print quality, and so on.
- Set the number of copies you want to print in the **Copies** spinner. Select the **Collate** check to print in sequence if there are more than one type of prints.

- Select the **Print to file** check box to print the model as a digital file. If you now click on the **Print** button from the cascading menu then the **Print to File** dialog box will be displayed; refer to Figure-35.

Figure-35. Print to File dialog box

- Select the **Print all colors as black** check box to print all the colored objects in Black and white format.
- Select the **Print as displayed** check box to print all the annotations and model as they are being displayed in the viewport.
- Select the **BGR Order** check box to print using Red, Blue, and Green color inks of printer.
- Click on the **Print** tool from the **Paper Print** cascading menu to print everything being displayed in the viewport. If you want to print specific portion of the viewport then click on the **Print Area** tool from the **Paper Print** cascading menu. You will be asked to define start point and end point of rectangle to define boundary of print area.
- Create the boundary enclosing the objects to be printed on paper; refer to Figure-36. The **Print Area** dialog box will be displayed; refer to Figure-37.

Figure-36. Creating boundary for print

Figure-37. Print Area dialog box

- You can modify the orientation, location, and scale factor for printing in the dialog box. If you want to modify the print area then click on the **X Y Range** button
- After setting desired parameters, click on the **OK** button from the dialog box. The model will be printed in the file or paper based on your selections in the **Paper Print** cascading menu.

Preparing File for 3D Printing

The options in the **3D Print** cascading menu are used to create file ready for 3D printing or post processing by 3D Printing software; refer to Figure-38. These options are discussed next.

Figure-38. 3D Print cascading menu

- Select desired option from the **File type** drop-down in the **3D Print Format** section. There are two formats available in this drop-down; 3MF and STL.
- Click on the **Options** button from the **Settings** section to define tolerance parameters related to conversion of model for 3D print ready model. The **Export Options** dialog box will be displayed; refer to Figure-39.

Figure-39. Export Options dialog box

- Select the check boxes from the **Export** area to define how elements of solid model will be exported in the 3D print ready model.
- Select desired radio button from the **Tolerance Options** area of the dialog box to define tolerance within which model will be converted to facets. Select **Fine** radio button to get better finish in converted model. Select the **Coarse** radio button to get faster conversion of model with lower finish. If you have selected the **Custom** radio button then you can manually define the conversion tolerance and surface plane angle.
- Set the other parameters as discussed earlier and click on the **OK** button.
- If you want to modify X, Y, and Z range of printer then specify the values in **Width**, **Depth**, and **Height** edit boxes of the **Printer Size** section and click on the **Apply** button.
- To further edit the 3D print model, click on the **3D Builder** button from the **Services** section. A message box will be displayed asking you to download the Microsoft 3D Builder app; refer to Figure-40. Note that 3D Builder is not supported in Windows 11.

Figure-40. Message box

- Click on the link in the message box to download the app. A web page will be displayed in default web browser. Click on the **Get** button from the web page to download and then click on the **Install** button.
- Now, if you click again on the **3D Builder** button from the **3D Print** cascading menu. The 3D Builder application will be displayed; refer to Figure-41.

Figure-41. 3D Builder application

- Click on the **Got It!** button and select the model to be modified. The arrow handles to change location of model on printer bed will be displayed. Drag the arrow handles to change location; refer to Figure-42. Using the tools in the toolbar, you can rotate and scale the model.

Figure-42. Tools and arrow handles in 3D Builder

- After modifying model, you can save the model for sharing using the **Save** button at the top in the application. If you want to 3D print model using your 3D printer then click on the **3D Print** button from the top in application window. The **Print 3D** window will be displayed.
- Set desired parameters and click on the **Print** button. Click on the **Close** button at top-right corner to exit the application.

Settings

The options in **Settings** cascading menu are used to customize Solid Edge as per user's requirement.

- Click on **Settings** option from **File** menu. The **Settings** cascading menu will be displayed; refer to Figure-43.

Figure-43. Settings cascading menu

Options

The **Options** tool in **Settings** cascading menu is used to define various default parameters for Solid Edge.

- Click on the **Options** tool from the cascading menu. The **Solid Edge Options** dialog box will be displayed; refer to Figure-44.

Figure-44. Solid Edge Options dialog box when no document is open

When a part model is already open in the software then **Solid Edge Options** dialog box will be displayed as shown in Figure-45. Some important options of Solid Edge are discussed next.

Figure-45. Solid Edge Options dialog box when a document is open

General Options

The options in the **General** page of **Options** dialog box are used to set parameters related to general functions of software like number of file that can be shown in recent list, reference plane size, paste behavior, and so on. Some important options in this page are discussed next.

- Select the **Show units in value fields** check box to display the units when you are typing value of a parameter in dynamic edit box.
- You can specify the maximum number of documents displayed in the recent list by specify respective value in the **Max number of recent files to show** edit box.

- The value specified in the **Maximum print file size** edit box is used to define the file size of print ready file of model.
- You can specify the maximum number of undo steps that can be stored in volatile memory in **Part and Assembly undo steps** edit box. Note that a higher value specified here will cause more usage of RAM.
- In the **Reference plane size** and **Feature origin size** edit boxes, you can specify the size of reference plane and origin being displayed when starting a new model file.
- The **Plane Name Display** button is used to change font, color, and text height of reference planes. On clicking this button, the **Plane Name Display** dialog box will be displayed where you can specify the parameters.
- Select the **Prompt for material in new model documents** check box to ask for selecting material when you are starting a new part model file.
- Select the **Check out-of-date properties in material table on file open** check box to check the model for an material data which has not been updated as per new material database.
- The options in the **Revise and Increment Name** area are used to define the sequence in which revision of model will be generated while collaborating with others on same file using the **Data Management** tools.
- Select the **Enable Dynamic Edit of profiles/sketches** check box to allow modification of sketches by simple mouse gestures like dragging an end point of a line to increase its length or location. On selecting this check box, two radio buttons below it will become active. Select the **Recompute continually during edit** radio button if you want to check the effect of dynamic operation in real-time. Select the **Recompute after edit** radio button if you want the effect of dynamic modification to be displayed after you have performed the operation. The later option is recommended if you do not have sufficient RAM and processing power.
- Select the **Recompute assembly during sketch edits** check box to automatically recompute the assembly when a sketch of part or assembly model is modified. The radio buttons available for this check box are same as discussed earlier.
- Select the **Enable value changes using the mouse wheel** check box to modify value in spinner edit box using the mouse wheel.
- Select the **Recompute Ordered part after Synchronous edit** check box to update the ordered part when it is modified by synchronous operation.
- Click on the **IntelliSketch Options** button from the dialog box to modify parameters related to automatic snapping and constraining of sketch. The **IntelliSketch** dialog box will be displayed; refer to Figure-46. By default, the **Relationships** tab is selected in the dialog box. Select check boxes for snap points to which relations will be applied automatically based on selection while creating sketch. You will learn more about this feature later in this book. Click on the **Auto-Dimension** tab from the dialog box and select the **Automatically create dimensions for new geometry** check box to automatically create dimensions while creating sketch; refer to Figure-47. After selecting the check box, select desired radio buttons to define when the dimensions will be created automatically. Click on the **Cursor** tab in the dialog box to define the size of cursor. After setting desired parameters, click on the **OK** button.

Figure-46. Intellisketch dialog box

Figure-47. Auto Dimension tab

- Select desired radio button from the **Preferred 2D Paste Behavior** section to define how 2D objects will be pasted after copying in Draft and Ordered sketch environment.
- Select the **Activate regions within sketches** check box from the **When Creating New Sketches** area to enable selection and creation of regions in sketches.
- Select the **Migrate geometry and dimensions when features are created (Synchronous sketches only)** check box to place the geometry and dimensions of features in respective categories after creating the feature.
- Select the **Indicate under-constrained profiles in PathFinder** check box to display whether profile is fully defined or under-defined.
- Select the **Enable "Undo All" for profile/sketch modifications** check box to display options for undoing all the modifications performed on sketch/profile.
- Set the other parameters as desired in the **General** page.

View Options

- Click on the **View** option from the left in the **Solid Edge Options** dialog box to modify view related parameters; refer to Figure-48.

Figure-48. View options

- Click on the **Quick View Cube Settings** button to modify location, color, and transparency parameters of Quick View Cube. The **Quick View Cube Settings** dialog box will be displayed. Set the parameters as discussed earlier.
- Select the **Show orientation triad** check box to display triad in viewport while orienting the model.
- Select the **Mouse wheel forward=zoom out, back=zoom in** check box to reverse mouse gestures for zoom in/out.
- Select the **3D input device** check box to enable use of 3D input devices for modeling.
- Set desired value in the **View transition** slider to define how fast Solid Edge transitions from one orientation to another.
- Select the **Culling** check box and using the slider specify how aggressive the software will skip small features in the large assembly or complex model.
- Select desired options in the **Auto-sharpen**, **Anti-aliasing (global)**, and **Floor reflection** drop-downs to set quality of model in viewport.
- Set desired value in the **Arc smoothness** drop-down to define how fine the arcs will be generated in sketch and model. Specify higher value in this drop-down if you are getting broken sketches in display.
- Select the **Use shading** check boxes to display shading on selection, highlight, reference planes, and so on.
- Similarly, you can set other view parameters.

Colors Options

In **Solid Edge**, you can use various color schemes for displaying the objects in application window. The process is discuss next.

- Click on the **Colors** option from left in the **Solid Edge Options** dialog box. The options in **Solid Edge Options** dialog box will be displayed as shown in Figure-49.

Figure-49. Colors option in Solid Edge Options dialog box

- Set desired colors in various drop-downs to define colors for respective objects and click on the **Apply** button.

Save Options

The options in the **Save** page of **Solid Edge Options** dialog box are used to specify auto save duration and related parameters; refer to Figure-50.

Figure-50. Save options in Solid Edge Options dialog box

- Select the **Automatically preserve documents by** check box to save or prompt you to save the active model. After selecting this check box, the related options will be displayed below it.
- Select the **Prompting me to save all documents every minutes** radio button and specify the time in minutes in related edit box so that software will prompt you to save file after specified duration.
- Select the **Creating uniquely named backup copies of unsaved documents every minutes** radio button to save backup copy of the document after specified duration.
- Select the **Backup model files** check box to backup model files. Select the **Backup Draft files** check box to create backup file of 2D drawings of the model.
- Select the **Prompt for File Properties on first save** check box if you want software to specify model properties also when you are saving the document for the first time.

File Locations

The **File Locations** tab is used to modify the locations of Solid Edge system files. The process is discuss next.

- Click on the **File Locations** option from the left area in the dialog box. The options in the dialog box will be displayed; refer to Figure-51.
- If you want to change any file then select it from the list and click on the **Modify** button. The **Select document** dialog box will be displayed; refer to Figure-52.

Figure-51. File Locations option in Solid Edge Options dialog box

Figure-52. Select Document dialog box

- Select desired document and click on the **Open** button from the dialog box.

Similarly, you can select folders for various objects like macros, reports, startup screens, and so on.

User Profile Options

The options in the **User Profile** page of the dialog box are used to specify information about user of the software; refer to Figure-53.

Figure-53. User Profile page in Solid Edge Options dialog box

- Specify desired name and initials in the **Name** and **Initials** edit boxes.
- In the **Mailing address** edit box, you can type your E-mail address.
- Select **Add my user profile information as properties to all Solid Edge Documents that I create or modify** check box to add the user profile information as properties to all the Solid Edge documents that created or modified.
- Select the **Display Solid Edge Discovery Center** check box to display Discovery at home page.
- Select the **Deploy Settings and Preferences during startup of Solid Edge** check box to deploy file of settings and preferences at startup.
- Select the **Capture Settings and Preferences during exit of Solid Edge** check box to deploy file of settings and preferences when you exit the software.
- You can modify the preference file by clicking on the **Modify** button.

Inter-Part Options

The options in **Inter-Part** page are used to activate and specify parameters related to inter-part linking; refer to Figure-54. You will learn about these options later in this book.

Figure-54. Inter-Part page of Solid Edge Options dialog box

Manage Options

The options in the **Manage** page are used to manage parameters related to vault and other data management systems; refer to Figure-55.

Figure-55. Manage page of Solid Edge Options dialog box

- In the **Vault Definition** tab, you can specify the local locations in your system to be shared with other data management users.
- In the **Custom Properties** tab, you can specify custom templates and properties to be used for sharing files using data management system.
- In the **Document Naming Rule** tab, you can specify default pattern to be used for generating name of revisions of files while sharing.
- In the **Life Cycle** tab, you can specify how files will be preserved or removed during life cycle of shared files.

- In the **Workflow** tab, you can specify the users allowed to access the files and define the priority given to various users for accessing the files.

Units Options

The options in the **Units** page are used to set the preference for units in Solid Edge. The process to set the units in Solid Edge is discussed next.

- Click on the **Units** option from the left area of the dialog box. The options to define units will be displayed in the dialog box; refer to Figure-56.

Figure-56. Units option in Solid Edge Options dialog box

- Select desired radio button from the **Units System** area to define unit system for creating model in Solid Edge.
- If you want to create custom unit system then set desired options in the **Base Units** and **Derived Units** tables.

Dimension Style

The options in the **Dimension Style** page of **Solid Edge Options** dialog box are used to define dimensioning style for drafting; refer to Figure-57.

Figure-57. Dimension Style page of Solid Edge Options dialog box

• Select the **Active dimension style** radio button and select desired option from the drop-down below it.

Shape Search Options

The options in the **Shape Search** page are used to define parameters for searching the models existing on specified server; refer to Figure-58. Using the shape search feature, you can find models identical to current model in 3D view. Note that the options in this page will be active when there is no file open in software.

Figure-58. Shape Search options

Helpers Options

The options in the **Helpers** page are used to set parameters for help options while working on the software; refer to Figure-59. These options are discussed next.

• Select desired radio button from the **Starting Solid Edge** area to define what type of starting page will be displayed in first interface of Solid Edge after starting.
• Select desired radio button from the **Start Part and Sheet Metal documents using this environment** area to define whether you want to start Part and Sheet Metal model in **Ordered** or **Synchronous** environment.

Figure-59. Helpers page of Solid Edge Options dialog box

- Click on the **Edit Creation Options** button from the dialog box to modify list of templates being displayed when you start new document. The **Template List Creation** dialog box will be displayed; refer to Figure-60. You can move the templates up/down or add new template in the list by using the options in this dialog box.

Figure-60. Template List Creation dialog box

- Click on the **Set Default Templates** button from the **Solid Edge Options** dialog box to define default templates for various documents. The **Default Templates** dialog box will be displayed; refer to Figure-61. Select the document type for which you want to change the template and click on the **Browse** button. The **Add Template** dialog box will be displayed where you can select a new template. After selecting template, click on the **OK** button from the **Add Template** dialog box and **Default Templates** dialog box.

Figure-61. Default Templates dialog box

- Select desired option from the **Application color scheme** drop-down to modify color scheme for interface of application.
- Select the **Make Part and Assembly PathFinder dockable** check box to display PathFinder at the bottom in interface.
- Select desired option from the **Steering Wheel, OrientXpress, and View Cube size** drop-down to define size of aforementioned objects.
- Select the **Show command tips** check box to display tips when you hover cursor on any command in interface.
- Select the **Show sensor indicator** check box to display sensor warnings in graphics area.
- Select desired radio button from the **Place new graphic window tabs** section to define where next document tab will be added against existing document tab when a new document is started.
- The options in the **Predict Commands** area are used to define how the **Predict Commands Toolbar** will be displayed in the interface. If you clear the **Use Predict Commands** check box then the toolbar will not be displayed in the interface.
- Similarly, you can set other parameters in this page.

Drawing Standards

The options in the **Drawing Standards** page are used to define standards for geometric design features; refer to Figure-62.

Figure-62. Drawing Standards page of Solid Edge Options dialog box

• Select desired radio button from the **Edge Condition** area to define what standards should be used for edges. Similarly, you can set standard for weld symbols and Limits & Fits.

You will learn about other settings later in the book when their respective tools will be used. Click on the **OK** button from the dialog box to apply settings.

Add-Ins

Add-Ins are used to add extra functions in Solid Edge like eAssistant. Before using Add-Ins, you need to download and install Add-Ins from Solid Edge certified developers. The procedure is discuss next.

• Click on the **Add-Ins** tool from the **Settings** cascading menu. The **Add-In Manager** dialog box will be displayed; refer to Figure-63.

Figure-63. Add In Manager dialog box

• Select desired check box from the dialog box to activate the Add-In and click on the **OK** button. The related tools will be added in the **Ribbon**.

Customize

The option in **Customize** dialog box are used to customize interface of Solid Edge. The procedure to customize is discussed next.

• Click on **Customize** tool from the **Settings** cascading menu of the **File menu**. The **Customize** dialog box will be displayed; refer to Figure-64.

Figure-64. Customize dialog box

Various options for customizing are discussed next.

Keyboard

The options in **Keyboard** tab are used to set keyboard shortcuts for various tools and options. The procedure to create keyboard shortcut is given next.

*   Select desired environment of Solid Edge in which you want to create custom shortcut key from **Environment to customize** drop-down.
*   Select desired option from **Choose commands from** drop-down to define the scope from which commands can be selected. The **Customize** dialog box will be displayed as shown in Figure-65.
*   Select desired tool from the **Command** column. Click on the **+** sign to expand the command groups.
*   Set desired modifier and key in the respective columns for selected command. Refer to Figure-66.

Figure-65. Customize Keyboard

Figure-66. Selecting button for keyboard shortcut

Quick Access

Quick Access->Environment to customize or Choose commands->select commands->Add

- Click on **Quick Access** tab in the dialog box to customize **Quick Access Toolbar**. The options in the dialog box will be displayed as shown in Figure-67.

Figure-67. Quick Access tab in Customize dialog box

- Select desired tool to be added in **Quick Access Toolbar** from the left area and click on the **Add** button. The tool will be added at the end of toolbar.
- Using the **Move Up** or **Move Down** tool to position the new tool in toolbar.

Ribbon
The process of customizing **Ribbon** has already been discussed earlier in this chapter.

Radial Menu
Radial menu->Environment to customise->Choose commands from->select commands and drag it

- **Radial Menu** is displayed on right-clicking in the drawing area. To customize the Radial Menu, click on the **Radial Menu** tab from **Customize** dialog box. The options to customize Radial Menu will be displayed; refer to Figure-68.

Figure-68. Radial Menu tab in Customize dialog box

- Expand desired node from which you want to include tool in **Radial menu**.
- Select desired tool and drag & drop to the **Radial menu** for replacing the tool; refer to Figure-69.

Figure-69. Radial menu

Layout

- Click on **Layout** tab in the dialog box to modify interface layout of software like location of CommandBar, tooltips, and so on. The options in the dialog box will be displayed as shown in Figure-70. Set desired options for repositioning the interface elements. After setting desired parameters, click on the **Close** button to exit the dialog box. You will be asked whether to save new scheme. Save the theme as desired.

Figure-70. Layout tab in Customize dialog box

Themes

Using the **Themes** tool, user can define the level of assistance provided by software related to tools and commands like displaying tips, help menus, and so on.

• On clicking the **Themes** option from **Settings** cascading menu of the **File menu**, the options to select a theme will be displayed; refer to Figure-71.

Figure-71. Themes menu

- Select desired theme option from the menu.

Tools

The option in **Tools** cascading menu are use to compare drawing file and convert the file from traditional to synchronous documents. The options in the **Tools** cascading menu are shown in Figure-72. You can also access models from various portals.

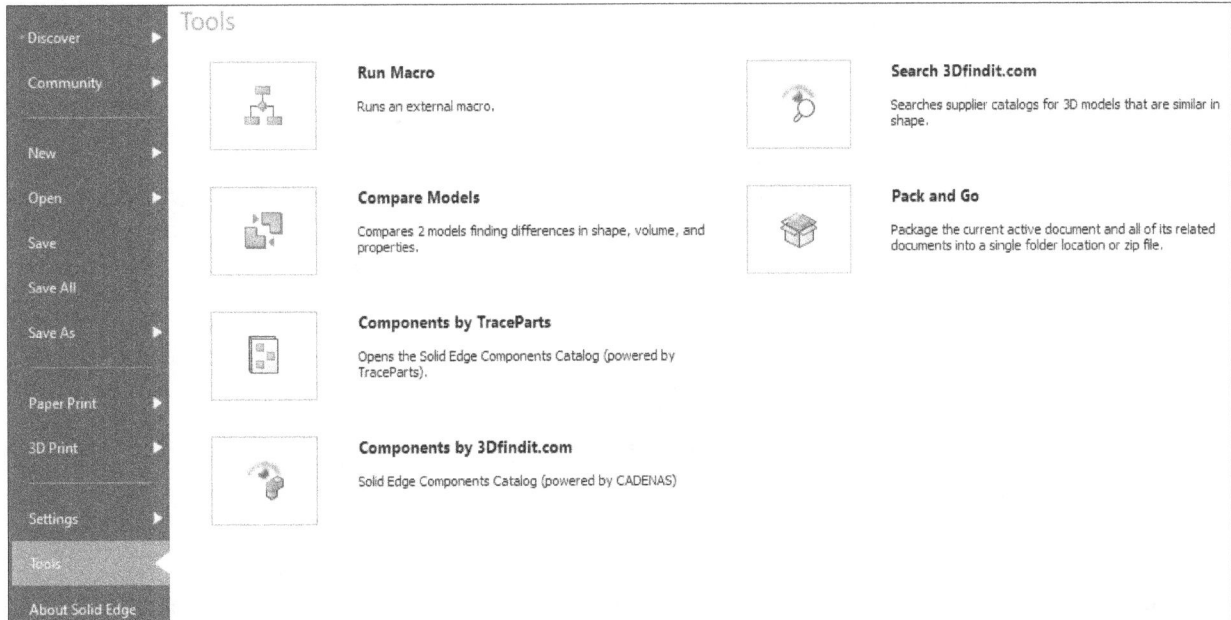

Figure-72. Tools cascading menu

Run Macro

Macros are scripts that automate tasks in software. The procedure to run a macro is given next.

- Click on the **Run Macro** tool from the **Tools** cascading menu of the **File menu**. The **Run Macro** dialog box will be displayed; refer to Figure-73.

Figure-73. Run Macro dialog box

- Select desired macro file and click on the **Open** button. The related macro dialog box will be displayed; refer to Figure-74. Use the macro as per related documentation.

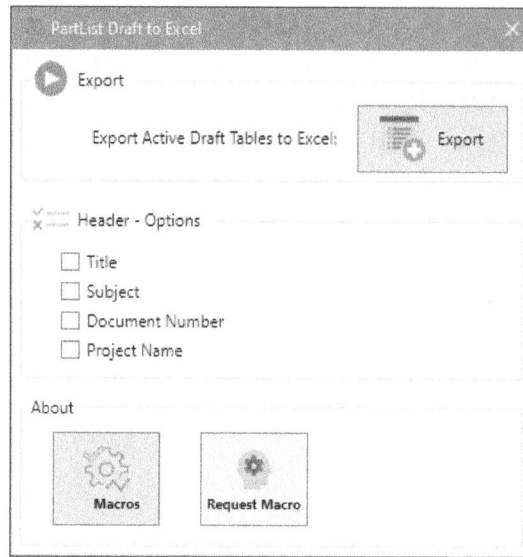

Figure-74. PartList Draft to Excel dialog box

Compare Models

The **Compare Models** option is used to compare the drawing files. The procedure is given next.

- Click on the **Compare Models** tool from **Tools** cascading menu. The **Compare Models** dialog box will be displayed; refer to Figure-75.

Figure-75. Compare Models dialog box

- Click on the **Browse** button for **Reference model** edit box and select desired file from **Open File** dialog box displayed; refer to Figure-76. Similarly, select the other file in **Working model** edit box.

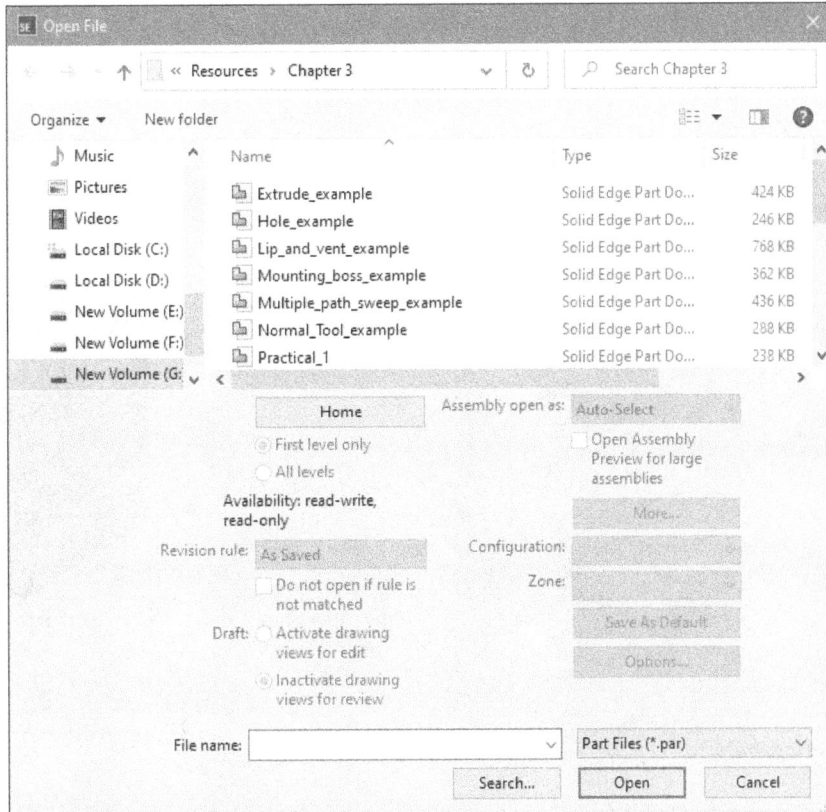

Figure-76. Open File dialog box

- Now, click on **Compare** button from **Compare** area of the dialog box. The differences between two files will be displayed; refer to Figure-77.

Figure-77. Comparing drawings

- Click on the **Report Options** button from the **Report** area to specify parameters related to report and then click on the **Generate Report** button. The **Save Comparison Report** dialog box will be displayed.
- Select desired format for report from the **Report type** drop-down and specify the path where you want to save file in the **Report folder path** edit box. Click on the **Save Report** button to save the file.
- Close the **Compare Models** dialog box using **Close** button at top-right corner.

Info

The options in the **Info** cascading menu of the **File menu** are used to modify document properties. The option in the **Info** cascading menu are shown in Figure-78 & Figure-79.

Figure-78. Info cascading menu when no document is open

Figure-79. Info cascading menu after opening new document

Material Table

The **Material Table** tool in the **Info** cascading menu is used to define and edit materials.

*   Click on the **Material Table** option from the **Info** cascading menu. The **Material Table** dialog box will be displayed; refer to Figure-80.

Figure-80. Material Table dialog box

- Select the material from the left list box, the preview and specifications of material are shown in the right window of dialog box.
- Modify the properties as desired and click on **Apply to Model** button.

File Units

The **File Units** is used to set the unit of active documents. It is same as we discuss earlier in **Options** section.

File Properties

The **File Properties** tool is used to show the properties of active document.

- Click on **File Properties** tool from **Info** cascading menu. The **Properties** dialog box will be displayed; refer to Figure-81.

Figure-81. Properties dialog box

You can check and modify various general properties of the model in the dialog box.

Property Manager

The **Property Manager** is used to modify Solid Edge document properties.

• Click on the **Property Manager** tool from **Info** cascading menu. The **Property Manager** dialog box will be displayed; refer to Figure-82. Select desired file to edit from the dialog box. You can edit any field for desired file by double-clicking.

Figure-82. Property Manager dialog box

• Click on the **OK** button from the dialog box to apply changes.

Where Used

The **Where Used** option in **Info** cascading menu is used to search for documents that use the selected documents.

• Click on the **Where Used** option from the **Info** cascading menu. The **Where Used** dialog box will be displayed; refer to Figure-83.

Figure-83. Where Used dialog box

- Select desired folder from left list box and click on **Add** button. The selected folders will be added in the list where software will search for use of current file.
- Click on the **Next** button from the dialog box. The **Where Used Results** dialog box will be displayed with list of files where the current document was used.

Edit Links

The **Edit Links** option in **Info** cascading menu is used to change and update links of particular document. The **Edit Links** option is active only when an assembly file with component is open.

- Click on **Edit Links** from the **Info** cascading menu. The **Links** dialog box will be displayed; refer to Figure-84.

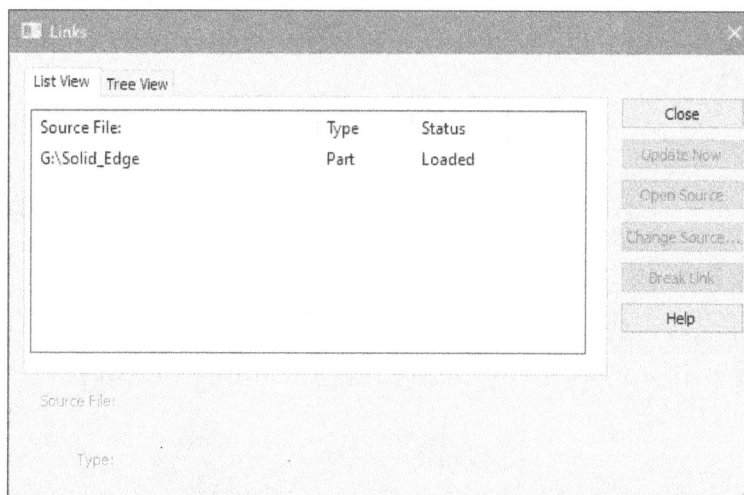

Figure-84. Links dialog box

- Select the file from the dialog box and select desired operation to be performed.

Binder

The **Binder** option in **Info** cascading menu is used to add linked or embedded objects to the document.

- Click on the **Binder** option from the **Info** cascading menu. The **Binder** dialog box will be displayed; refer to Figure-85.

Figure-85. Binder dialog box

- Click on the **Add** button to bind a new document. The **Insert Object** dialog box will be displayed; refer to Figure-86. Set desired parameters and click on the **OK** button.

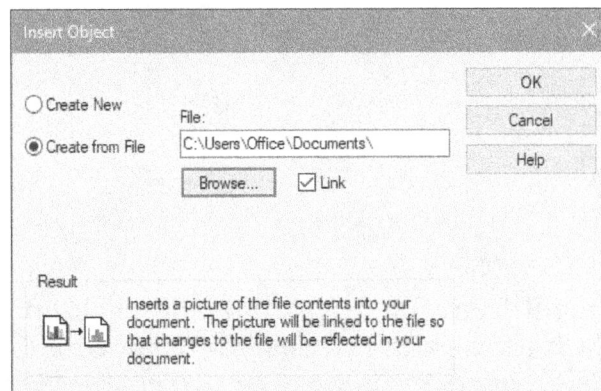

Figure-86. Insert Object dialog box

- If you have selected the **Create from File** radio button then click on the **Browse** button. The **Browse** dialog box will be displayed; refer to Figure-87.

Figure-87. Browse dialog box

- Select desired document from the dialog box and click on **Open** button.
- After that click on **OK** button from the **Insert Object** dialog box.
- If you want to create a new document then select the **Create New** radio button. The dialog box will be displayed as shown in Figure-88.

Figure-88. Create New option in Insert Object dialog box

- Select desired option from the **Object Type** list and click on the **OK** button. A new document will be displayed in respective application.
- Similarly, you can add more documents. Click on the **Close** button from the dialog box.

Exit Solid Edge

Click on the **Exit Solid Edge** tool from the **File menu** to exit the Solid Edge software.

MOUSE BUTTON FUNCTION

Rotate view
- To rotate the model in viewport: Press the Middle mouse button and drag.

Pan
- Hold down **SHIFT** key from keyboard and drag with middle mouse button.

Zoom In/Out
- For Zoom In/out scroll the middle mouse button up/down or hold down **CTRL** key and drag up/down with middle mouse button pressed.

WORKFLOW IN SOLID EDGE

The first step in Solid Edge is to create a sketch. After creating sketch of desired feature, create solid or surface model from that sketch. After doing desired operations on the solid/surface model, we go for assembly or analysis. After, we are satisfied with the assembly/analyses, we create the engineering drawings from the model to perform manufacturing of the model into a real world object.

SELF ASSESSMENT

1. The templates whose names end with part.par are used for creating assembly files. (T/F)

2. The templates whose names end with sheetmetal.psm are used to start sheetmetal design. (T/F)

3. The options in **Open** cascading menu are used to open CAD files in Solid Edge. (T/F)

4. The options in **Tools** cascading menu are used to compare drawings. (T/F)

5. To create new document, which of the following option can be used?
    (a) **New** tool
    (b) CTRL + N
    (c) SHIFT + N
    (d) Both a & b

6. Which of the following templates is used to start new drawing file?
    (a) draft.dft
    (b) part.par
    (c) assembly.asm
    (d) weldment.asm

7. Hold down **SHIFT** and drag with middle mouse button to
    (a) Rotate view
    (b) Pan
    (c) Zoom In/Out
    (d) Fit

8. The .... button display the application menu, which provide access to all documents level functions, templates, and standards.

9. The **PromptBar** shows the .... you have selected.

10. What is the purpose of Add-Ins in Solid Edge?

11. Write down the steps to create new documents?

12. What type of modeling styles does Solid Edge support?

A) Parametric only
B) Synchronous only
C) Both Parametric and Synchronous
D) Neither

13. Which keyboard shortcut can be used to create a new document in Solid Edge?

A) Ctrl + N
B) Alt + N
C) Shift + N
D) Ctrl + M

14. What type of file does the template ending with .asm create?

A) Part file
B) Assembly file
C) Sheet metal design
D) Drawing file

15. What option should you select to record audio while capturing video in Solid Edge?

A) Record Video
B) Record audio check box
C) Audio Settings
D) Enable Audio

16. What does the Quick Access Toolbar allow you to do?
A) Access all files on your computer
B) Access common tools like Save, Undo, Redo
C) Customize keyboard shortcuts
D) Change application settings

17. How can you upload a recorded video to YouTube from Solid Edge?
A) Use the Upload button in the Record Video dialog
B) Use the Share option in the File menu
C) Drag the file to the YouTube website
D) Email the video to YouTube

18. What does the Predict Commands Toolbar do in Solid Edge?

A) Displays all available commands
B) Suggests tools based on user history
C) Records user actions
D) Changes the interface layout

19. What is the primary purpose of the Save As Image tool?

A) To save a file in image formats like JPEG and TIFF
B) To export the model in IGES format
C) To create a 3D print-ready file
D) To print the model on paper

20. Which option in the Save As cascading menu is used to store a copy of the current active model with a new name?

A) Save As Image
B) Save Copy As
C) Save Model As
D) Save As Translated

21. What is the function of the Save As Translated tool?

A) To export the model in a non-native CAD format
B) To save the model as an image
C) To print the model
D) To create a simplified model without design history

22. What does selecting the 'Brep solid as NURBS' option in the IGES Export Wizard do?

A) Converts the model into basic analytic surfaces
B) Exports the model as a skeleton with lines and arcs
C) Converts solid faces into B-spline surfaces
D) Exports the model as a black-and-white image

23. In the Paper Print cascading menu, which option allows you to print a digital file of the model?

A) Print to file
B) Print as displayed
C) Print all colors as black
D) Print Area

24. Which 3D printing file formats are available in the 3D Print Format section?

A) JPEG and PNG
B) IGES and TIFF
C) 3MF and STL
D) PDF and DOCX

25. What is the function of the 3D Builder tool in the 3D Print cascading menu?

A) To rotate and scale the model for printing
B) To change the color of the model
C) To print the model directly on paper
D) To save the model as a JPEG image

26. What happens when you select the 'Save the remappings to the original configuration file' option in the IGES Export Wizard?

A) A new configuration file is created
B) The original configuration file is updated with the new settings
C) The configuration file is exported to PDF
D) The model is printed on paper

27. Which tool is used to save a model for use on tablet devices?

A) Save As Image
B) Save for Tablet
C) Save As Translated
D) Save Model As

28. The 3D Print cascading menu in Solid Edge allows users to prepare files for:
A) 2D printing only
B) Image export
C) 3D printing or post-processing
D) Archival storage

29. What does the "Max number of recent files to show" option in General Options control?
A) Maximum files that can be saved
B) Maximum recent files displayed in the list
C) Maximum files that can be printed
D) Maximum files that can be opened in one session

30. In View Options, the "Mouse wheel forward=zoom out, back=zoom in" check box:
A) Inverts zoom functionality for the mouse wheel
B) Disables zooming completely
C) Zooms in only
D) Zooms out only

31. The File Locations option in Solid Edge is used to:
A) Set file export formats
B) Customize the locations of system files
C) Define recent files list
D) Set up user profile information

32. What purpose does the "Enable 'Undo All' for profile/sketch modifications" check box serve?
A) Enables undo for all operations
B) Allows batch undo for profile/sketch modifications
C) Prevents undo for profile modifications

D) Restricts undo to recent actions only

33. The Manage Options in Solid Edge primarily allow users to manage:
A) View settings
B) Vault and data management parameters
C) Sketch parameters
D) Printing options

34. Which option in the General Options page allows you to automatically create dimensions while sketching?
A) IntelliSketch Options
B) Prompt for material
C) Enable Dynamic Edit of profiles/sketches
D) Auto-Dimension

35. What does the "Enable value changes using the mouse wheel" option allow in Solid Edge?
A) Rotates the model when using the mouse wheel
B) Modifies parameter values in spinner edit boxes with the mouse wheel
C) Zooms in and out using the mouse wheel
D) Changes window size using the mouse wheel

36. In View Options, the Culling slider controls:
A) The number of backup files created
B) The shading style of selected elements
C) The aggressiveness of skipping small features in complex models
D) The maximum number of documents displayed

40. The "Enable Dynamic Edit of profiles/sketches" option in General Options allows users to:
A) Automatically save sketches after each modification
B) Use mouse gestures to dynamically modify sketches
C) Lock sketches after completion
D) Automatically apply dimensions to sketches

41. The "Activate regions within sketches" option in the General Options page is used to:
A) Allow color customization within sketches
B) Enable selection and creation of regions in sketches
C) Lock sketch regions to prevent accidental edits
D) Automatically save each region as a separate file

42. The Workflow tab in the Manage Options page allows users to:
A) Set default print settings
B) Define user access and file handling priorities
C) Adjust color schemes
D) Set model backup locations

Chapter 2

Sketching

Topics Covered

The major topics covered in this chapter are:

- *Basic for Sketching*
- *Synchronous Sketching Environment*
- *Ordered Sketching Environment*
- *View Toolbar*
- *Sketch Creation Tools*
- *Relate Tools*
- *Dimensions*
- *Practical 1*
- *Practical 2*
- *Practice*

BASICS FOR SKETCHING

In engineering, sketching does not mean drawing birds or animals. It refers to sketches based on the real dimensions of real-world objects. You will now work with geometric entities like lines, arcs, circles, rectangles, ellipses, and more, which you have likely used in your school and engineering courses. However, this time you will be using software instead of a pencil, ruler, paper, and other geometry tools. Note that sketching is the foundation for most 3D models, so you should aim to become proficient in it. To start with sketching, you must have a good understanding of planes in **Solid Edge**.

PLANES

In Solid Edge, the planes are displayed in the same orientation as shown in Figure-1. To check the planes of Solid Edge, click on the planes (Front plane, Top plane, Right plane) in **Pathfinder**. You can also select the **Show** button before **Base Reference Planes** option in **Pathfinder** to display all the planes; refer to Figure-2.

Figure-1. Plane

Figure-2. Planes in Pathfinder

- Select the plane on which you want to create sketch.

Relation between sketch, plane and 3D model

Sketch has a direct relationship with plane and the outcome which is generally a **3D** model; refer to Figure-3. In this figure, rectangle is created on the XY plane which is also called **Top** plane. The circle is created on the YZ plane which is also called **Right** plane. The polygon is created on the XZ plane which is also called **Front** plane. In a 3D model, the geometry seen from the Top view should be drawn on the **TOP** plane. Similarly, geometry seen from the Right view should be drawn on **Right** plane and geometry seen from the Front view should be drawn on the **Front** plane.

Figure-3. Sketches created on different planes

STARTING SKETCH

In Solid Edge, sketching can be performed in **Synchronous** and **Ordered** environments. To change from Synchronous to Ordered environment, right-click in the empty area of drawing area. A shortcut menu will be displayed; refer to Figure-4.

Figure-4. Right click shortcut menu

- Select the **Transition to Ordered** option from the shortcut menu to activate ordered sketching.

Sketching Environment and Tools

In this section, we will discuss different type of sketching tools which are used in sketch creation.

Synchronous Sketching Environment

The tools to perform sketching in Synchronous environment are available in **Home** tab of **Ribbon**; refer to Figure-5. You can directly use these tools and later select the plane where you want to create sketch entities.

Figure-5. Synchronous Sketch Environment

Ordered Sketching Environment

You need to switch to Ordered environment first before using the sketching tools of Ordered environment. The procedure to switch environments has been discussed earlier. The Ordered environment will be displayed as shown in Figure-6.

Figure-6. Ordered Sketching Environment

The procedure to create sketch in Ordered environment is given next.

- Click on the **Sketch** tool from the **Sketch** panel in the **Home** tab of **Ribbon**. The **Sketch** command box will be displayed and you will be asked to select a sketching plane; refer to Figure-7.

Figure-7. Sketch command box

- Select desired plane from the drawing area or **PathFinder**. The tools to create sketch will be displayed along with **Line and Arc** dialog box.

VIEW TOOLBAR

The **View Toolbar** contains tools to change the view and orientation of the model; refer to Figure-8.

Figure-8. View Toolbar

1. **Command Finder** : The **Command Finder** tool is used to search tools and options in Solid Edge. For example, if you can not find **Line** tool then type **Line** in **Command Finder** and press **ENTER**. The tools with Line in their name will be displayed; refer to Figure-9. Hovering the cursor on option in the list will show its location in user interface. Clicking the tool/option from the list will activate it.

Figure-9. Command Finder dialog box

2. **Zoom Area** : The **Zoom Area** tool is used to display a specific area in the viewport zoomed to the full extent. To use this tool, click on it. The cursor will change to a zoom box selection cursor and you will be asked to create a boundary box surrounding the entities which you want to zoom in. Click to specify the starting point of the zoom box and then drag to the point till where you want to complete the zoom box. The area in the box will zoom automatically. Shortcut key for Zoom Area is **Alt + MMB**.

3. **Zoom** : The **Zoom** tool is used to Zoom-in and zoom-out of the drawing area. After clicking this tool, drag cursor upward to zoom-out and drag cursor downward to zoom-in. Shortcut key for Zoom is **Ctrl + MMB**.

4. **Fit** : The **Fit** tool is used to display all the objects created in the viewport. To use this tool, click on the tool once. The objects will automatically fit in the current viewport.

5. **Pan** : The **Pan** tool is used to changes the viewed area by dragging the entities. Shortcut key is **Shift + MMB**.

6. **Rotate** : The **Rotate** tool is used to rotate the entities 360 degree. After selecting this tool, drag the cursor in desired direction to rotate. Shortcut key is **MMB**.

7. **Sketch View** : The **Sketch View** tool is used to make the sketching plane parallel to screen. Shortcut key is **Ctrl + H**.

8. **View Orientation** : The **View Orientation** tool is used to change the view orientation of model. When you click on this tool, a toolbox will be displayed as shown in Figure-10. These buttons orient the model as per their shapes and annotations.

Figure-10. View orientation toolbox

9. **View Styles** : These tools are used to display the model in different styles like shaded, hidden, no hidden, and so on.

SKETCH CREATION TOOLS (SYNCHRONOUS ENVIRONMENT)

The tools to create sketch are available in **Home** tab of **Ribbon**. These tools are discussed next.

Line Tools

There are four tools in the **Line** drop-down for Synchronous as shown in Figure-11 and three line tools for Ordered as shown in Figure-12.

Figure-11. Line tools in Synchronous Environment

Figure-12. Line tools in Ordered Environment

Line

The **Line** tool is used to create each type of lines in creating sketch. The procedure to create line is discussed next.

* Click on the **Line** tool from the **Line** drop-down. The **Line and Arc** command box will be displayed as shown in Figure-13. Specify the origin point of line and drag the mouse. The preview is shown in Figure-14 & Figure-15.

Figure-13. Line and Arc command box

Figure-14. Line created

Figure-15. Line arc created

- Select the type of sketch entity that you want to create from **Type** area in the command box.
- Specify desired length of line in the **Length** edit box from **Properties** rollout.
- Specify desired orientation angle of line in the **Angle** edit box from **Properties** rollout. Zero degree angle is horizontal to the x-axis.
- Specify desired radius value and sweep angle in the **Radius** edit box and **Sweep** edit box of the command box, respectively. These options are active only in **Arc** mode.
- Select desired style of entity from **Style** drop-down in the **Style** rollout.
- Select desired color of entity from **Color** drop-down.
- Specify desired width of entity in **Width** edit box of **Style** rollout.
- Specify desired line type from **Type** drop-down in the **Style** rollout. Select **More** button from **Type** drop-down, the **Custom Line Type** dialog box will be displayed; refer to Figure-16. Specify desired parameters in the dialog box and click on the **Close** button.

Figure-16. Custom Line Type dialog box

- After setting desired parameters in the command box, click at desired location to specify the start point. You will be asked to specify end point of the line.
- If you move the cursor horizontally or vertically then respective constrain will be applied. Note that if there is another sketch entity then cursor will automatically snap to center, intersection, midpoint, or endpoints when moving cursor over the entity. Also, if you hold **SHIFT** key while moving the cursor then it move in increment of 15 degree.
- Click at desired location to specify end point of line. You can also specify the length and angle values in dynamic edit boxes displayed with line to define end point location. Press **TAB** to switch between dynamic edit boxes.
- After specifying end point of first line, you will be asked to specify end point of next line. Specify the end point parameters or press **ESC** to exit the tool. Note that you can press **A** key from Keyboard to switch to tangent arc creation and **L** key to switch back to line creation when this tool is active.

Point

Points are generally used as references for creating sketches entities. To create point, click on the **Point** tool from **Line** drop-down in the **Draw** panel of **Home** tab in the **Ribbon**. You will be asked to specify the location of point and **Point** command box will be displayed; refer to Figure-17. Click at desired location to place point. You can also specify coordinates in the **X** and **Y** edits boxes and press **Enter** to create points.

Figure-17. Point command box

Draw

The **Draw** tool is used to create precise geometry by sketching freehand with the help of pen, finger, or mouse. This tool is available in synchronous sketching environment only. On clicking the **Draw** tool from the **Line** drop-down in the **Home** tab of **Ribbon**, the **Draw** command box will be displayed as shown in Figure-18.

Figure-18. Draw command box

Creating Geometry

Select the **Create Geometry** button to dynamically create geometries. Create geometry by dragging your pen (if using digitizer), fingertip (if using touch screen device) or mouse.

Trim Geometry

Select the **Trim Geometry** button to trim geometry while dragging the cursor. Trim or delete by dragging your pen, finger, or mouse as discussed earlier.

Freesketch

When you select **FreeSketch** from **Line** drop-down, then you can draw rough sketch of basic entities like arcs, lines, circles, rectangles, and so on. Based on drag motion, the respective sketch entities will be created. This tool is available in both Synchronous and Ordered Sketching Environment. On selecting **FreeSketch** tool, the **FreeSketch** command box will be displayed as shown in Figure-19.

Figure-19. FreeSketch command box

- **Adjust On** : **Adjust On** button is active by default. It means when you draw any entities roughly. The software recognise the orientation of entities and convert it into the suitable entities.

- **Adjust Off** : When you select this button, the software do not recognise the orientation of entities and keep the orientation the same as you draw them.

- You should try to draw entities with **Adjust On** and **Adjust Off** buttons to find out which suits better for creating an entity.
- Press **ESC** to exit the tool.

Circle Tools

The **Circle** tool is used to draw circles. In **Circle** drop-down, there are five tools. Three tools are used to create circles and two tools are used to create ellipses; refer to Figure-20.

Figure-20. Circle drop down

Circle by Center Point

The **Circle by Center Point** tool is the widely used tool to draw circle. The procedure to use this tool is given next.

- Click on the **Circle by Center Point** tool from the **Circle** drop-down. The **Circle** command box will be displayed; refer to Figure-21.

Figure-21. Circle command box

- Specify diameter/radius of circle in the **Diameter/Radius** edit boxes of command box, respectively and click at desired location to place the circle; refer to Figure-22.

Figure-22. Circle by center point

Circle by 3 Points

The **Circle by 3 Points** tool is used to draw circle but it is different from using **Circle by Center Point** tool. As the name suggest, you need to specify three points to draw circle. After specifying first and second point of circle, the cursor gets attached to third point of circle. Click at desired location to specify third point of circle; refer to Figure-23.

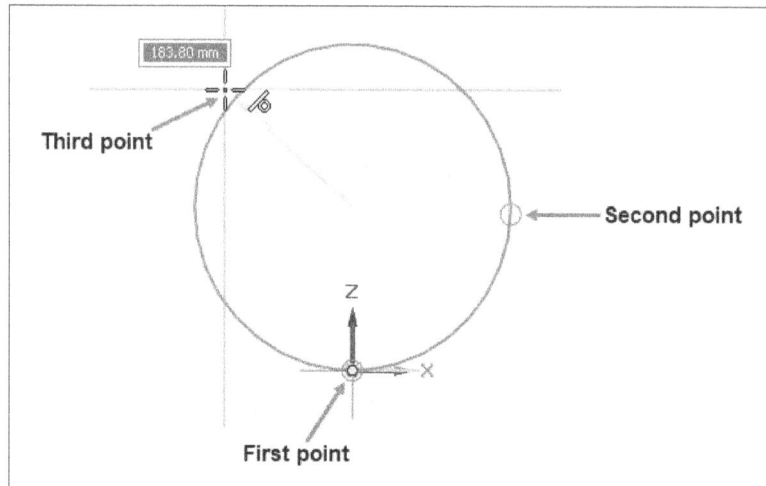

Figure-23. Circle by 3 points

Tangent Circle

The **Tangent Circle** tool is used to draw circle tangent to the existing one or two entities. To draw this type of circle, select the **Tangent Circle** tool from the **Circle** drop-down. To draw tangent circle, click on the two entities to which circle should be tangent or click on one entity and specify radius/diameter; refer to Figure-24.

Figure-24. Tangent circle

Ellipse Tools

There are two tools in **Circle** drop-down to create ellipses which are:-

• Ellipse by Center Point
• Ellipse by 3 Point

Ellipse by Center Point

The **Ellipse by Center Point** tool is used to draw ellipse by specifying center point, major axis, and minor axis. The procedure is given next.

- Click on the **Ellipse by Center Point** tool from **Circle** drop-down. The **Ellipse** command box will be displayed; refer to Figure-25. You will be asked to specify the centre of the ellipse.

Figure-25. Ellipse command box

- Click to specify the center. You will be asked to specify the radius along the major axis.
- Click to specify the radius of major and minor axes or specify the values in the **Primary** and **Secondary** edit boxes of the command box. You can also specify angle in the **Angle** edit box of command box. The ellipse will be created; refer to Figure-26.

Figure-26. Ellipse by center point

Ellipse by 3 Point

The **Ellipse by 3 Point** tool is used to create ellipse by three points. As the name suggest, you need to specify three points of ellipse. The procedure is given next.

- Click on the **Ellipse by 3 Point** tool from **Circle** drop-down. The **Ellipse by 3 Point** command box will be displayed similar to Figure-25.

- Specify the first, second, and third point of ellipse. When you specify the first and second point, the annotations are displayed on the ellipse. Select desired location to place third point of ellipse or specify parameters in the command box. You can also create ellipse by providing dimensions of primary and secondary axes in command box of ellipse. The preview is shown in Figure-27.

Figure-27. Ellipse by 3 point

Rectangle Tools

In **Rectangle** drop-down, there are four tools out of which three are used to draw rectangles and one of them is used for creating polygon. The tools of **Rectangle** drop-down are shown in Figure-28. These tools are discussed next.

Figure-28. Rectangle drop down

Rectangle by Center

The **Rectangle by Center** tool is used to draw rectangle by specifying center point and a diagonal corner point. The procedure is discussed next.

- Click on the **Rectangle by Center** tool from **Rectangle** drop-down. The **Rectangle** command box will be displayed; refer to Figure-29.

Figure-29. Rectangle command box

- Specify the center point and diagonal point of rectangle by clicking at desired locations. Refer to Figure-30.

Figure-30. Rectangle by center

- Press **ESC** to exit the tool.

Rectangle by 2 Points

The **Rectangle by 2 Points** tool is used to draw rectangle by specifying two diagonal corner points. The procedure is discuss next.

- Click on the **Rectangle by 2 Points** tool from **Rectangle** drop-down. The **Rectangle** command box will be displayed similar to Figure-31.

Figure-31. Rectangle by 2 points

Rectangle by 3 Points

The **Rectangle by 3 Points** tool is used to draw rectangle by three corner points. The procedure is discuss next.

- Click on the **Rectangle by 3 Points** from **Rectangle** drop-down. The **Rectangle** command box will be displayed similar to Figure-29.
- Specify the first, second, and third point of rectangle. Refer to Figure-32.

Figure-32. Rectangle by three points

Polygon by Center

The **Polygon by Center** tool is used to create regular polygon by defining number of sides and size of polygon. In Solid Edge, you can create polygon with number of sides ranging from 3 to 200.

- Click on the **Polygon by Center** tool from **Rectangle** drop-down. The **Polygon by Center** command box will be displayed; refer to Figure-33.

Figure-33. Polygon by Center command box

- **By Vertex** : Select the **By Vertex** button from the command box to create polygon using vertices; refer to Figure-34.

Figure-34. Polygon by vertex

- **By Midpoint** : Select the **By Midpoint** button from the command box to create polygon using mid points of sides of polygon; refer to Figure-35.

Figure-35. Polygon by midpoint

- Click at desired location to specify the center point of polygon and then specify the mid point or vertex point of polygon side.
- After creating polygon, press **ESC** to exit the tool.

Fillet Tools

There are two tools in **Fillet** drop-down to create fillets and chamfers; refer to Figure-36.

Figure-36. Fillet drop down

Fillet

The **Fillet** tool is used to create fillets or rounds at the corners formed by the intersection of two entities. A fillet is sometimes also referred to as a round. Generally, it is advised to use this tool after the rest of the sketch has been created, and when no significant changes are needed later, as it can otherwise distort the entire sketch or cause constraint overlaps. The procedure to create fillet is given next.

* Click on the **Fillet** tool from **Fillet** drop-down in **Draw** panel of **Home** tab in the **Ribbon**. The **Fillet** command box will be displayed; refer to Figure-37.

Figure-37. Fillet command box

* When **No trim** check box in **Fillet** command box is not selected then trim takes place while performing fillet. Refer to Figure-38.

Figure-38. Fillet trim

* When the **No Trim** check box is selected in the Fillet command box, no trimming occurs while creating the fillet. Refer to Figure-39.

Figure-39. Fillet no trim

Chamfer

The **Chamfer** tool is used to create a sloped edge on an object at the sharp intersection. You can define the chamfer by specifying the chamfer angle, the distance from the first side, and the distance from the second side. The procedure to create chamfer is given next.

* Click on the **Chamfer** tool from the **Fillet** drop down in the **Home** tab of **Ribbon**. The **Chamfer** command box will be displayed; refer to Figure-40.

Figure-40. Chamfer command box

* Specify desired parameters like chamfer angle, distance A, and distance B in respective edit boxes.
* Click on two intersecting lines, where you want to apply chamfer. The preview of chamfer will be displayed and you will be able to switch A and B sides by cursor movement.
* Click at desired side of centerline to define A and B sides of chamfer. The chamfer will be created; refer to Figure-41.

Figure-41. Chamfer applied to sketch

Arc Tools

There are three tools available in **Arc** drop-down of **Draw** panel in **Ribbon** to create different types of arcs; refer to Figure-42. These tools are discussed next.

Figure-42. Arc drop down

Tangent Arc

The **Tangent Arc** tool is used to create an arc starting tangentially to selected sketch entity. The procedure to use this tool is given next.

*   Click on the **Tangent Arc** tool from the **Arc** drop-down. The **Arc** command box will be displayed; refer to Figure-43.

Figure-43. Arc command box

*   Select the end point of the entity to which the arc should be tangent.
*   Click to specify the end point of the arc. Note that the tangent constrained is automatically applied at the arc. The preview of tangent arc is shown in Figure-44.

Figure-44. Tangent arc

Arc by 3 Points

The **Arc by 3 Points** tool is used to create an arc by specifying start point, end point, and an intermediate point in sequence. The procedure is given next.

- Click on the **Arc by 3 Points** tool from **Arc** drop-down. The **Arc** command box will be displayed as earlier.
- Click to specify the start point of the arc.
- Click to specify the end point of the arc.
- Click to specify a point on the arc to set the radius of the arc. The preview of arc by 3 points is shown in Figure-45.

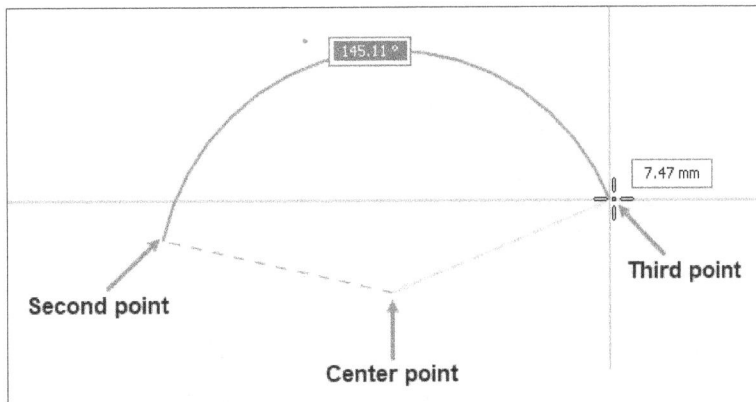

Figure-45. Arc by center point

Arc by Center Point

The **Arc by Center Point** tool is used to create an arc by specifying center point, start point and end point of the arc. The procedure is given next.

- Click on the **Arc by Center Point** tool from **Arc** drop-down. The **Arc** command box will be displayed as earlier.
- Click to specify the center point of the arc.
- Click to specify the start point of the arc.
- Click to specify the end point of the arc. The preview of arc by center point is shown in Figure-45. You can also specify the radius and sweep angles in command box after specifying start point.

Curve Tools

There are two tools available in **Curve** drop-down of **Draw** panel in **Ribbon**; refer to Figure-46. These tools are discussed next.

Figure-46. Curve drop down

Curve Tool

The **Curve** tool in **Curve** drop-down is used to draw spline curves. The procedure is given next.

- Click on the **Curve** tool from **Draw** panel. The **Curve** command box will be displayed; refer to Figure-47. You are asked to specify the points through which the curve should pass.

Figure-47. Curve command box

- One by one click in the drawing area to specify the points of the curve.
- Press **ESC** key from the keyboard to exit the tool. The curve will be created; refer to Figure-48. Select the recently created curve, the **Edit Definition** command box will be displayed with all the buttons activated; refer to Figure-49. Some of the important buttons in the command box are discussed next.

Figure-48. Curve created

Figure-49. Edit Definition command box

- Select **Add/Remove Points (Alt+Click)** check box from **Modify Curve** rollout to add or remove the points on the curve.
- Select desired option from **Edit** drop-down to edit the shape of curve using edit points or control vertex points. Select **Local** option to modify the section of curve by dragging respective point of curve. Select **Shape** option to modify shape of full curve by dragging a point of curve.
- Select desired option from **Relationship Mode** drop-down to control the shape of the curve using relationships, such as connects, tangency, and dimensions.
- Specify desired number of degrees for the curve from **Degree** drop-down to define smoothness of the curve.
- Specify desired tolerance value in the **Tolerance** edit box to be applied to the selected curve and click on the **Apply** button to simplify the curve by reducing the number of points in the curve. The number of edit points and control vertices will be displayed in the respective edit boxes of **Original** and **Current** area of the **Simplify Curve** rollout.

• Select desired check boxes from **Show/Hide** rollout to show the control polygon, curvature comb, edit points, and control vertex points. Select the **Close Curve** check box to draw the close curve; refer to Figure-50.

Figure-50. Close curve

• After specifying desired parameters, click on the **Close** button from the command box.

Conic Curve

The **Conic** tool in **Curve** drop-down is used to create conical arc of specified rho value. The procedure to use this tool is given next.

• Click on the **Conic** tool from the **Curve** drop-down in the **Draw** panel of the **Ribbon**. The **Conic** command box will be displayed; refer to Figure-51 and you will be asked to specify starting point for conic arc.

Figure-51. Conic command box

• Click in the graphics area to specify start point, end point, and mid point of the conic arc.
• If you want to create a hyperbola, parabola, or elliptical arc then select respective option from the **Shape** drop-down of **Properties** rollout.
• Specify desired value of rho in the **Rho** edit box of command box or dynamic input box; refer to Figure-52. The value can be 0.05 to 0.95.

Figure-52. Specifying rho for conic curve

*   Specify other parameters as discussed earlier and press **ESC** to exit the tool.

Trim Tools

There are two tools available in **Trim** drop-down of **Draw** panel in **Ribbon**; refer to Figure-53. These tools are discussed next.

Figure-53. Trim drop down

Trim Tool

The **Trim** tool is used to delete unwanted segment of an entity in sketch. Suppose you have created a sketch and small segment of an entity is extending out of the sketch then you can use this tool to remove that extra segment. The procedure to use this tool is given next.

*   Click on the **Trim** tool from the **Draw** panel in the **Home** tab of **Ribbon**. You will be asked to select the entity to be trimmed.
*   Click at desired side of entity to be removed by trimming or you can create a freehand curve to trim desired side of curve; refer to Figure-54.

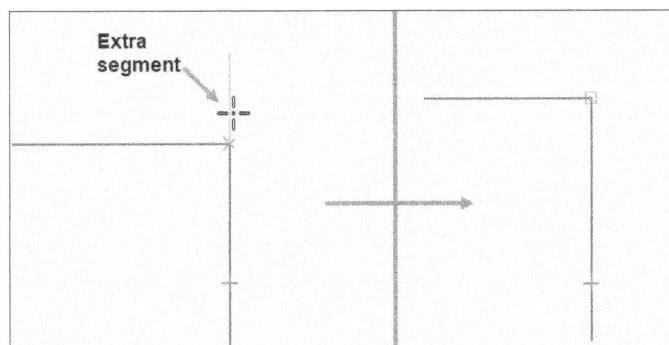

Figure-54. Trimming extra segment

Trim Corner Tool

The **Trim Corner** tool is used to trim or extend intersecting entities to form a corner, simultaneously. For example, you have created a sketch and you want their corners to be extend to next corner as shown in Figure-55. The procedure to use this tool is given next.

Figure-55. Trim corner

- Click on the **Trim Corner** tool from the **Draw** panel in the **Home** tab of **Ribbon**. You will be asked to select the first entity.
- Select the first entity and then hover the cursor on second entity. Preview of trim will be displayed; refer to Figure-56.

Figure-56. Trimming corners

- If the preview is as desired then click on the entity. The corner will be created by trimming.

Split drop-down

There are two tools available in **Split** drop-down; **Split** and **Extend to Next**; refer to Figure-57. These tools are discussed next.

Figure-57. Split drop down

Split

The **Split** tool is used to divide sketch entity at selected point. The procedure to use this tool is given next.

- Click on the **Split** tool from the **Split** drop-down in the **Draw** panel of **Home** tab in the **Ribbon**. You will be asked to select the sketch entity to be split.
- Select the object which you want to split. An arrow will be displayed showing the direction of split.
- Click at desired location on sketch entity where you want to split the entity. A round circle is created on the object that indicates that the object has split; refer to Figure-58. The number of circles show number of splits created. Press **ESC** to exit the tool.

Figure-58. Splitting line

Extend To Next

The **Extend to Next** tool is used to extend the line to the next object. If you have drawn a line but there is certain gap between it then you can use this tool to fill the gap. The procedure to use this tool is given next.

- Click on the **Extend to Next** tool from the **Split** drop-down in the **Draw** panel of **Home** tab in the **Ribbon**. You will be asked to select entity to be extended.
- Hover the cursor at desired side of sketch entity. Preview of extended sketch will be displayed; refer to Figure-59. If the sketch extension is as desired then click on the entity to create extended entity. Press **ESC** to exit the tool.

Figure-59. Extend to next

Offset Tools

There are two tools available in the **Offset** drop-down, **Offset** and **Symmetric Offset**; refer to Figure-60. These tools are discussed next.

Figure-60. Offset drop down

Offset

The **Offset** tool is used to create copy of the selected entities at specified distance from them. If you are user of **Autocad** then you would be quite familiar with this tool. The procedure to use this tool is given next.

*   Click on the **Offset** tool from the **Offset** drop-down in the **Draw** panel of **Home** tab in the **Ribbon**. The **Offset** command box will be displayed; refer to Figure-61.

Figure-61. Offset command box

*   Select desired option from the **Selection Type** drop-down to define whether you want to select single curve for offset or a chain of curves. Select the **Single** option from the drop-down to select single curves and select the **Chain** option from the drop-down to select chain of connected curves.
*   Specify the distance by which you want to offset curves in the **Distance** edit box of the command box.

- You can modify the line color, width, and type using the respective buttons in the command box as discussed earlier.
- Select the curves to be offset from the drawing area and click on the **Accept** button from the command box or right-click in the empty area. Preview of offset will be displayed with arrow showing direction of offset; refer to Figure-62.

Figure-62. Offset created

- If you want to modify the offset distance then you can do so by specifying value in the **Distance** edit box of command box. Click on desired side of original curves to create offset copy.
- Press **ESC** to exit the tool.

Symmetric Offset

The **Symmetric Offset** tool is used to create symmetric offset on both sides of selected entities. The procedure to use this tool is given next.

- Click on the **Symmetric Offset** tool from the **Offset** drop-down in the **Draw** panel of **Home** tab in the **Ribbon**. The **Symmetric Offset Options** dialog box will be displayed; refer to Figure-63.

Figure-63. Symmetric Offset Options dialog box

- Specify desired value of distance between two symmetric offset copies in the **Width** edit box.
- Select the **Apply radii if fillet radius = 0** check box to apply fillet at the corner even if not radius is applied in original.
- Specify desired value of fillet radius in the **Radius** edit box.

- By default, **Line** radio button is selected in the dialog box and so a line connects the open end points of offset copies. You can also specify fillet at the capped corners by specifying value in the **Cap fillet radius** edit box.
- If you want to create a semi circular cap then select the **Arc** radio button. Similarly, select the **Offset Arc** radio button to create semi-circular cap at offset distance equal to radius of arc.
- After setting desired parameters, click on the **OK** button from the dialog box. The **Symmetric Offset** command box will be displayed asking you to select the entities to be offset.
- Set desired parameters in the command box as discussed earlier and select the entities.
- Click on the **OK** button from the command box or right-click to create symmetric offset copy; refer to Figure-64.
- Press **ESC** to exit the tool.

Figure-64. Symmetric offset copy created

Move Tools

The tools in the **Move** drop-down are used to move and rotate selected sketch entities; refer to Figure-65. These tools are discussed next.

Figure-65. Move drop down

Move

The **Move** tool is used to move the entities from one position to another position. You can move the entities either by specifying coordinates or by dragging the entities. The procedure to use this tool is given next.

- Click on the **Move** tool from the **Move** drop-down in the **Draw** panel of **Home** tab in the **Ribbon**. The **Move** command box will be displayed; refer to Figure-66.

Figure-66. Move command box

- **Copy button Off** :- When **Copy** button is off, then the entities can be moved by selecting and dragging the element from one location to another. In this case, the original entities will move. Refer to Figure-67.

Figure-67. Copy button off

- **Copy button On** :- When the **Copy** button is **On**, then the original entities remain at their location and copies of entities are moved from one location to another by selecting and dragging the entities. Refer to Figure-68.

Figure-68. Copy button on

Rotate

The **Rotate** tool is available in the **Move** drop-down. This tool is used to rotate the entities by specified angle or by dragging.

- Click on the **Rotate** tool from the **Move** drop-down. The **Rotate** command box will be displayed as shown in Figure-69.

Figure-69. Rotate command box

- Select the entity that you want to rotate. You will asked to specify center point for rotation.
- Click at desired location/point to specify center point. You will be asked to specify start point for rotation.
- Specify the angle and the value at which the object will rotate.
- **Copy button Off** :- When the **Copy** button is **Off**, then the entities can be moved by selecting and rotating the element from one location to another. Refer to Figure-70.

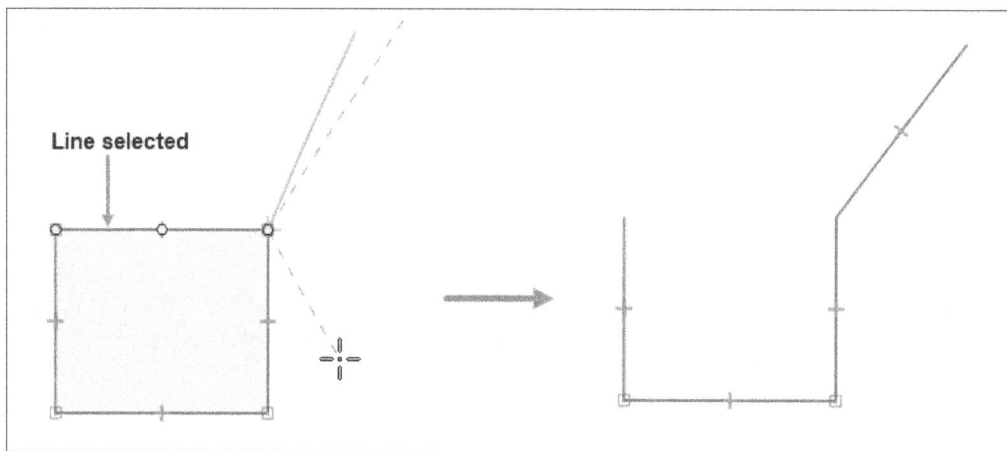

Figure-70. Copy button off

- **Copy button On** :- When the **Copy** button is **On**, then the entities are moved and copied from one location to another by selecting and rotating the entity. Refer to Figure-71.

Figure-71. Copy button on

Mirror Tools

In **Mirror** drop-down, there are three tools; **Mirror**, **Scale**, and **Stretch**; refer to Figure-72.

Figure-72. Mirror drop down

Mirror

The **Mirror** tool is used to create mirror copy of the selected entities with respect to a reference called mirror line. The procedure to create mirror entities is given next.

*   Click on the **Mirror** tool from **Mirror** drop-down in **Draw** panel of the **Home** tab in the **Ribbon**. The **Mirror** command box will be displayed; refer to Figure-73.

Figure-73. Mirror command box

*   Select the entity/entities that you want to mirror copy while holding **CTRL** key (if there are multiple entities).
*   Click on the reference line about which you want to mirror the entities. You can create multiple mirror copies by selecting different reference lines one by one.
*   **Copy button Off** :- When the **Copy** button is **Off**, then the entities can be mirror without being copied by selecting the entities and reference line. The base entity will be deleted after mirroring. Refer to Figure-74.

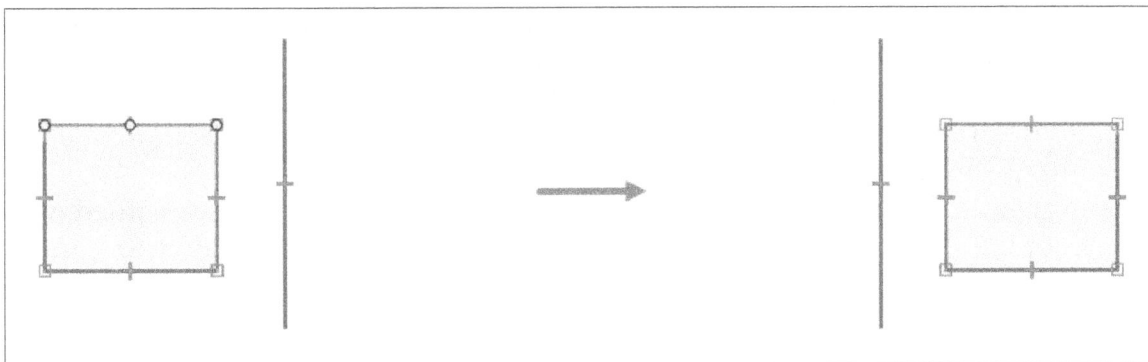

Figure-74. Copy button off

*   **Copy button On** :- When the **Copy** button is **On**, then the entities can be mirror copied to other location by selecting the entities and reference line. Refer to Figure-75.

Figure-75. Copy button on

Scale

The **Scale** tool is available in the **Mirror** drop-down. This tool is used to increase or decrease the size of an entity by specifying scale value. The procedure to use this tool is given next.

- Click on the **Scale** tool from **Mirror** drop-down. The **Scale** command box will be displayed as shown in Figure-76.

Figure-76. Scale command box

- Select the entities that you want to scale up or scale down. You can select multiple entities by pressing and holding the **CTRL** key during selection. After selecting entities, you will be asked to specify center point for scaling.
- Select desired base point about which you want to scale the entities.
- Specify desired scale value in the **Scale** edit box of command box or use cursor movement for scaling.
- **Copy button Off** :- When the **Copy** button is **Off**, then the entities are scaled without being copied. In that case, the original entities will be deleted. Refer to Figure-77.

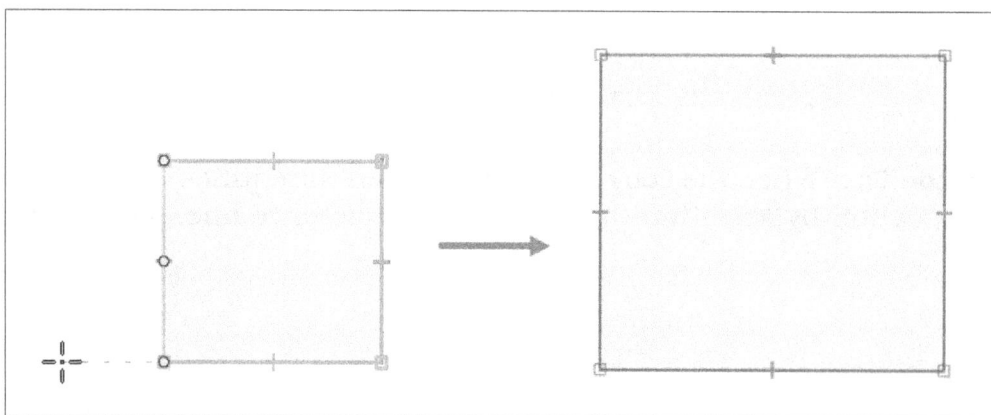

Figure-77. Copy button off

- **Copy button On** :- When the **Copy** button is **On**, then the entities are scaled and copied. In that case, the selected entities will remain unchanged. Refer to Figure-78.

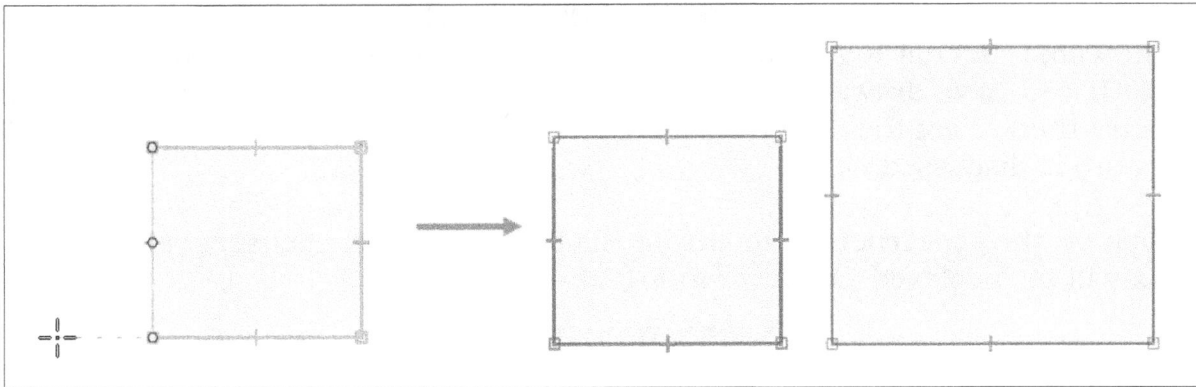

Figure-78. Copy button on

Stretch

The **Stretch** tool is available in the **Mirror** drop-down. This tool is used to stretch any sketched entity which is constrained at one side. The procedure to use this tool is given next.

- Click on the **Stretch** tool from **Mirror** drop-down in **Draw** panel of **Home** tab in the **Ribbon**. The **Stretch** command box will be displayed as shown in Figure-79.

Figure-79. Stretch command box

- By using the cross-rectangle selection, select the portion of entities that you want to stretch. You will be asked to select point for moving entities.
- Select the base point on the sketch. The selected portion of sketch will get attached to cursor.
- Move the cursor to desired position and click to stretch the entities.
- You can also stretch the entity by specifying desired values in the **X** and **Y** edit boxes of command box for selected point; refer to Figure-80.

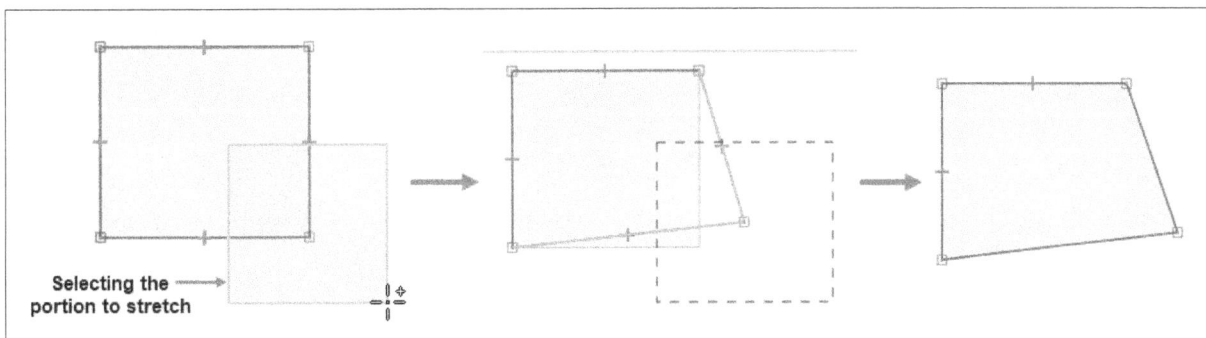

Figure-80. Stretching entities

Project to Sketch Tool

The **Project To Sketch** tool is used when you need the projection of any face, edge, or sketch entity in another sketch. This tool will be discussed in Chapter 3.

Construction Tool

The **Construction** tool is used to switch the entities between construction or profile. If you have drawn any sketch and you want to convert it in construction geometry then select the sketch entities and click on the **Construction** button. The procedure is discussed next.

- Click on the **Construction** tool from the **Draw** panel. The **Construction Command box** will be displayed as shown in Figure-81.

Figure-81. Construction command box

- Select the **Single** option from the **Select** drop-down to select single curve at a time for conversion. Select the **Chain** option to select consecutively connected curves as a chain.
- Select the entities from drawing area which you want to convert in construction geometry; refer to Figure-82 and Figure-83.

Figure-82. Single construction

Figure-83. Chain construction

Create As Construction Tool

The **Create As Construction** tool is used to draw construction entities. If you want to draw construction entities, you need to turn **"ON"** **Create As Construction** tool to create construction entities. If it is **"OFF"** then profile entities will be drawn. The procedure is discussed next.

- Click on the **Create As Construction** tool from the **Draw** panel in the **Home** tab of **Ribbon** to turn on the construction mode.
- Select desired sketch creation tool from the **Draw** panel and draw the entities. Refer to Figure-84.

Figure-84. Created as construction

Clean Sketch

The **Clean Sketch** tool is used to remove duplicate elements, clean B-spline curves, and generate sketch elements identical to imported data. The procedure to use this tool is given next.

- Click on the **Clean Sketch** tool. The **Clean Sketch** command box will be displayed; refer to Figure-85.

Figure-85. Clean Sketch command box

- Click on the **Clean Sketch Options** button from the command box to define clean sketch tool settings. The **Clean Sketch Options** dialog box will be displayed; refer to Figure-86.

Figure-86. Clean Sketch Options dialog box

- Select the **Clean B-Spline curves** check box to remove extra points from spline and repair it.
- Select the **Delete elements** radio button to delete identical elements, point elements and lines or arcs with size less than specified values.
- Select the **Move elements to** radio button and select desired layer from the next drop-down to define the layer on which identical elements, point elements and lines or arcs with size less than specified values will be moved.
- After setting desired parameters, click on the **OK** button. You will be asked to select the imported sketch entities to be cleaned.
- Select desired sketch entities and click on the **Finish** button from the command box to perform the operation.

Auto-Scale Sketch

The **Auto-Scale Sketch** tool is a toggle button used to scale whole sketch based on specified driving dimension. Generally, this button is kept toggled **ON** so that all the sketch entities are in visible range.

You will learn about other tools in this panel later in the book.

RELATE TOOLS

The tools in the **Relate** panel are used to apply different type of relations between entities; refer to Figure-87. Relations are used to constrain the sketch entities geometrically. These constraints are used to constrain the shape/position of sketch entities with respect to other entities. The constrains available in **Relate** panel are discussed next.

Figure-87. Relate tools

Connect

The **Connect** tool is used to connect the two entities at specified points. In other words, this tool makes two selected points coincident. To invoke this tool, select the **Connect** tool from **Relate** panel in the **Home** tab of **Ribbon**. Select the entities on their snap points where they will be connected; refer to Figure-88.

Figure-88. Connecting points

Horizontal / Vertical

Horizontal - The horizontal constraint makes one or more selected lines or center line to becomes horizontal. You can select external entity such as an edge, plane, axis, or sketch curve on external sketch that will act as a reference to apply this constraint. You can also make two or more points horizontally aligned using the Horizontal constraint. A point can be a sketch point, a center point, an endpoint, a control point of a spline, or an external entity such as origin, vertex, axis, or point in the external sketch. To apply this constraint, select the **Horizontal/Vertical** tools from **Relate** panel and select desired entities.

Vertical - The vertical constraint makes one or more selected lines or center lines to become vertical. You can force two or more point to become vertically aligned by using Vertical Constraint. To apply this constraint, select the **Horizontal/Vertical** tools from the relate groups; refer to Figure-89.

Figure-89. Horizontal and vertical constraints

Tangent

The **Tangent** constraint makes selected arc, circle, spline, or ellipse to become tangent to other arc, circle, spline, ellipse, line, or edge. The procedure to apply constraint is given next.

* Click on the **Tangent** button from the **Relate** panel in the **Home** tab of **Ribbon**. The **Tangent** command box will be displayed; refer to Figure-90.

Figure-90. Tangent command box

* Select **Tangent** option from **Type** drop-down to tangent the two entities. You will be asked to select the first entity.
* Select the first entity and then select second entity. You can change the position of entities by selecting and dragging them if the sketch is not fully constrained; refer to Figure-91.

Figure-91. Tangent constraint

* Select **Tangent+Equal Curvature** option from **Type** drop-down to tangent the two entities with an equal radius of curvature; refer to Figure-92.

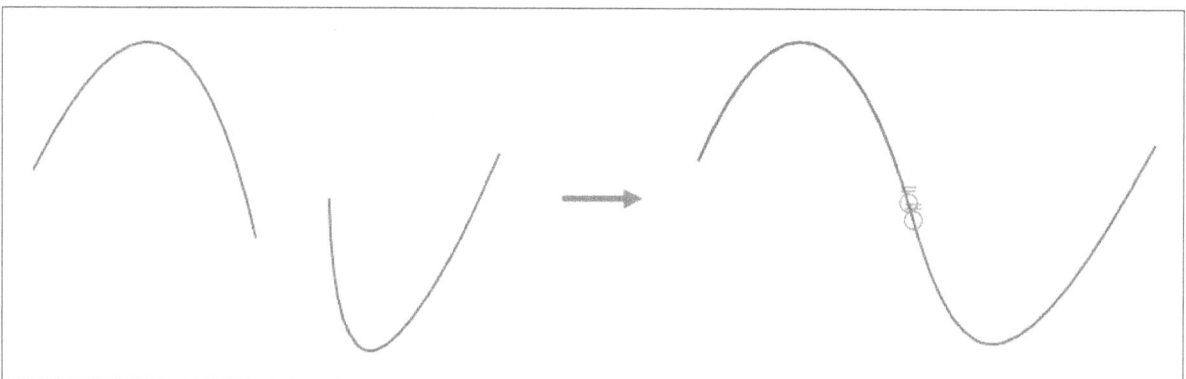

Figure-92. Tangent + Equal Curvature constraint

* Select **Parallel Tangent Vectors** option from **Type** drop-down to tangent the vectors parallel for the two entities; refer to Figure-93.

Figure-93. Parallel Tangent Vectors constraint

- Select **Parallel Tangent Vectors+Equal Curvature** option from **Type** drop-down to tangent the vectors parallel for the two entities with an equal radius of curvature; refer to Figure-94.

Figure-94. Parallel Tangent Vectors + Equal Curvature constraint

Parallel

The **Parallel** constraint makes selected lines parallel to each other. The procedure to apply this constraint is given next.

- Click on the **Parallel** tool from the **Relate** panel in the **Home** tab of **Ribbon**. You will be asked to select the entities.
- Select first line and then select second line. The constraint will be applied; refer to Figure-95.

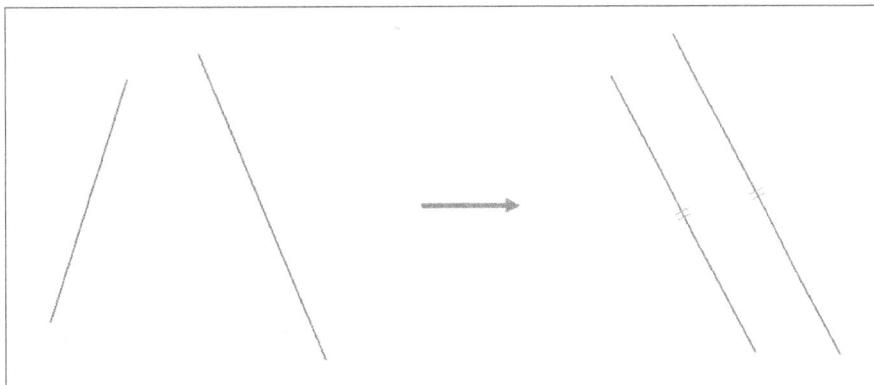

Figure-95. Parallel constraint

Equal

= The **Equal** constraint makes selected lines to have equal length and the selected arcs, circles, or arc and circle to have equal radii. The procedure to apply this constraint is given next.

- Click on the **Equal** tool from the **Relate** panel in the **Home** tab of **Ribbon**. You will be asked to select the entities.
- Select first entity and then select second entity. The constraint will be applied; refer to Figure-96.

Figure-96. Equal constraint

Symmetric

⌗ The **Symmetric** constraint makes two selected entities mirror copy of each other about selected center line. Note that the objects selected for making symmetric should be of same type. The procedure is given next.

- Click on **Symmetric** tool from the **Relate** panel in the **Home** tab of **Ribbon**. You will be asked to select the reference line.
- Select the reference line. You will be asked to select the first entity.
- Select first entity and then select second entity. The constraint will be applied; refer to Figure-97.

Figure-97. Symmetric constraint

Concentric

◎ The **Concentric** constraint makes selected arc or circle to share the same center point with other arc, circle, point, vertex or circular edge. The procedure to use this constraint is given next.

- Click on the **Concentric** tool from the **Relate** panel in the **Home** tab of **Ribbon**.
- Select the first entity and then select the second entity to be made concentric. The constraint will be applied; refer to Figure-98.

Figure-98. Concentric constraint

Perpendicular

The **Perpendicular** constraint is used to make one entity perpendicular to another. It can make line perpendicular to circle, ellipse, arc, or other entities. The procedure is given next.

- Click on the **Perpendicular** tool from **Relate** panel in the **Home** tab of **Ribbon**. You will be asked to select the entities.
- Select first entity which must be a line.
- Select second entity (line, arc, circle, ellipse, or other curve). The constraint will be applied; refer to Figure-99.

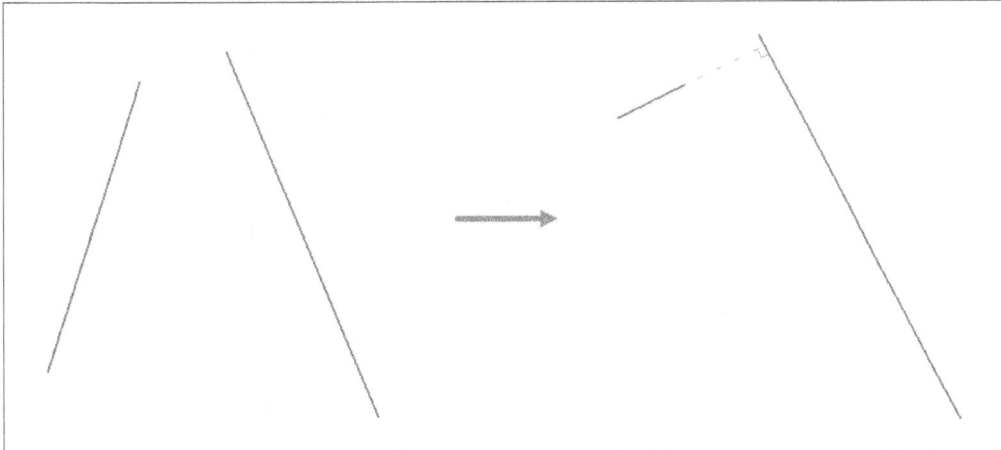

Figure-99. Perpendicular constraint

Collinear

The **Collinear** constraint makes the selected lines to lie on the same imaginary infinite line. The procedure is given next.

- Click on the **Collinear** from **Relate** panel in the **Home** tab of **Ribbon**. You will be asked to select the entities.
- Select first entity and then select the second entity. The constraint will be applied; refer to Figure-100.

Figure-100. Collinear constraint

Lock

The **Lock** tool is used to fix movement of entities and key points so that they are not modified accidently. Once a key point is locked, it cannot modified without unlocking it. A locked key point/entity can be moved by manipulation commands like Move/Rotate and will be fixed at the new location after manipulation. You can apply multiple lock command at the same time.

Rigid Set

The **Rigid Set** tool is available in **Relate** panel. This tool is used to create rigid sketch which means you can add multiple sketch entities to a rigid set and they will behave as single object. To procedure to apply this tool is given next.

*   Select the **Rigid Set** tool from the **Relate** group in the **Home** tab of **Ribbon**. The **Rigid Set** command box will be displayed; refer to Figure-101 and you will be asked to select the entities.

Figure-101. Rigid Set command box

*   Select multiple entities that you want to add in a single set and click **Accept** button from the command box. The entities will be added to the set; refer to Figure-102. Now, if you drag one element of the set then the complete set will move.

Figure-102. Rigid set

Maintain Relationships

The **Maintain Relationships** tool is used to activate or deactivate the automatic recognition of geometric constraints while creating the sketch. If this tool is turn **"ON"** then the application or software automatically detects the geometric constraint. It depend upon the user to turn **"ON"** or turn **"OFF"** this option. You can delete constraint by selecting it and pressing **DELETE** key any time.

Relationship Handles

The **Relationship Handles** tool is available in the **Relate** panel. This tool is used to display or hide constraints applied on the model. Note that hiding a constraint will not remove the constraint. If this tool is turn **"ON"**, it show the geometric constraints which are applied to the sketch. Refer to Figure-103.

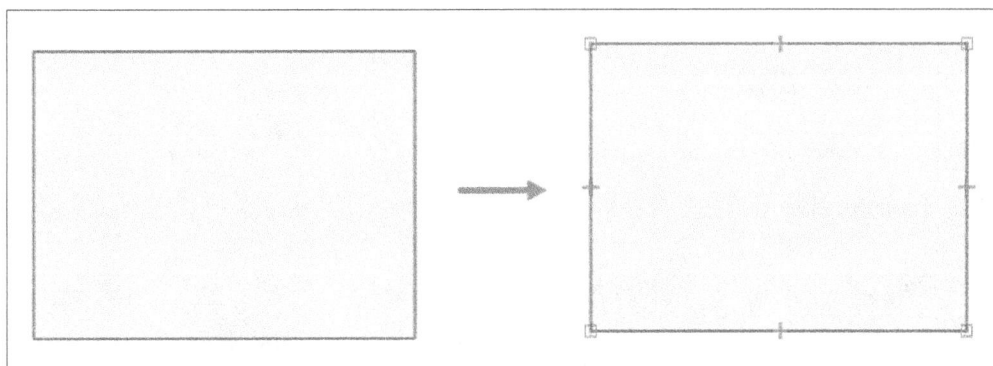

Figure-103. Relationship handle

Relationship Assistant

The **Relationship Assistant** tool is used to apply dimensions to selected entities with respect to selected horizontal and vertical references. The procedure to use this tool is given next.

- Click on the **Relationship Assistant** tool from the **Relate** panel in the **Home** tab of **Ribbon**. You will be asked to select the entities to be dimensioned and **Relationship Assistant** command box will be displayed; refer to Figure-104.

Figure-104. Relationship Assistant command box

- One by one click on the elements that you want to be dimensioned and click **Accept** button from the command box. You will be asked to select horizontal dimension reference.
- Select desired line or edge to define horizontal reference. You will be asked to specify vertical reference.
- Select desired line to define vertical reference. The dimensions will be created automatically; refer to Figure-105.

Figure-105. Sketch dimensioned by relationship assistant

- Press **ESC** to exit the tool.

DIMENSIONS

Dimensions are used to specify size of the sketch entities. For example; specifying length of the line, specifying diameter of the circle, and so on. The tools to dimension sketch entities are available in the **Dimension** panel. These tools are discussed next.

Smart Dimension

The **Smart Dimension** tool is used to dimension various entities like line, arc, circle, and so on in the sketch. The procedure to use **Smart Dimension** tool is discussed next.

- Click on the **Smart Dimension** tool from the **Dimension** panel in the **Home** tab of **Ribbon**. The **Smart Dimension** command box will be displayed as shown in Figure-106.

Figure-106. Smart dimension command box

The procedure to dimension various entities using the **Smart Dimension** tool is given next.

Dimensioning Lines

- After selecting **Smart Dimension** tool, select the line from the middle. The dimension will get attached to cursor; refer to Figure-107.

Figure-107. Dimensioning a line

- Click at desired location to place the dimension. A dynamic input box will be displayed for specifying value of dimension.
- Enter desired value in input box if you want to change the length of line or click in the empty area to lock the displayed dimension.

Dimensioning Inclined Lines

- After selecting **Smart Dimension** tool, select the line from the middle. An inclined dimension will get attached to cursor.
- Click at desired location to place the dimension. A dynamic input box will be displayed for modifying the length of line; refer to Figure-108.

Figure-108. Dynamic input box for dimensioning

- Enter desired value or click in the empty area to lock the dimension.

Now, we need to define angle dimension for inclined line.

- Make sure the **Smart Dimension** tool is still active. Click on the inclined line and then reference line/edge from which you want to dimension angle; refer to Figure-109. You will be asked to select a key point for specifying 3 point angle.

Figure-109. Selecting lines for angle dimension

- Click at the point where reference line and inclined line intersects. The angle dimension will get attached to cursor; refer to Figure-110.

Figure-110. Placing inclined dimension

- Click at desired location to place dimension. A dynamic input box will be displayed for modifying the dimension.
- Enter desired value in the input box. The dimension will be applied.

Dimensioning a Circle

- Click on the **Smart Dimension** tool from the **Dimension** panel in the **Home** tab of **Ribbon** if tool is not activated yet.
- Click on the circumference of circle. The diameter dimension will get attached to cursor. Note that depending on the location of cursor, the single arrow or double arrow line will be displayed for diameter dimensioning; refer to Figure-111.

Figure-111. Dimensioning a circle

- Click at desired location to place the dimension and enter desired value in the dynamic input box.

Dimensioning an Arc

- After activating the **Smart Dimension** tool, click on the circumference of arc. The radius dimension will get attached to cursor.
- Click at desired location to place the dimension and enter the value of radius in dynamic input box.
- To specify arc angle, click on the circumference of arc while **Smart Dimension** tool is active and then press **A** from keyboard. The arc angle dimension will be displayed; refer to Figure-112. If you again press **A** from keyboard then arc length dimension will be displayed; refer to Figure-113. You can also switch between various arc dimensioning styles by selecting the respective button from **Smart Dimension** command box; refer to Figure-114.

Figure-112. Angular arc dimension

Figure-113. Arc length dimension

Figure-114. Buttons for changing dimension style

- Click at desired location to place the dimension and specify dimension value in the dynamic input box.

Dimensioning a Chamfer

- After activating the **Smart Dimension** tool, select the inclined line of chamfer and place horizontal or vertical dimension to specify length of chamfer by holding **SHIFT** key from keyboard while placing the dimension; refer to Figure-115.

Figure-115. Placing horizontal dimension

- Enter desired value in the dynamic input box to specify length of chamfer.
- To specify angle, select the inclined line again and click on the **Smart Dimension - Angle** button from the command box or press **A** key. The angle dimension will get attached to cursor; refer to Figure-116.

Figure-116. Angle dimension

- Place the dimension at desired location and enter the angle value in dynamic input box.

Various options of the **Smart Dimension** dialog box are discussed next.

Format Options

- Select the **Enable dimension style mapping** check box to apply dimension style of selected template. Clear this check box if you want to use a different dimension style from the **Dimension Style** drop-down.
- Select the **Driving Dimension** radio button if the current dimension changes the size of related entity. Select the **Driven Dimension** radio button if current dimension does not change the size of entity but it is used as reference to report size or location.
- Specify desired precision value (digits after separator . in value) in the **Round Off** drop-down for dimension.
- Expand the **Advanced Options** node to define precision for dimension value and tolerance.

Properties Options

- By default, the **Automatic** option is selected in the **Orientation** drop-down of **Properties** section in the **Smart Dimension** command box so orientation of dimension is automatically decided based on selection. Select the **Horizontal/Vertical** option from the drop-down to create only horizontal or vertical dimensions. Select the **By 2 Points** option from the drop-down to define orientation of dimensions based on line connected selected two points on the entity. Select the **Use Dimension Axis** option from the drop-down to use selected line for orientation of dimensions. After selecting this option from the drop-down, click on the **Dimension Axis** button next to drop-down and select desired reference line. After setting this option, all new dimensions will be oriented according to selected reference; refer to Figure-117.

Figure-117. Dimension oriented using Dimension axis

- After selecting an entity from the graphics area to be dimensioned, you can select desired radio button from the **Properties** section of the command box to define the type of dimension to be created; refer to Figure-118. Click on **Show More Options** button in the command box to check all the available radio buttons.

Figure-118. Options to define dimension type

- Select desired option from the **Dimension Type** drop-down to define the type of machining tolerances to be applied to current dimension. Select the **Nominal** option from the drop-down if you want to create dimension without tolerance. Select the **Unit Tolerance** option from the drop-down if you want to specify numerical values of upper and lower tolerances. Select the **Alpha Tolerance** option from the drop-down if you want to specify tolerance in alphanumerical value. Select the **Class** option from the drop-down to apply hole-shaft class tolerances. Select the Limit option from the drop-down to specify upper and lower limits of dimension. Select the **Basic** option from the drop-down if you do not want to define numerical tolerance for the dimension and leave the tolerance value to GD&T symbols. Select the **Reference** option from the drop-down if you want to create a reference dimension. Select the **Feature Callout** option from the drop-down if you want to create a callout using GD&T symbol. Select the **Blank** option from the drop-down if you create dimension lines without value. After selecting desired option from the drop-down, specify desired values in the **Tolerance** section of the dialog box.
- Select the **Inspection Dimension** check box to make current dimension as inspection dimension.
- Select the **Prefix** check box to apply desired text before dimension.

Hybrid Annotation

The options in the **Hybrid Annotation** section of the dialog box are used to attach Geometric Dimensioning and Tolerance symbols in the dimension. Various options of this section are discussed next.

- Select the **Feature Control Frame** check box to apply desired GD&T symbol to current dimension. After selecting check box, click on the **Hybrid Feature Control Frame** button next to check box. The **Feature Control Frame Properties** dialog box will be displayed. The options of this dialog box are discussed later in the book.
- Set the other parameters as discussed earlier. After creating dimensions, press **ESC** to exit the tool.

Creating Distance Between Dimension

The **Distance Between** tool is used to give linear distance between the elements or key points. This tool is available in the **Dimension** panel. Click on the **Distance Between** tool, the **Distance Between** command box will be displayed which is same as **Smart Dimension** command box. Select the edge and corresponding edges to give dimension and click at desired location to place the dimension; refer to Figure-119. You can specify the dimension value in the same way as discussed earlier.

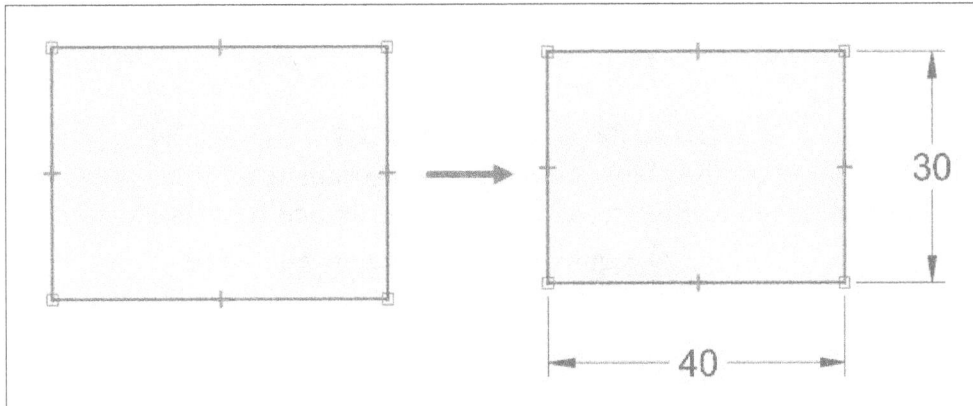

Figure-119. Distance between

Dimensioning an Ellipse

- After activating **Distance Between** tool, one by one select two opposite quadrant points of the ellipse. The dimension will get attached to cursor; refer to Figure-120.

Figure-120. Selecting points of ellipse

- Click at desired location to place the dimension and enter the value in dynamic input box.
- Similarly, you can dimension other side of ellipse.

Creating Angle Between Dimension

The **Angle Between** tool is used to dimension angle between elements or key points. On clicking this tool, the respective command box will be displayed. When you select the elements or key points, the dimension is placed in one of four quadrants. As you move the cursor, the dimension dynamically changes to another quadrant. You can place angular dimensions in stacked or chained dimension groups; refer to Figure-121.

Figure-121. Angle between

Coordinate Dimension Tools

There are two tools available in **Coordinate Dimension** drop-down of **Dimension** panel in the **Ribbon**; refer to Figure-122. These tools are discussed next.

Figure-122. Coordinate Dimension drop-down

Creating Coordinate Dimension

The **Coordinate Dimension** tool is used to dimension entities by their coordinates relative to selected reference origin. The procedure to create coordinate dimensions is given next.

- Click on the **Coordinate Dimension** tool from **Coordinate Dimension** drop-down in the **Dimension** panel of the **Home** tab in the **Ribbon**. The **Coordinate Dimension** command box will be diplayed which is same as **Smart Dimension** command box and you will be asked to select reference origin for dimensioning.
- Click at desired point for specifying origin. The **0** reference dimension will get attached to cursor; refer to Figure-123.

Figure-123. Reference dimension attached to cursor

- Click at desired location to place zero reference. You can create horizontal reference or vertical reference based on defining the location.

- One by one click on the lines which you want to dimension. The coordinate dimensions aligned to selected reference dimension will be displayed; refer to Figure-124. You can specify the value of dimension as discussed earlier. If you want to change the location of dimension then press **ESC**, select the dimension and drag it using the key point displayed on dimension. Note that by default all the dimensions will move by dragging. To move a single dimension, press **ALT** key while dragging.

Figure-124. Creating coordinate dimensions

- After you have specified the dimensions in one direction. Press **ESC** key and then restart the tool.
- Now, select the next zero references and repeat the procedure. Figure-125 shows a sketch dimensioned by coordinates.

Figure-125. Coordinate dimension created

Changing Origin of Coordinate Dimensions

The **Change Coordinate Origin** tool is used to change the origin dimension location for selected ordinate dimensions. The procedure to use this tool is given next.

- Click on the **Change Coordinate Origin** tool from the **Coordinate Dimension** dropdown in the **Dimension** panel of **Home** tab in **Ribbon**. You will be asked to select the dimension to be set as origin.
- Select the dimension from coordinate dimension set to set it as origin. Selected dimension will be set as **0** point for other coordinate dimensions; refer to Figure-126.

Figure-126. Changing origin of coordinate dimensions

Angular Coordinate Dimension

The **Angular Coordinate Dimension** tool is used to dimension the angle with coordinate values. The procedure is same as discussed earlier for coordinate dimensioning. Refer to Figure-127.

Figure-127. Angular coordinate dimension

Symmetric Diameter

The **Symmetric Diameter** tool is used to specify symmetric diameter between two element or key points from the center line. The procedure of this tool is discussed next.

- Click on the **Symmetric Diameter** tool from the **Dimension** panel in the **Home** tab of **Ribbon**. You will be asked to select the reference origin and the **Symmetric Diameter** command box will be displayed.
- Select the reference point or center line. You will be asked to select entities to be dimensioned.
- Select the entities to be dimensioned. The dimensions will be created; refer to Figure-128.

Figure-128. Symmetric diameter dimensions created

• Press **ESC** to exit the tool.

Maintain Alignment Set

By default, the **Maintain Alignment Set** toggle button is selected so you can move complete set of dimensions by moving one dimension of the set. The alignment set is automatically created when you create horizontal or vertical dimensions collinear; refer to Figure-129. Clear this toggle button if you do not want to create alignment sets for dimensions.

Figure-129. Dimensions in alignment set

Removing dimension from Alignment Set

The **Remove from Alignment Set** tool is used to remove selected dimension from the alignment set. The procedure to use this tool is given next.

• Click on the **Remove from Alignment Set** tool from the **Dimension** panel in **Home** tab of **Ribbon**. The **Remove from Alignment Set** command box will be displayed; refer to Figure-130.

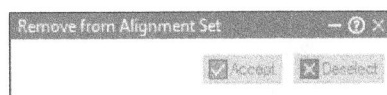

Figure-130. Remove from Alignment Set command box

- Select the dimensions to be removed from the alignment set and click on the **OK** button.

ORDERED TOOLS

When we transition from **Synchronous** to **Ordered** then some more tools are displayed for sketching in the **Ribbon**. These tools are discussed next.

The procedure to transition from **Synchronous** to **Ordered** has been discussed earlier. After changing from **Synchronous** to **Ordered** modeling, the **Ribbon** will be displayed as shown in Figure-131. In order to start a sketch, we have two type of option:-

Figure-131. Ordered Ribbon

1. Sketch (For 2D sketch)
2. 3D Sketch.

In this topic, we will discuss about 2D sketch. 3D sketch will be discussed in next chapter.

Starting 2D Sketch in Ordered Environment

In **Synchronous** environment, you can select sketching tools and directly draw on a plane. In **Ordered** environment, you need to select a sketching plane first then you can use the sketching tools to draw sketch. The procedure to start a 2D sketch is given next.

- Click on the **Sketch** tool from **Sketch** drop-down in the **Sketch** panel of **Home** tab in the **Ribbon**. The **Sketch** command box will be displayed; refer to Figure-132 and you will be asked to select a sketching plane.

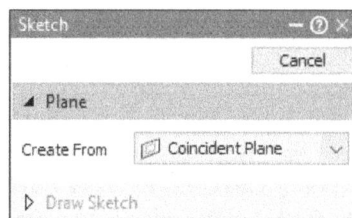

Figure-132. Sketch command box

- Select desired plane or face of solid. The selected sketching plane will become parallel to screen. The interface to draw 2D sketch will be activated and the **Ribbon** will be displayed as shown in Figure-133.

Figure-133. 2D Ribbon

After creating 2D sketch, click on the **Close Sketch** button from **Close** panel in the **Home** tab of **Ribbon** to exit sketching mode.

Tools In Draw Group

Most of the tools in **Draw** panel have been discussed earlier like Line, Rectangle, Circle, Arc, Curve, Fillet, and so on. Here, we will discuss rest of the tools available in the **Draw** panel like Convert to Curve, Fill, and so on.

Convert To Curve

The **Convert to Curve** tool is used to convert analytic geometry into a B-spline curves while using analytic creation tools. Suppose, you have drawn a line in a model but you want a curve in place of line for modifications. Instead of deleting the line and creating the curve using respective tool, you can turn line into curve by invoking **Convert to Curve** tool. The procedure is discussed next.

- Select the entities to be converted and click on the **Convert to Curve** tool from the **Draw** panel in the **Home** tab of **Ribbon**. The tool will be applied to the entities.
- Press **ESC** key to exit the tool.
- Now, if you select the entity then drag points will be displayed on entities and **Edit Definition** command box will be displayed as shown in Figure-134.

Figure-134. Converted to curve

Fill

The **Fill** tool is used to hatch closed region. The procedure is discussed next.

- Click on the **Fill** tool from the **Draw** panel in the **Home** tab of **Ribbon**. The **Fill** command box will be displayed; refer to Figure-135.

Figure-135. Fill command box

You can change color, angle, and spacing of grid lines using the command box.

- Click in the closed region where you want to apply hatching. The fill will be created; refer to Figure-136.

Figure-136. Hatch fill created

Frame Origin ▢

The **Frame Origin** tool is used to set default origin for frame design objects. You will learn about frame design objects later in chapter related to assembly. The procedure to specify location of frame origin is given next.

- Click on the **Frame Origin** tool from the **Draw** panel in the **Home** tab of the **Ribbon**. The **Frame Origin** command box will be displayed; refer to Figure-137.

Figure-137. Frame Origin command box

- Click on desired point from the sketch to define frame origin. You will be asked to specify default orientation of the frame.
- Select desired line segment to be used for orienting the frame and right-click in the drawing area to define frame origin.

Show Grid ▸⊞

The **Show Grid** tool is used to display grids in sketching environment; refer to Figure-138. These grid lines can be used as reference for creating entities.

Figure-138. Show Grid

Snap to Grid

The **Snap to Grid** tool is used to snap cursor to the grid. To invoke this tool, click on the **Snap to Grid** tool from the **Draw** panel in the **Home** tab of **Ribbon**; refer to Figure-139.

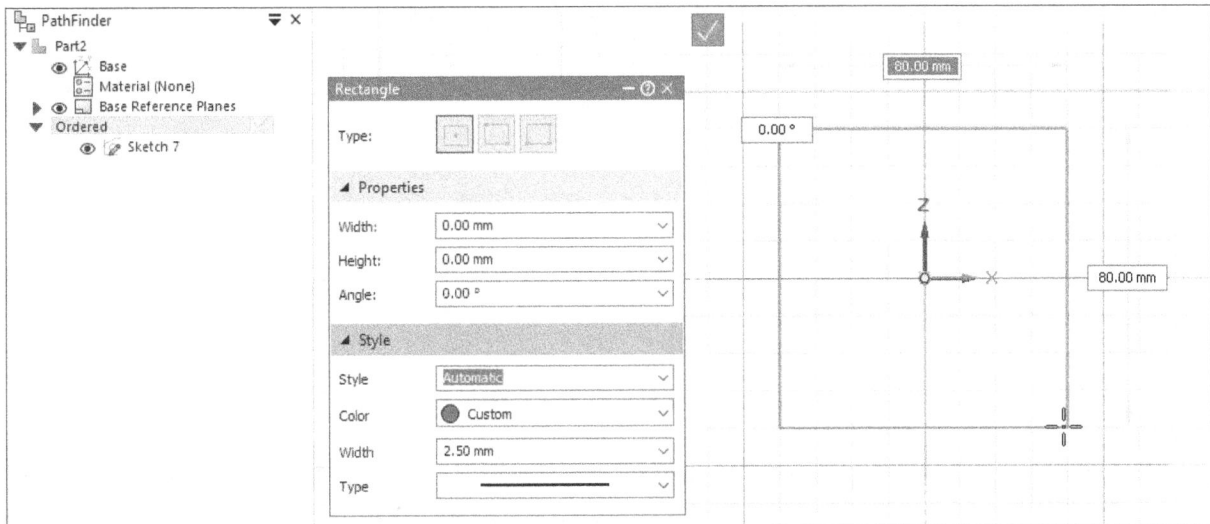

Figure-139. Snap to Grid

XY Key-in

Select the **XY Key-in** toggle button from **Draw** panel in **Home** tab of **Ribbon** to activate input boxes for specify X and Y coordinates when placing sketch objects; refer to Figure-140.

Figure-140. XY input boxes

Repositioning Origin

The **Reposition Origin** tool is used to change position of grid origin at specified location. The procedure to use this tool is given next.

* Click on the **Reposition Origin** tool from the **Draw** panel in the **Home** tab of **Ribbon**. You will be asked to specify location for new grid origin.
* Click at desired location to reposition the grid origin. The grids will move accordingly.

Zero Origin

The **Zero Origin** tool in **Draw** panel is used to place origin at origin of sheet/sketch. In simple words, this tool positions origin to its default location in sheet/sketch.

Grid Options

The **Grid Options** tool is used to set the grid appearance. The options of this dialog box are discussed next.

- Click on the **Grid Options** tool from the **Draw** panel in the **Home** tab of **Ribbon**. The **Grid Options** dialog box will be displayed; refer to Figure-141.

Figure-141. Grid Options dialog box

- Select the **Show readouts** check box to display coordinates while creating sketch entities.
- Select the **Show grid** check box and then select desired radio button below the check box to display lines or points of grid.
- Select the **Show alignment lines** check box to display horizontal and vertical lines on the cursor for alignment while creating sketch entities.
- Select the **Snap to grid** check box to automatically snap cursor to grid and then select desired radio button below the check box to define whether you want to snap cursor to grid lines or grid points.
- Select the **Enable key-ins (X,Y)** check box to display the **X** and **Y** coordinates while creating sketch entities; refer to Figure-142.

Figure-142. Displaying coordinates

- Specify desired value in **Angle** edit box to rotate grid lines to specified angle.
- Specify desired value in **Major line spacing** edit box to define gap between two consecutive major grid lines.
- Specify desired value in **Minor spaces per major** edit box to define how many minor grid lines are to be created between two consecutive major lines. You can specify value from minimum 1 to maximum 255 in this edit box.
- Select desired colors for grid lines in the **Major line color** and **Minor line color** drop-downs.
- After setting desired parameters, click on the **OK** button to apply settings.

Symmetry Axis 🔳

The **Symmetry Axis** button in the **Relate** panel of **Home** tab is used to convert selected line into symmetry axis. After converting a line to symmetry axis if you use **Symmetric** button to apply the constraint then this converted line will be automatically used as symmetry axis.

IntelliSketch Tools

In this panel, you will find options used to apply constraints in sketch intelligently while creating the sketch. Note that using Intellisketch options, you can snap cursor to key points of other sketches and solid model edges as well. The options in this panel are discussed next.

IntelliSketch Options

The **IntelliSketch Options** tool is used to set parameters related to intellisketch interface. The procedure to use this tool is given next.

- Click on the **IntelliSketch Options** tool from the **IntelliSketch** panel in the **Home** tab of **Ribbon**. The **IntelliSketch** dialog box will be displayed; refer to Figure-143.

Figure-143. IntelliSketch dialog box

- Select the check boxes from the **Relationships** tab to activate respective snaps as discussed earlier.
- Click on the **Auto-Dimension** tab to specify parameters related to auto dimensioning; refer to Figure-144. The options in this tab have been discussed earlier.

Figure-144. Auto Dimension tab

- Click on the **Cursor** tab in the dialog box to set size of cursor and snap points in the drawing; refer to Figure-145.

Figure-145. Cursor tab

- After setting desired parameters in the dialog box, click on the **OK** button.

The usage of various IntelliSketch constraints are discussed next.

End point

Select the **End point** check box to enable display and snapping of end points of sketch entities while creating sketch; refer to Figure-146.

End points displayed

Figure-146. Endpoints displayed

Midpoint

Select the **Midpoint** check box to enable display and snapping of mid points of sketch during sketch creation; refer to Figure-147.

Figure-147. Midpoint displayed

Point on element

Select the **Point on element** check box to display and snap to points on sketch entities or edges of other solids; refer to Figure-148.

Figure-148. Point on element

Center point

The **Center point** check box is used to display and snap to center of the circle, arc, and other round entities; refer to Figure-149. Note that center point is displayed when you hover cursor over the curve.

Figure-149. Center point

Silhouette point

The **Silhouette point** check box is used to display and snap to silhouette points on the arc, circle, ellipse, and other curves; refer to Figure-150.

Figure-150. Silhouette point

Intersection point

The **Intersection point** check box is used to display and snap to intersection point of two entities; refer to Figure-151.

Figure-151. Intersection point

Edit Point

This check box is used to display and snap to the dynamic edit point of the curves while creating other sketch entities. Refer to Figure-152.

Figure-152. Edit points displayed on curve

Curve Control Vertex

The **Curve Control Vertex** tool is used to display and snap to the curve control vertices of curves while creating sketch; refer to Figure-153.

Figure-153. Curve control vertex

Parallel

The **Parallel** tool is used to automatically recognize when a line is parallel to earlier created line while creating sketch; refer to Figure-154.

Figure-154. Parallel constraint recognition

Perpendicular

The **Perpendicular** tool is used to recognize when a line or curve is perpendicular to another line/curve. Refer to Figure-155.

Figure-155. Perpendicular line recognition

Tangent

The **Tangent** check box is used to recognize when a line/curve is tangent to another curve while creating; refer to Figure-156.

Figure-156. Tangent line recognition

Horizontal or vertical ▯

The **Horizontal or vertical** check box is used to recognize when a line is either Horizontal or Vertical with respect to X axis of sketch. Refer to Figure-157.

Figure-157. Horizontal vertical line recognition

Relationship Colors

The **Relationship Colors** tool is used to display sketch element in colors that indicate relationship status like sketch defined and so on. The colors can be modified in the **Color** page of the **Solid Edge Options** dialog box. To show color indications for relationships, you need to turn "**ON**" this toggle button.

Alignment Indicator

The **Alignment Indicator** tool is used to indicate horizontal and vertical relationship while creating sketch entities.

Creating and Using Dimension Axis

The **Dimension Axis** tool is used to set reference for other dimensions in Ordered environment. The procedure to use this tool is given next.

* Click on the **Dimension Axis** tool from the **Dimension** panel in the **Home** tab of **Ribbon**. You will be asked to select key points or line for defining dimension axis.
* Select desired line or two key points. A green line will be displayed for an instance showing that reference dimension axis has been created.
* Now, click on the **Smart Dimension** or **Distance Between** tool to create dimensions. The respective command box will be displayed.

- Select the **Use Dimension Axis** tool from the **Smart Dimension - Orientation** drop-down in the **Properties** rollout of command box. Preview of dimension axis will be displayed; refer to Figure-158.

Figure-158. Use Dimension Axis option

- One by one select the entities to be dimensioned and place the dimensions at desired locations; refer to Figure-159.

Figure-159. Dimensions aligned to dimension axis

Activating Auto Dimensioning

The **Auto-Dimension** tool is used to toggle auto-dimensioning scheme in sketch. When this button is toggle **ON** then dimensions will be automatically created on specifying value while creating sketch entities. You will learn about other dimension toggle buttons later in this book.

Styles

The **Styles** tool is used to manage the styles of text, dimensions, hatching, and so on. The procedure to use this tool is discussed next.

- Click on the **Styles** tool from the **Dimension** panel in the **Home** tab of **Ribbon**. The **Style** dialog box will be displayed as shown in Figure-160.

Figure-160. Style dialog box

- Select desired object from the **Style type** list box whose properties are to be modified. The list of default available styles available for the object will be displayed in the **Styles** section of dialog box and details of style will be displayed in the **Description** area of dialog box.
- To modify properties of selected style, click on the **Modify** button from the dialog box. The related modify dialog box will be displayed. The procedures to modify styles of different objects are given next.

Modifying Style of Dimension

- After selecting a dimension style, click on the **Modify** button from the **Style** dialog box. The **Modify Dimension Style** dialog box will be displayed; refer to Figure-161.

Figure-161. Modify Dimension Style dialog box

- Specify desired name of style in the **Name** edit box. If the style is based on any other style then you can select it from the **Based on** drop-down. The properties of style selected in drop-down will be copied.

Units tab

- Click on the **Units** tab to modify unit parameters of the property. The options will be displayed as shown in Figure-162.

Figure-162. Units tab

- Select desired option from the **Units** drop-down to define unit for dimensions.
- In the **Round-off** drop-down, you need to specify up to what accuracy the dimension will be created. You can create a dimension up to 7th place after decimal.
- In the **Unit label** edit box, you can specify characters to be used as unit label.
- In the **Subunit label** edit box, you can specify the characters to be used for subunit label of dimension.
- In the **Fraction separator** edit box, you can specify up to 5 characters by which whole value will be separated from the fraction value in dimension. If no value is specified in this edit box then fraction will be separated by single space like 15 2/3.
- In the **Maximum subunits** edit box, you can specify number of subunits available in one primary unit. For example, if primary unit is Feet then you can specify 12 as subunit which are Inches.
- You can set the above discussed parameters in the **Linear**, **Linear Tolerance**, **Angular**, and **Angular Tolerance** areas of the dialog box.
- Select desired radio button from the **Round-up** area to define how dimensions will be rounded to nearest accuracy level.
- Select desired radio button from the **Delimiter** area to define decimal delimiter. You can select period, comma, or space as delimiter.
- In the **Zeroes** area, you can select **Leading** and **Trailing** check boxes to specify whether you want to add zeros before decimal and after decimal if there is no number before and after decimal, respectively.
- Select the **Zero inches for ft-in** check box from the **Zeroes** area to display zero if there is no inch value in the dimension.

Secondary Units tab

The options in the **Secondary Units** tab are used to specify properties of secondary unit if applied in dimension. The options in this tab are same as discussed for **Units** tab.

Text tab

The options in the **Text** tab are used to modify text font and other related parameters.

*   Click on the **Text** tab from the **Modify Dimension Style** dialog box. The options will be displayed as shown in Figure-163.

Figure-163. Text tab in Modify Dimension Style dialog box

*   Select desired font from the **Font** drop-down. The dimensions will be displayed in selected font.
*   Select desired option from the **Symbol font** drop-down to define font for symbols used in dimension.
*   Select desired option from the **Font style** drop-down to define the style of font.
*   In the **Font size** edit box, you can specify the size of dimension text.
*   Select desired option from the **Orientation** drop-down to define whether text will be written horizontal, vertical, parallel to dimension line, or perpendicular to dimension line.
*   Select desired option from the **Position** drop-down to define where to place the dimension.
*   Select the **Override pulled-out text** check box if you want to set different parameters for pulled-out text of dimension.
*   Specify desired value in the **Size** edit box to define reduction in size for tolerance text.

- Select desired option from the **Limit Arrangement** drop-down to define how limits will be placed for tolerance in dimension.
- Set desired position and alignment parameters for tolerance in the **Position** and **Align to** drop-downs.
- Select the **Use tolerance text size for combined tolerance values** check box to display upper and lower tolerance values as combined if they are same.
- Select the **Display degree symbol after numeric angular tolerance values** check box to display degree symbol after angular tolerance.
- Select the **Use 45 degree character** check box to display 45 degree value for chamfer.
- Select the **Use lower case multiplication symbol "x"** check box to add "x" between chamfer value and angle value.
- Select the **Inhibit display of 0.0 values for automatic fit tolerances** check box to not display 0.0 value in tolerance when fit tolerance is generated automatically.
- Similarly, set other parameters in the dialog box.

Lines and Coordinate tab

The options in the **Lines and Coordinate** tab are used to modify parameters related to lines and coordinates; refer to Figure-164.

Figure-164. Lines and Coordinate tab

- Select the **Connect** check box if you want dimensions lines to automatically extend up to terminators of dimension.
- Set desired option in the **Display** drop-down to define which side of dimension lines will be displayed for a given dimension.
- In the **Width** edit box, you can specify the width of dimension line.

- Set desired value in the **Stack pitch** edit box to define gap between two stacked dimensions.
- Set desired value in the **Initial stack distance** edit box to define gap between two innermost stacked dimensions.
- Set desired value in **Break line** edit box to define length of break line in case of stack dimensions.
- Select the **Apply Break Line gap** check box if stacked text is oriented horizontally and you want to apply gap between stacked dimensions.
- Select desired option from the **Center line type** drop-down to define what type of line will be used for showing centerline in the drawing/sketch.
- Similarly, you can specify the line extension, center mark size and other parameters in **Center Line/Mark** area.
- Select the **Alternate text positions** check box in **Diameter** area to place dimension text alternatively left and right when dimensions are stacked.
- Select the **Underline symbol and prefix** check box to underline symbols and prefixes in the dimension.
- The options in the **Projection Line** area are used to modify projection lines created for projected views in the drawing.
- The options in the **Coordinate** area are used to set parameters for displaying coordinate system in the drawing.

Spacing tab

The options in **Spacing** tab are used to modify various gaps in dimensions like text clearance gap from dimension line, vertical gap between dual dimensions, and so on; refer to Figure-165. Set desired values in the edit boxes to modify gaps.

Figure-165. Spacing tab

The other options in this dialog box will be discussed later in chapter related to drafting.

Increase PMI Font

The **Increase PMI Font** tool is used to increase the font size of all the text and dimensions in the drawing.

Decrease PMI Font

The **Decrease PMI Font** tool is used to decrease the font size of all the text and dimensions in the drawing.

PRACTICAL 1

Create the sketch as shown in Figure-166. Also, dimension the sketch as per the figure.

Figure-166. Practical 1

Steps to be performed

- Start a new part file and switch to Ordered environment of designing.
- Select a sketching plane and activate sketching mode.
- Create the sketch using **Line** tool.
- Apply the dimensions using the **Smart Dimension** tool.
- Save the file.

Starting Sketching Environment

- Start Solid Edge if not started already.
- Click on the **New** button from the **Application** menu and click on the **New** tool from the cascading menu. The **New** dialog box will be displayed; refer to Figure-167.

Figure-167. New Solid Edge document dialog box

- Select the **ANSI Metric** option from the Standard Templates list at the left in the dialog box and double-click on the **ansi metric part.par** template at the right in the dialog box. The part environment of Solid Edge will be displayed; refer to Figure-168. You can draw sketch in **Synchronous** or **Ordered** sketch environment as discussed earlier. In this tutorial, you will create the sketch in Ordered environment.

Figure-168. Solid Edge Application window

- Right-click in the empty area of application window and select the **Transition to Ordered** option from the shortcut menu displayed. The Ordered environment will become active.
- Click on the **Sketch** tool from the **Sketch** panel in the **Home** tab of the **Ribbon**. You will be asked to select a sketching plane.
- Expand the **Base Reference Planes** node from the **Path Finder** at the left in the application window and select the **Front (xz)** plane. You can also check the default planes by hovering cursor on the coordinate system. On selecting plane, various tools to create sketch will be displayed in the **Ribbon**.

- Click on the **Line** tool from the **Draw** panel in the **Home** tab of the **Ribbon**. The **Line** tool will become active and you will be asked to specify the start point of line.
- Specify the start point of line and draw a line of length **100** at **0** degree angle. Refer to Figure-169.

Figure-169. Starting creation line

- Move the cursor upwards and draw a line of **20** with **90** degree angle.
- Move the cursor left and draw a line of **33** with **180** degree angle.
- Move the cursor upwards and draw a line of **50** with **90** degree angle.
- Again, move the cursor left and draw a line of **34** with **180** degree angle.
- Move the cursor downwards and draw a line of **50** with **-90** degree angle or make equal to adjacent line.
- Move the cursor left and draw a line of **33** with **180** degree angle.
- Move the cursor downwards and draw a line of **20** with **-90** degree angle. After creating the sketch, press **ESC** to exit the tool.
- Note that sketch is not fully defined yet. To make sketch fully defined, we need to connect the middle point of line of **100** length with origin. Then the sketch will be fixed at its place and will become fully defined.
- Click on the **Connect** tool from the **Relate** panel in the **Home** tab of the **Ribbon**. You will be asked to select the entities to be connected.
- Select the middle point of the line and origin. The middle point of line will get connected with origin; refer to Figure-170.

Figure-170. Connecting mid point of line with origin

- Press **CTRL+S** from keyboard and save the file at desired location.

PRACTICAL 2

In this practical, we will create the sketch as shown in Figure-171.

Figure-171. Practical 2

Steps to be performed

- Start a new part file and switch to Synchronous environment of designing.
- Select a sketching plane and activate sketching mode.
- Create the sketch using **Line**, **Circle**, **Fillet**, and **Arc** tools.
- Apply the dimensions using the **Smart Dimension** tool.
- Save the file.

Starting New Document

- Start Solid Edge if not started already.
- Click on the **New** button from the **Application** menu and click on the **New** tool from the cascading menu. The **New** dialog box will be displayed.
- Select the **ANSI Metric** option from the Standard Templates list at the left in the dialog box and double-click on the **ansi metric part.par** template at the right in the dialog box. The part environment of Solid Edge will be displayed. You can draw sketch in **Synchronous** or **Ordered** sketch environment as discussed earlier. In this tutorial, we will create the sketch in synchronous environment which is default selected when starting a new document.

Creating circle

- Click on the **Circle by Center Point** tool from the **Circle** drop-down in the **Draw** panel of the **Sketching** tab in the **Ribbon**. You will be asked to specify the center point.
- Click on the origin to specify center point of circle. Specify the diameter of circle as **13**. Similarly, draw a circle with same center point of diameter **26**. Refer to Figure-172.

Figure-172. Creating circles

- Click on the **Sketch View** button from the **View** toolbar at the bottom right corner of the application window or press **CTRL + H** from keyboard to orient sketching plane parallel to screen.

Creating Lines

- Click on the **Line** tool from the **Line** drop-down in the **Draw** panel of **Sketching** tab in the **Ribbon**. You will be asked to specify the start point of the line.
- Select the **Silhouette** point of the circle as a start point; refer to Figure-173. Move the cursor downward and draw a line of **50** with **90** degree angle.

Figure-173. Creating line

- Move cursor towards right and draw a line of **20** with **00** degree angle.
- Move cursor downwards and draw a line of **10** with **90** degree angle.
- Move cursor towards left and draw a line of **33** with **180** degree angle.
- Click on the **Create as Construction** tool from the **Draw** panel in the **Ribbon** and draw a construction line upto center point of circle; refer to Figure-173. Press **ESC** to exit the tool. Deselect the **Create as Construction** tool to exit construction mode.

Creating Mirror

- Click on the **Mirror** tool from the **Mirror** drop-down in the **Draw** panel of the **Ribbon**. Select all the lines recently drawn except construction line while holding **CTRL** key. You will be asked to select line for mirror reference.
- Select the construction line as reference line for mirror. The mirror copy will be created; refer to Figure-174. Press **ESC** to exit the tool.

Figure-174. Mirror copy generated

Applying Fillet

- Click on the **Fillet** tool from the **Fillet** drop-down in the **Draw** panel of the **Ribbon**. You will be asked to select two elements between which fillet will be applied and the **Fillet Command box** will be displayed.

- Specify the radius of fillet as **5** in the **Radius** edit box of the **Command box**. Select the lines between which fillets are to be created; refer to Figure-175. Press **ESC** to exit the tool.

Figure-175. Fillet applied

Trim

- Click on the **Trim** tool from the **Draw** panel in the **Sketching** tab of **Ribbon** and trim the lower portion of circle with diameter **26**; refer to Figure-176.

Figure-176. Trim circle

Applying Constraints and Dimensions

- Click on the **Tangent** tool from the **Relate** panel in the **Sketching** tab of **Ribbon** to make the lines tangent to the trimmed circle so that the sketch should be fully defined.

- Click on the **Smart Dimension** tool from the **Dimension** panel in **Ribbon** and dimension all the entities as shown in Figure-177.

Figure-177. Tangent relation and dimensions applied

PRACTICE 1

In this practice session, you will create a sketch for the drawing shown in Figure-178.

Figure–178. Practice 1

PRACTICE 2

In this practice, you will create a sketch for drawing shown in Figure-179.

Figure-179. Practice 2

PRACTICE 3

In this practice session, you will create a sketch for the drawing given in Figure-180.

Figure-180. Practice 3

PRACTICE 4

In this practice session, you will create a sketch for the drawing given in Figure-181 when viewing from the direction shown in figure.

Figure-181. Practice 4

PRACTICE 5

In this practice session, you will create a sketch for the drawing given in Figure-182 when viewing from the direction shown in figure.

Figure-182. Practice 5

PRACTICE 6

In this practice session, you will create a sketch for the drawing given in Figure-183.

Figure-183. Practice 6

SELF ASSESSMENT

1. In engineering, what does "sketching" primarily refer to?

A. Drawing animals and scenery
B. Creating detailed portraits
C. Drawing geometrically accurate representations of real-world objects
D. Sketching in 3D software with a freehand approach

2. Which of the following tools is NOT commonly used for sketching in engineering software?

A. Pencil and ruler
B. PathFinder
C. Line tool
D. Command Finder

3. In Solid Edge, the plane on which a circle is typically drawn when viewed from the Right view is called the:

A. Top plane (XY plane)
B. Front plane (XZ plane)
C. Right plane (YZ plane)
D. Sketch plane

4. Which shortcut key combination is used to zoom into a specific area in Solid Edge?

A. Alt + MMB
B. Ctrl + MMB
C. Shift + MMB
D. MMB only

5. What does the "View Orientation" tool in Solid Edge allow you to do?

A. Change the style of the model display
B. Rotate the entities 360 degrees
C. Change the view orientation of the model
D. Zoom in and out of the drawing area

6. In Solid Edge's Circle drop-down menu, which tool requires specifying three points to create a circle?

A. Circle by Center Point
B. Circle by 3 Points
C. Tangent Circle
D. Circle by Diameter

7. In the Rectangle drop-down menu, which tool allows you to draw a rectangle by specifying its center and one corner point?

A. Rectangle by 2 Points
B. Rectangle by 3 Points
C. Rectangle by Center
D. Polygon by Center

8. What is the purpose of the "Adjust On" option in the FreeSketch tool in Solid Edge?

A. To turn off snapping for better manual control
B. To allow the software to recognize and adjust the orientation of sketch entities
C. To enable freehand drawing without corrections
D. To adjust the width and color of entities

9. The Fillet tool in Solid Edge is mainly used to:

A. Create a sloped edge at an intersection
B. Delete sketch entities
C. Draw freehand shapes with curves
D. Add rounded corners to intersections of two entities

10. What do you need to specify when using the "Arc by 3 Points" tool in Solid Edge?

A. Start point and end point only
B. Center point, radius, and angle
C. Start point, end point, and an intermediate point
D. Major and minor axis lengths

11. Which tool is used to draw spline curves in the Draw panel?
A) Conic Tool
B) Curve Tool
C) Trim Tool
D) Mirror Tool

12. In the Curve Tool, which option from the Edit drop-down allows modification of a section of the curve by dragging a point?
A) Global
B) Shape
C) Local
D) Degree

13. What is the purpose of the Conic Tool in the Curve drop-down?
A) To create a spline curve
B) To create a conical arc with a specified rho value
C) To delete unwanted sketch segments
D) To create a mirror copy of entities

14. Which tool can be used to delete an unwanted segment of an entity in a sketch?
A) Trim Tool
B) Split Tool
C) Mirror Tool
D) Conic Tool

15. The Trim Corner Tool is used to:
A) Remove duplicate sketch elements
B) Stretch an entity at a constrained end
C) Trim or extend intersecting entities to form a corner
D) Offset entities symmetrically

16. What does the Split Tool do?
A) Extends a line to the next object
B) Divides a sketch entity at a selected point
C) Rotates an entity around a center point
D) Scales an entity

17. Which tool from the Offset drop-down allows you to create a copy of selected entities at a specified distance on both sides?
A) Offset Tool
B) Symmetric Offset Tool
C) Move Tool
D) Rotate Tool

18. The Move tool's "Copy button On" function allows you to:
A) Delete the original entities
B) Rotate the entities to a new position
C) Move the original entities
D) Move and copy the entities

19. Which tool would you use to create a mirrored copy of an entity across a selected reference line?
A) Stretch Tool
B) Mirror Tool
C) Symmetric Tool
D) Scale Tool

20. What does the Scale Tool do when the Copy button is Off?
A) Copies the scaled entities
B) Deletes the scaled entities
C) Scales the selected entities without copying them
D) Creates symmetric offsets

21. The Symmetric Tool in the Relate panel is used to:
A) Make selected entities the same length
B) Make two entities mirror images of each other
C) Connect two entities at selected points
D) Make lines lie on the same line

22. In the Relate panel, which tool makes an arc or circle share the same center point with another entity?
A) Concentric Tool
B) Equal Tool
C) Horizontal Tool
D) Tangent Tool

23. Which tool from the Relate panel can be used to align a line perpendicular to an arc or circle?
A) Tangent Tool
B) Horizontal Tool
C) Parallel Tool
D) Perpendicular Tool

24. What is the function of the Collinear Tool in the Relate panel?
A) Align two lines on the same infinite line
B) Set two points at the same level
C) Create a mirror copy of an entity
D) Scale a line to match another

25. What does the Lock tool primarily do in a sketching environment?
A) Fixes entities to prevent accidental modification
B) Converts lines to curves
C) Fills closed regions with color
D) Creates rigid sets

26. Which tool in the Relate panel is used to group multiple sketch entities so they behave as a single object?
A) Maintain Relationships
B) Lock
C) Rigid Set
D) Convert to Curve

27. If the Maintain Relationships tool is turned "ON," what does it do?
A) Locks the entities
B) Displays the geometric constraints
C) Automatically detects geometric constraints
D) Hides the geometric constraints

28. Which tool allows you to display or hide constraints applied on the model?
A) Maintain Relationships
B) Relationship Handles
C) Relationship Assistant
D) Smart Dimension

29. Which tool in the Dimension panel allows you to specify dimensions for lines, arcs, circles, and other entities?
A) Rigid Set
B) Distance Between
C) Smart Dimension
D) Fill

30. The Symmetric Diameter tool is used to:
A) Set the grid origin
B) Specify symmetric diameter between two elements or key points
C) Dimension inclined lines
D) Remove dimensions from an alignment set

31. The Distance Between tool is primarily used to:
A) Measure angular distances between elements
B) Measure linear distances between elements or key points
C) Convert lines into curves
D) Maintain alignment of dimensions

32. To change the origin dimension location for selected ordinate dimensions, which tool should be used?
A) Change Coordinate Origin
B) Maintain Alignment Set
C) Reposition Origin
D) Zero Origin

33. When using the Smart Dimension tool, pressing the "A" key while selecting an arc's circumference will:
A) Activate alignment options
B) Toggle between radius and diameter dimensions
C) Toggle between arc angle and arc length dimensions
D) Lock the dimension

34. What is the purpose of the Grid Options tool?
A) Define alignment of dimensions
B) Set appearance and behavior of grid lines
C) Specify dimension tolerances
D) Apply symmetry axis

35. To create a sketch where dimensions are automatically added whenever a value is specified for an element, you would:
A) Enable the Auto-Dimension tool in IntelliSketch Options.
B) Select the Dimension Axis tool in the Home tab of Ribbon.
C) Use the Smart Dimension tool and manually add dimensions.
D) Apply the Relationship Colors to preview dimensioning values.

36. When setting up dimensions with the Modify Dimension Style dialog box, you would use the Alternate text positions check box in the Diameter area to:

A) Force text to align with a specified dimension line orientation.

B) Alternate text placement on stacked dimensions.

C) Add additional tolerance values to dimension text.

D) Place the dimension text only to the left of the dimension line.

37. If you want the sketch entities to appear in colors indicating defined or undefined constraints, which tool would you enable?

A) Alignment Indicator

B) IntelliSketch Options

C) Relationship Colors

D) Curve Control Vertex

Chapter 3

3D Sketch and Solid Modeling

Topics Covered

The major topics covered in this chapter are:

- *3D Sketch*
- *Solid Modeling tools*
- *Planes and Coordinate System*
- *Modifying tools*
- *Practical 1*
- *Practical 2*
- *Practice*

INTRODUCTION

In previous chapter, we have discussed about the 2D sketch and related tools. In this chapter, we will discuss about 3D sketch and modeling tools.

3D SKETCH

The sketch creation tools are same in both **2D** and **3D** but in 3D sketch, you are using all three coordinates **X, Y, Z** simultaneously for defining positions of entities. The procedure to create a 3D sketch is discussed next.

- Click on the **New 3D Sketch** tool from the **New Sketch** panel in the **3D Sketching** tab of **Ribbon** in case of **Synchronous** environment or click on the **3D Sketch** tool from the **Sketch** panel in the **Home** tab of **Ribbon** in case of **Ordered** environment. If you are using the **Synchronous** environment then a new sketch will be added in the **Path Finder** and you need to right-click on the sketch and activate it for editing. Here, we will discuss the procedure for **Ordered** environment.
- After selecting **3D Sketch** tool from the **Ribbon**, the tools in **Ribbon** will be displayed as shown in Figure-1.

Figure-1. 3D Sketch Ribbon

- Select desired sketch creation tool from the **3D Draw** panel in the **Home** tab of **Ribbon** and draw sketch as desired. Note that the sketch entity will be automatically aligned to nearest axis while drawing sketch. You can use the **View Cube** to change orientation while drawing sketch.
- If you want to create sketch entity on a plane or along an axis then select it from the **OrientXpress** after activating the tool; refer to Figure-2.

Figure-2. OrientXpress

- After selecting desired orientation, create the sketch as desired; refer to Figure-3.

Figure-3. 3D Sketch creation

- After creating sketch, click on the **Close 3D Sketch** button from the **Close** panel in the **Ribbon** to exit.

Tips:
- You can use 3D sketch tools to connect various edges of 3D model using lines, arcs, and so on.
- The relations available in the **3D Relate** panel are applicable on external sketches, edges, vertices, and faces of 3D models as well.
- Using the **Routing Path** tool, you can create guidelines for piping, electrical wiring, and other systems.

EXTRUDE

The **Extrude** tool is used to create a solid volume by adding height to the selected sketch. In other words, this tool adds material in the direction perpendicular to the plane of sketch while dragging boundaries of sketch. There are two ways to create extrude feature in Solid Edge: In first method, select the **Extrude** tool and then make sketch by selecting a plane. In second method, sketch is created first and then **Extrude** tool is activated. After activating tool, you need to select the sketch for creating extrude feature. The procedure to create extrude features by second method is given next.

- Create a 2D sketch with non intersecting curves and forming one close loop chain; refer to Figure-4. Note that there can be multiple closed regions inside the loop.

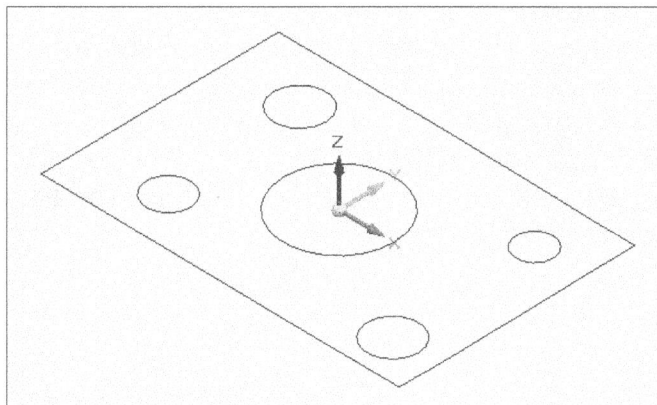

Figure-4. Sketch created for extrude feature

- Click on the **Extrude** tool from the **Solids** panel in the **Home** tab of **Ribbon**. You will be asked to select plane or planar face. Select the **Select from Sketch** option from the **Extrude - Create-From Options** drop-down in the **Sketch Step** cascading menu of **Extrude** command box as shown in Figure-5.

Figure-5. Extrude Create From Options drop down

- Select the sketch from the drawing area. Note that you can select multiple loops in the sketch by selecting them one by one. After selecting sketch, right-click in the drawing area or click on the **OK** button from the command box. The top face of extrude feature will get attached to cursor and the **Extrude** command box will be displayed as shown in Figure-6.

Figure-6. Extrude command box

- Specify desired distance in the **Distance** edit box of command box and left click to define side where you want to create the extrude feature with respect to sketching plane. The extrude feature will be created; refer to Figure-7. Note that if you want to modify any step of extrude creation then you can select respective button from the command box. After setting parameters, click on the **Finish** button from the command box to create the feature.

Figure-7. Extrude creation

- Select **Extrude - Automatic** button from the **Operation** area to add or remove the material of the model based on the material detection. Select **Extrude - Add** button to add the material of the model. Select **Extrude - Cut** button to remove the material of the model. Select **Extrude - Add Body** button to add a new body of the model.

Various options in the **Extent** rollout of **Extrude** command box are discussed next.

Extrude - One-sided Extent :- Select the **One-sided Extent** button to extend the feature in one direction only; refer to Figure-8.

Figure-8. Extrude One sided Extent

- **Extrude - Non-symmetric Extent :-** The **Non-symmetric Extent** button is used to extrude the sketch on both sides of sketching plane but you can specify different heights for both sides individually; refer to Figure-9.

Figure-9. Extrude Non Symmetric Extent

- **Extrude - Symmetric Extent :-** Select this button to extrude the sketch symmetrically on both sides of sketching plane; refer to Figure-10.

Figure-10. Extrude Symmetric Extent

- **Extrude - Through All :-** Select this button when you want to extrude the sketch through all of the solids earlier created. Note that if there is no face in perpendicular direction of sketching plane then this option will not generate any feature. Refer to Figure-11.

Figure-11. Extrude Through All

- **Extrude - Through Next :-** Select this button to extrude the sketch upto the next face/surface in direction perpendicular to sketching plane. Refer to Figure-12.

Figure-12. Extrude Through Next

- **Extrude - From/To Extent :-** This button is used to extrude the sketch from the selected reference location to another location in perpendicular direction. After selecting this button from command box, select the face to define "**From Surface**" point and then select another face to define "**To Surface**" point for extrude feature; refer to Figure-13.

Figure-13. Extrude From Or To Extent

• **Extrude - Finite Extent :-** Select the **Finite Extent** button when you want to extrude the sketch up to the finite extent means user will define the limit of extrude; refer to Figure-14.

Figure-14. Extrude Finite Extent

After specifying extent of extrude feature, expand the **Treatment** rollout from the command box. The options in command box will be displayed as shown in Figure-15.

Figure-15. Extrude command box with Treatment Step options

Various options in the **Extrude** command box at **Treatment** rollout are discussed next.

• Select the **Treatment Parameters Prompt** check box if you want the software to ask for specifying the treatment parameters when using this tool.
• **Extrude - No Treatment -** Select this toggle button if you do not want to apply any treatment like drafting or round.
• **Extrude - Draft :-** Select this button if you want to create taper side faces in the extrude feature; refer to Figure-16. After selecting this button, specify desired value of draft angle in the **Angle** edit box of the command box. If you want to change the direction of draft then select the **Flip Side** toggle button at the end in the command box.

Figure-16. Draft applied in extrude feature

- **Extrude - Crown :-** Select this button to create curved side faces of the extrude feature. On selecting this button, the parameters will be displayed along with arrow displaying direction of crown on the model; refer to Figure-17. Select desired option from the **Direction 1** drop-down to define how crown will be created. Select the **No Crown** option from the drop-down if you do not want to create crown on the side faces. Select the **Radius** option if you want to specify the radius of crown on side faces; refer to Figure-18. Select the **Radius and take-off** option from the drop-down if you want to specify start angle along with radius for creating crown; refer to Figure-19. Select the **Offset** option from the drop-down if you want to specify the offset distance from side walls up to which crown walls will span. You can use the **Offset and take-off** option from the drop-down to specify take-off value as well. You can flip the direction and curvature of crown by using the **Flip Side** and **Flip Curvature** toggle buttons, respectively from the dialog box.

Similarly, you can specified desired parameters in the Direction 2 area in the command box.

Figure-17. Crown Parameters dialog box

Figure-18. Radius parameter specified for crown

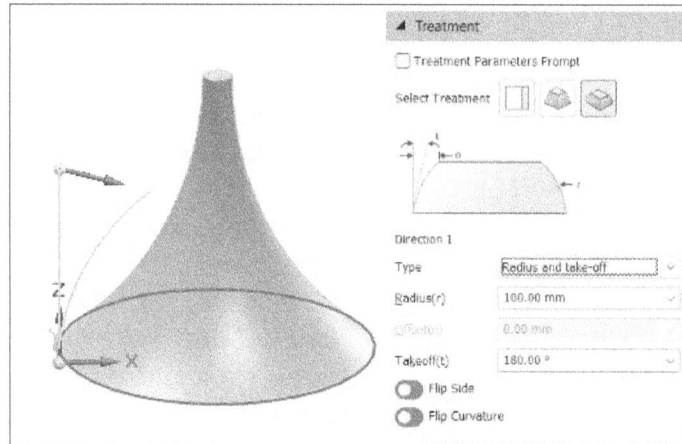

Figure-19. Radius and takeoff parameters specified for crown

- Right-click in the empty area to check the preview of extrude feature. If the preview of feature is created as desired then right-click again to create the feature.

Project To Sketch

The **Project To Sketch** tool is used when you need the projection of any face, edge, or sketch entity in another sketch. In this way, you can create the sketch entities from the projection of other features. The tool is available in the **Draw** group of Synchronous environment. The procedure is discussed next.

- Click on **Project To Sketch** tool from the **Draw** panel in the **Home** tab of **Ribbon** after activating sketch in Ordered environment or from the Synchronous environment. The **Project to Sketch** dialog box will be displayed; refer to Figure-20.

Figure-20. Project To Sketch dialog box

- Select the **Project with offset** check box to specify offset distance after projecting edges/faces/other sketch entities.
- Select the **Project internal face loops** check box to project internal loops as well when a face is selected for projection.
- Select the **Allow locate of peer assembly parts and sketches** toggle button to include entities of peer assemblies as well for projection.
- Select **Maintain associativity when projection geometry** toggle button to automatically keeps the link between the sketch projection and the source geometry.
- Select desired option from the **Selection Type** drop-down in the **Select** rollout to define selection filter. For example, if you want to select complete loops when clicking on one entity of loop then select the **Loop** option from the drop-down.
- Click on the **Select Element** button and select desired edge to display the name of the selected element.
- Set desired offset distance in the **Distance** edit box if needed.
- After setting desired parameters, select desired edges/faces/curves external to current sketch. After selection, right-click or click on the **OK** button from the command box to create projection; refer to Figure-21. Press **ESC** to exit the tool.

Figure-21. Project to Sketch creation

Copying or Moving Sketch Elements to New Sketch

The **Tear-Off** tool is used to copy/move sketch elements from one sketch to a new sketch. The procedure to use this tool is given next.

- Click on the **Tear-Off** tool from **Sketch** drop-down in the **Sketch** panel of **Home** tab in the **Ribbon**. The **Tear-Off Sketch** command box will be displayed; refer to Figure-22 and you will be asked to select the plane/face on which you want to copy/move the other sketch elements.

Figure-22. Tear Off Sketch command box

- Select desired face/plane where you want to create new sketch. You will be asked to select sketch elements to be copied/moved.
- Select the entities to be copied/moved. Preview of copying will be displayed; refer to Figure-23.

Figure-23. Preview of copying the entity

- Click on the **Options** button from the command box. The **Tear-Off Sketch Options** dialog box will be displayed; refer to Figure-24.

Figure-24. Tear Off Sketch Options dialog box

- Select the **Copy elements (Associative)** radio button if you want the copied elements to be modified automatically if original sketch elements are modified. Select the **Copy elements (Non-associative)** radio button if you want the copied elements to be not linked with original elements after copying. Select the **Move elements** radio button if you want the original elements to move at new plane selected for sketch. Note that you need to break the blocks before copying/moving by this method. (You will learn about blocks later). After setting parameters, click on the **OK** button from the **Tear-Off Sketch Options** dialog box.
- Right-click in the empty area to perform copy/move of elements. A new sketch will be created and you will be asked to specify name of the sketch in command box.
- Specify desired name in the **Name** edit box and click on the **Finish** button from the command box.
- Press **ESC** to exit the tool.

PERFORMING EXTRUDE CUT

The extrude cut operation is performed to remove material from the model. It can create simple hole and other cuts using the close loop sketch sections. The procedure to create a cut is discussed next.

- After activating the **Extrude** tool and select the **Cut** button from the **Operation** section in the **Extrude** dialog box displayed; refer to Figure-25. The options in the command box will be same as discussed earlier.

Figure-25. Cut option in Extrude command box

- Select plane or features face to create a sketch which you want to use for removing material. Note that if you are using plane to create sketch then the plane should be parallel to the face through which you want to cut the material.
- After creating sketch, click on the **Close Sketch** button from the **Close** panel in the **Home** tab of **Ribbon**. Preview of cut feature will be displayed with extent of feature attached to cursor and the command box will be displayed as shown in Figure-26.

Figure-26. Extrude command box with preview of cut feature

- Set desired parameters in the command box. The feature preview will be displayed; refer to Figure-27.

Figure-27. Preview of cut creation

Various options in the command box for Extent are discussed next.

- **Through All :-** If this option is chosen then the cut is created through all part. Refer to Figure-28.

Figure-28. Cut Through All

- **Through Next :-** If this option is selected then the cut is made through next face of the model; refer to Figure-29.

Figure-29. Cut Through Next

- **From/To Extent :-** Select this option if you want to define the start point and end point of cut feature by selecting faces, planes, or other references. Note that the selected references should be in parallel planes for defining extents. After selecting the **From/To Extent** button, select the start reference and then end reference. The preview of feature will be displayed; refer to Figure-30.

Figure-30. Cut From / To Extent

- **Finite Extent :-** If this option is selected then you need to define the distance up to which the cut will be created in the **Distance** edit box; refer to Figure-31.

Figure-31. Cut Finite Extent

The other options in the command box have been discussed earlier.

Before we further discuss the 3D modeling tools, it is important to understand the use of reference planes and Coordinate System in 3D Modeling.

CREATING PLANES AND COORDINATE SYSTEM

The tools in **Planes** panel are used to create planes and coordinate system; refer to Figure-32. Various tools in this panel are discussed next.

Figure-32. Planes tool

Creating Coincident Plane

The **Coincident Plane** tool is used to create plane coincident with selected face or reference plane. The procedure to use this tool is given next.

* Click on the **Coincident Plane** tool from the **Planes** panel of **Home** tab in the **Ribbon**. You will be asked to select face/reference plane at which new plane will be created.
* Select desired face/plane. The plane will be created and move handles will be displayed; refer to Figure-33.

Figure-33. Creating coincident plane ordered

* Using the handles, you can translate or rotate the plane as need. After changing orientation and position of plane, click in the empty area of screen. The plane will be created.

More Planes drop-down

The tools to create planes are available in **More Planes** drop-down. In **Synchronous** environment, only three tools are available in **More Planes** drop-down to create planes; refer to Figure-34. In **Ordered** environment, several other tools are available which are shown in Figure-35.

Figure-34. More Planes drop down in Synchronous environment

Figure-35. More Planes drop down in Ordered environment

Creating Parallel Plane

The **Parallel** tool in **More Planes** drop-down is used to create plane parallel to selected face or plane at specified distance. The procedure to use this tool is given next.

- Click on the **Parallel** tool from the **More Planes** drop-down in the **Planes** panel of **Home** tab in the **Ribbon** in **Ordered** environment. You will be asked to select a planar face or reference plane.
- Select desired planar face/plane. Preview of plane will be displayed with **Parallel** command box; refer to Figure-36.

Figure-36. Preview of parallel plane

- Specify desired value of distance in the **Distance** edit box of command box and press **ENTER**. You will be asked to define side on which plane will be created with respect to selected plane/face.
- Click on desired side to create the plane.

Creating Plane at Angle

The **Angled** tool in **More Planes** drop-down is used to create plane at an angle to selected face/reference plane. The procedure to use this tool is given next.

- Click on the **Angled** tool from the **More Planes** drop-down in the **Planes** panel of **Home** tab in the **Ribbon**. You will be asked to select a reference plane/face.
- Select desired face/plane with respect to which angle will be specified for new plane. You will be asked to select face/edge/plane to used as base for new plane.

- Select the edge (recommended), face, or plane intersecting with previous selected face/plane. You will be asked to click near the end point of selected edge to define orientation of plane.
- Click at desired side of selected edge. The plane will get attached to cursor and you will be asked to specify the value of angle; refer to Figure-37.

Figure-37. Specifying angle for plane

- Specify desired value in **Angle** edit box of command box and press **ENTER**. You will be asked to specify the side at which you want to create the plane.
- Click on desired side of selected face to create plane above or below the selected face.

Creating Perpendicular Plane

The **Perpendicular** tool in **More Planes** drop-down is used to create plane perpendicular to selected planar face/reference plane. The procedure to use this tool is given next.

- Click on the **Perpendicular** tool from the **More Planes** drop-down in the **Planes** panel of **Ribbon**. You will be asked to select a plane/planar face.
- Select desired planar face/plane. You will be asked to select the base of plane which can be face, edge, or plane.
- Select desired plane/edge/face. Rest of the procedure is same as discussed for plane at angle.

Similarly, you can use the other tools in **More Planes** drop-down. The **Coincident by Axis** tool is used to create plane coincident to selected intersecting planes or plane and edge. The **Normal to Curve** tool in **More Planes** drop-down is to create plane perpendicular to selected curve or edge. The **By 3 Points** tool is used to create a plane using 3 selected points; refer to Figure-38. The **Tangent** tool in **More Planes** drop-down is used to create plane tangent to selected curved face.

Figure-38. Creating plane using 3 points

Creating Mid Plane

The **Mid** tool in **More Planes** drop-down is used to create a reference plane exactly halfway between two selected parallel reference planes or planar faces. The procedure to use this tool is discussed next.

- Click on the **Mid** tool from the **More Planes** drop-down in the **Planes** panel of **Home** tab in the **Ribbon** of **Ordered** environment. You will be asked to select a planar face or reference plane.
- Select desired two parallel planar faces/planes. The plane will be created; refer to Figure-39.

Figure-39. Creating the mid plane

Creating Coordinate System (Ordered Environment)

The **Coordinate System** tool is used to create custom coordinate system at desired location. The procedure to use this tool is given next.

- Click on the **Coordinate System** tool from the **Planes** panel in the **Home** tab of **Ribbon**. The **Coordinate System Options** dialog box will be displayed; refer to Figure-40.

Figure-40. Coordinate System Options dialog box

- Select desired option from the dialog box to specify how the coordinate system's location and orientation will be defined. Select the **Key-in** radio button to create coordinate system in reference to another coordinate system. Select the **Geometry** radio button if you want to place the coordinate system at desired point and specify its orientation. We will discuss the **Geometry** option here, you can apply it to **Key-in** option as well.

- Select the **Geometry** radio button from the dialog box and click on the **OK** button. The **Coordinate System** command box will be displayed; refer to Figure-41 and you will be asked to specify the key point of placing coordinate system.

Figure-41. Coordinate System command box

- Click at desired location or specify the coordinates in **X**, **Y**, and **Z** edit boxes of command box to place the coordinate system. You will be asked to define the direction references.

- Select desired edge to define direction of **X** axis in coordinate system and press **ENTER**. An arrow will be displayed attached to cursor denoting the direction of **X** axis.

- Click on desired side of selected location to define the **X** axis direction. You will be asked to specify the **Y** axis direction. Set the **Y** axis direction in the same way.

- On specifying the direction, preview of coordinate system will be displayed; refer to Figure-42.

Figure-42. Preview of custom Coordinate System

- Click on the **Finish** button to create the coordinate system and press **ESC** to exit the tool.

REVOLVE

The **Revolve** tool is used to create revolve protrusion features. This tool is available in the **Solids** panel of **Home** tab in the **Ribbon**. The procedure to create a revolve is discussed next.

- Click on the **Revolve** tool from the **Solids** panel in the **Home** tab of **Ribbon**. The **Revolve Command box** will be displayed; refer to Figure-43 and you will be asked to select a plane for creating base sketch.

Figure-43. Revolve command box

- If you have a sketch earlier created then you can use it as discussed earlier or you can select a plane and create the sketch; refer to Figure-44.

Figure-44. Sketch created for revolve feature

- After creating sketch, click on the **Close Sketch** tool from the **Close** panel in the **Home** tab of **Ribbon**. The preview of feature will be displayed with **Revolve** command box; refer to Figure-45.

Figure-45. Preview of revolve feature

- If you want to create a full 360 revolve feature then select the **Revolve 360°** button from the command box. If you want to specify the angle span for feature then select the **Finite Extent** button and specify the value in **Angle** edit box of command box; refer to Figure-46.

Figure-46. Revolve creation

- After setting desired parameters, click on the **Finish** button from the command box or right-click on empty area to create the feature. Press **ESC** to exit the tool.

REVOLVE CUT

The revolve cut operation is performed to remove material from the model by revolving close loop sketch section about an axis. Click on the **Revolve - Cut** button from the **Operation** section in the **Revolve** dialog box to perform the operation; refer to Figure-47. The procedure is same as we have discussed for the **Revolve** tool. The difference is that in case of revolve, we add material and in case of revolve cut, we remove material from the model. Refer to Figure-48.

Figure-47. Cut option in Revolve command box

Figure-48. Revolve cut creation

HOLE TOOLS

There are three tools available in the **Hole** drop-down: **Hole**, **Thread**, and **Slot**; refer to Figure-49. These tools are discussed next.

Figure-49. Hole drop-down (Ordered Environment)

HOLE

The **Hole** tool is used to create holes that comply with the real machining tools available in market. Solid Edge has library of standard holes that can be created in the solid model. You can use this standard library or you can create a customized hole by using the tool. The procedure to use this tool is given next.

*   Click on **Hole** tool from **Hole** drop-down in the **Solids** panel of **Home** tab in the **Ribbon**. The **Hole** command box will be displayed; refer to Figure-50 and you will be asked to select plane/face for creating hole.
*   Select desired face/plane on which you want to create the hole. The sketching environment will be displayed along with **Hole Circle** command box; refer to Figure-51.

Figure-50. Hole command box

Figure-51. Hole Circle command box

*   **Hole Options :-** This button is used to select the type of hole. On clicking this button, the **Hole Options** dialog box will be displayed; refer to Figure-52. Using the options in this dialog box, you can define the shape and size of the holes. You can create five types of holes using the options in this dialog box. The methods to define properties of these holes are discussed next.

Figure-52. Hole Options dialog box

Creating Simple Hole

- Select the **Simple Hole** button from the dialog box. The dialog box will be displayed; refer to Figure-52.
- Select desired option from the **Standard** drop-down to define database standard of holes.
- Select desired option from the **Sub type** drop-down to define how hole size will be selected. There are three options in this drop-down viz. **Drill Size**, **Dowel**, and **General Screw Clearance**.
- Select desired option from the **Size** and **Fit** drop-downs to define size of holes.
- Select desired button from the **Hole extents** area to define the depth of holes. If you have selected the **Finite Extent** button then you can create V type bottom of holes by selecting the **V bottom angle** check box. After selecting the check box, specify desired angle value in the edit box below it.
- Specify desired diameter and depth of hole in the edit boxes displayed.
- Expand the **Chamfers** section of the dialog box and select desired check boxes to create start and end chamfers on the hole. Preview of chamfer will be displayed in the preview area of the dialog box.
- After setting desired parameters in the dialog box, click on the **OK** button from the dialog box. You will be asked to specify the location of centers of holes.
- Click at desired locations to specify the hole positions; refer to Figure-53.

Figure-53. Selecting hole location

- After specifying the hole locations, click on the **Close Sketch** button from the **Close** panel of **Ribbon**. The preview of holes will be displayed with the command box; refer to Figure-54. Major options of **Hole Options** dialog box are discussed next.

Figure-54. Hole creation

Creating Threaded Hole

- Select the **Threaded** button from the **Hole Options** dialog box. The options in the dialog box will be displayed as shown in Figure-55.
- Set desired standard and sub-type for defining thread standard and then select the size of hole with related pitch. Preview of hole will be displayed in the dialog box.
- Expand the **Threads** section of the dialog box and select desired radio button to define the manufacturing parameter for creating hole on machine. Set desired diameter value in the respective edit box.
- Specify the value of pitch and thread depth in respective edit boxes of the **Threads** section in the dialog box.
- Rest of the options in this dialog box are same as discussed for standard hole. Click on the **OK** button and specify location for threaded holes as discussed earlier.

Figure-55. Hole Options dialog box for threaded hole

Creating Counterbore Holes

- Select the **Counterbore** button from the dialog box. The options in dialog box will be displayed as shown in Figure-56.

Figure-56. Hole Options dialog box for counterbore hole

- The options in this dialog box have been discussed earlier. After specifying parameters of hole, click on the **OK** button from the dialog box and place the hole centers at desired locations as discussed earlier.

Similarly, you can create **Countersink** and **Tapered** holes.

THREAD

The **Thread** tool is used to create thread in existing hole or on a cylindrical surface. This tool is available in the **Hole** drop-down of **Solids** panel in the **Ribbon**. The procedure to create thread using this tool is discussed next.

- Click on the **Thread** tool from **Hole** drop-down in **Solids** panel of **Home** tab in the **Ribbon**. The **Thread Options** dialog box will be displayed; refer to Figure-57. Select desired option from the **Standard** and **Type** drop-downs to specify type of threads.

Figure-57. Thread Options dialog box

- Select desired radio button from the **Internal Thread (Holes)** area of the dialog box to specify diameter of internal threads. If you are creating external threads then nominal diameter of threads will be selected automatically based on shaft diameters.

- After setting desired parameters, click on the **OK** button. The **Thread** command box will be displayed; refer to Figure-58 and you will be asked to select cylindrical surface.

*Figure-58. Select Cylinder Step in
Thread command box*

- Select the cylindrical surface and then select the bottom edge of cylinder. The command box will be displayed as shown in Figure-59. Specify desired parameters in the command box and click on the **Finish** button. The threads will be created; refer to Figure-60.

Figure-59. Parameters Step in Thread command box

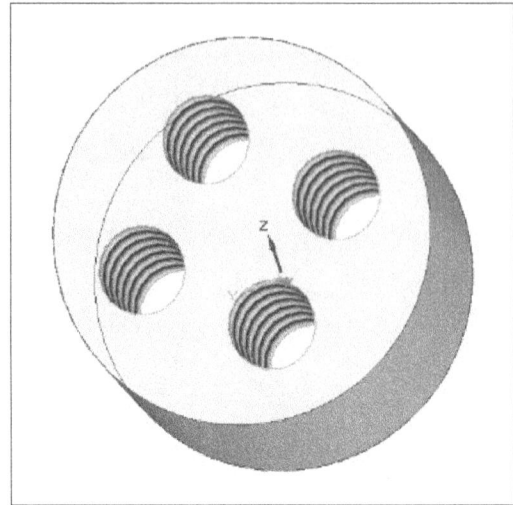

Figure-60. Thread creation

- Press **ESC** to exit the tool.

SLOT

The **Slot** tool is used to create slots using tangent continuous sketch elements. This tool is available in the **Hole** drop-down list of **Solids** panel in the **Ribbon**. The procedure to use this tool is discussed next.

- Click on the **Slot** tool from the **Hole** drop-down in the **Solids** panel of the **Home** tab in the **Ribbon**. The **Slot** command box will be displayed; refer to Figure-61.
- Click on the **Options** button from the **Slot** command box to define slot width, path offset, depth offset, and so on. The **Slot Options** dialog box will be displayed; refer to Figure-62.

Figure-61. Slot command box

Figure-62. Slot Options dialog box

- Select the **Counterbore** check box if you want to create counterbore slot of specified parameters. You can create two types of counterbore slots; **Recessed** and **Raised** by selecting respective radio button below the selected check box.
- Specify desired value of slot width in the **Slot width** edit box. If you have opted to create counterbore slot then specify desired values in **Path offset** and **Depth offset** edit boxes.

- By default, the **Flat end** radio button is selected so rectangular shaped cuts are created in the slot. If you want to create slot with semi-circular shaped ends then select the **Arc end** radio button.
- After setting desired parameter in the dialog box, click on the **OK** button. You will be asked to select tangent continuous sketch elements.
- Select the sketch elements if you want to use earlier created elements for slot creation or select desired plane or face on which you want to create sketch for slot.
- After selecting plane/face, create a sketch on that plane and click on **Close Sketch** button. Preview of the slot will be displayed with command box as shown in Figure-63.

Figure-63. Slot creation

- Specify desired value of depth in the **Distance** edit box or set desired extent as discussed earlier. After specifying the value, click in the drawing area to create the slot.
- Click on the **Finish** button from the command box to create the feature. Press **ESC** to exit the tool.

Recognizing Holes

The **Recognize Holes** tool is available in **Synchronous** environment. This tool is used to automatically recognize circular cut features (created by sketch) as holes if possible. The procedure to use this tool is given next.

- Click on the **Recognize Holes** tool from the **Hole** drop-down in the **Solids** panel of the **Home** tab in the **Ribbon**. The **Hole Recognition** dialog box will be displayed; refer to Figure-64.

Figure-64. Hole Recognition dialog box

- If you want to change parameter for any hole for your reference then click on the button from the **Select Alternate Hole Type** column in the table for respective hole. The **Hole Options** dialog box will be displayed as discussed earlier. Set the parameters as desired and click on the **OK** button from the dialog box. The preview of selected hole will be displayed in different color.
- Select the check boxes for the holes to be recognized from the **Recognize** column.
- Click on the **OK** button from the **Hole Recognition** dialog box to recognize the holes. The hole features will be created and displayed in the **Path Finder**.

ROUND (ORDERED ENVIRONMENT)

The **Round** tool is used to apply fillets at sharp edges or face of solid model. The **Round** tool is available in the **Solids** panel. The function of this tool is similar to fillet in sketch. The difference is that here we are creating round in solid model. The procedure to use this tool is discuss next.

- Click on the **Round** tool from the **Solids** panel in the **Home** tab of **Ribbon**. The **Round** command box will be displayed; refer to Figure-65.

Figure-65. Round command box

- Select desired button from the **Options** area to define the shape and type of fillet; refer to Figure-66.

Figure-66. Round Options dialog box

Creating Constant Radius Round

- Select the **Round - Constant radius** button if you want to create fixed radius value fillet.
- Click on the **Round Parameters** button from the **Round** command box to define how rounds will be created on corners and tangent edges; refer to Figure-67.

Figure-67. Round Parameters rollout

- Select the **Roll across tangent edges** toggle button to create tangential edges of the fillet at the base.
- Select the **Force roll along at blend ends** toggle button to round sharp edges of the blend ends.
- Select the **Round - Cap Sharp Edges** button from the **Sharp Edges** area to make fillets without sharp edges. Note that you should select **Curvature Continuous** option from command box before selecting this check box to get best finish. Select the **Round - Roll Along Sharp Edges** button to round sharp edges using a roller ball of size equal to round radius.

- Select the **Round - Miter at Corner** button from the **Corners** area to remove extra fillet material at the corners while creating the feature. Select the **Round - Roll Around Corners** radio button to create round base at the corner.

- Select desired option from the **Shape** area to define how the surface continuity will be followed. Select the **Round - Tangent Continuous** option if you want to create rounds with edges tangent to connected surfaces. Select the **Round - Curvature Continuous** option want the round to follow curvature of connected faces/surfaces.

- Select desired option from the **Selection Type** drop-down to set selection filter for selecting entities to create round; refer to Figure-68.

Figure-68. Selection Type drop down

- Select desired edges/faces/features to be used for creating round. Preview of round will be displayed; refer to Figure-69.

Figure-69. Round created

- Specify desired value of radius in the **Radius** edit box and click on the **OK** button from the command box.

Creating Variable Radius Round

- Select the **Round - Variable Radius** button from the **Options** area then the command box will be displayed; refer to Figure-70.

Figure-70. Round command box for variable radius

- Select desired edges on which you want to apply variable radius round and then click on the **OK** button from the command box. You will be asked to select a vertex on which you want to specify radius for round.
- Select desired vertex/point and specify the radius in the command box; refer to Figure-71. Click on the **OK** button from the command box. You will be asked to select next vertex for specifying radius.

Figure-71. Specifying radius at a vertex

- Repeat the previous step until you have specified radius at desired vertices of selected edges; refer to Figure-72.

Figure-72. Variable radius specified on vertices

- Click on the **Preview** button to check the preview of round and click in the **Name** edit box from the **Soften Corner** area and specify desired name of the variable radius then click on the **Finish** button from the command box to create the feature.

Creating Blend Round

- Select the **Round - Blend** radio button from the **Options** area then the command box will be displayed; refer to Figure-73. You will be asked to select faces/surfaces.

Figure-73. Round command box for blend round

- Select the two faces which you want to blend using the round. Preview of round feature will be displayed; refer to Figure-74.

Figure-74. Preview of blend round

- Select desired option from the **Shape** drop-down to define the shape of blend. The **Tangent continuous** and **Curvature continuous** options have been discussed earlier. Select the **Constant width** option to create round of specified width value. Select the **Chamfer** option from the drop-down to create chamfer at blending edge of specified setback. Note that chamfers will be created at 45 degree by default. Select the **Bevel** option from the drop-down to chamfer the edge with different setbacks for both sides. Note that the parameter specified in **Value** edit box defines the ratio of bevel length of both sides. If **1** is specified in the **Value** edit box then both sides will be equal. Select the **Conic** option from the drop-down to create conical shaped blend. The shapes of blend for different options are shown in Figure-75.

Figure-75. Shapes for different types of rounds

- After setting desired parameters, click on the ☑ button to create the round.

Creating Surface Blend

- Select the **Round - Surface Blend** button from the **Options** area. You will be asked to select two surfaces for creating surface blend.
- Select desired surfaces and create the round as discussed earlier for blend rounds. Note that sometimes you need to correct the direction of blend for creating this feature.

CHAMFER

The **Chamfer** tool is used to bevel the sharp edges of the solid model. This tool works in the same way as the sketch chamfer do. The procedure to use this tool is discussed next.

- Click on **Chamfer** tool from **Round** drop-down in the **Solids** panel of **Home** tab in the **Ribbon**. The **Chamfer** command box will be displayed; refer to Figure-76.

Figure-76. Chamfer command box

There are three ways in which you can create chamfer: By specifying equal setbacks on both sides, by specifying chamfer angle and length, or by specifying different length for both sides of chamfer. The procedure to create chamfer by all three options is almost similar. Here, we will use the **Chamfer - Angle and setback** option. You can apply the same procedure for other types.

*   Select the **Chamfer - Angle and setback** button from the dialog box. You will be asked to select the face containing edges to be chamfered.
*   Select desired face and click on the **OK** button from the command box. The **Chamfer** command box will be displayed as shown in Figure-77. Note that the angle and distance will be measured from the selected face.

Figure-77. Setback and Angle selection

*   Select the edges of the face to be chamfered and specify desired parameters in the command box. Preview of chamfer will be displayed; refer to Figure-78.

Figure-78. Chamfer created

*   Click on the **OK** button from the dialog box to create chamfers. Press **ESC** to exit the tool.

DRAFT

The **Draft** tool is used to apply taper to the faces of a solid model. This tool is mainly useful when you are designing components for molding or casting. The taper applied using **Draft** tool allows easy and safe ejection of part from the dies. The procedure to use this tool is given next.

- Click on the **Draft** tool from the **Solids** panel in the **Home** tab of the **Ribbon**. The **Draft** command box will be displayed; refer to Figure-79 and you will be asked to select a face/plane to be used reference for applying draft angle to other faces/surfaces.

Figure-79. Draft command box

- Select desired face/plane. You will be asked to select side faces to be used for applying draft angle.
- Select desired faces for applying draft angle. The command box will be displayed as shown in Figure-80.

Figure-80. Specifying draft angle

- Specify desired angle value for draft in the **Draft angle** edit box. After setting desired value, click on the **OK** button from the command box or press **ENTER**.
- If you want to add more faces for draft then you can select them now. After that click on the **OK** button from the command box and click **RMB** in an empty area. You will be asked to specify the direction of draft angle.
- Click on desired side of draft faces to define direction; refer to Figure-81. Draft will be created and draft angle annotation will be displayed on the model; refer to Figure-82.

Figure-81. Draft direction handle displayed

Figure-82. Draft created

• Click on the **Close** button from the command box to create feature and press **ESC** to exit the tool.

THIN WALL DROP-DOWN

The tools in **Thin Wall** drop-down (refer to Figure-83) are used to create various structure supporting features like rib, web network, and so on. The Thin Wall, Thin Region, and Vent tools are used to remove material from the solid.

Figure-83. Thin Wall drop down

Various tools available in this drop-down are discussed next.

Thin Wall (Shell)

The **Thin Wall** tool is similar to **Shell** tool in other 3D CAD software. This tool is used to make solid part hollow and remove one or more selected faces. The procedure to use this tool is discussed next.

• Click on the **Thin Wall** tool from the **Solids** panel in the **Home** tab of the **Ribbon**. The **Thin Wall** command box will be displayed; refer to Figure-84.

Figure-84. Thin Wall command box

- Specify desired value of thickness in the **Common thickness** edit box. The walls of solid will have this amount of thickness after applying thin wall feature.
- Select desired button from the command box to define which side the material will be offset for creating walls. Select the **Thin Wall - Offset Outside** button to create thin wall of specified thickness outside the faces of current solid. Select the **Thin Wall - Offset Inside** button to create wall inside the faces of current solid. Select the **Thin Wall - Symmetrical** button to use faces of solid as midplane for creating thin wall.
- After setting desired parameters, press **ENTER**. You will be asked to select faces to be removed from shell body and the command box will be displayed as shown in Figure-85.

Figure-85. Thin Wall Preview command box

- Select desired faces to be removed; refer to Figure-86. Note that specified thickness will be applied to all the faces even if you do not remove any face of the model.

Figure-86. Face selected for removing

- After selecting the face(s), right-click in the empty area or press **ENTER** to confirm selection.

- If you want to specify unique thickness of any face then click on the **Unique Thickness** button from the command box. You will be asked to select the faces.
- Select desired face(s) and enter the thickness value in **Unique Thickness** edit box.
- After setting all the parameters, click on the **Preview** button to check the preview of feature; refer to Figure-87.

Figure-87. Thin Wall created

- Click on the **Finish** button from the command box. The thin wall feature will be created.

THIN REGION

The **Thin Region** tool is used to make a closed portion of the part as thin feature. This tool is available in **Thin Wall** drop-down. The procedure to use this tool is discussed next.

- Click on the **Thin Region** tool from the **Thin Wall** drop-down of **Solids** panel in the **Home** tab of **Ribbon**. The **Thin Region** command box will be displayed as shown in Figure-88.

Figure-88. Thin Region command box

- Select desired enclosed face chain in the model to be converted to thin feature and specify the thickness value in the **Common Thickness** edit box; refer to Figure-89.

Figure-89. Faces selected for thin region

- Click on the **OK** button from the command box after specifying parameters. You will be asked to select faces to be removed from the feature.
- Select desired faces as discussed earlier. Click on the **Capping Faces** button from the command box and select the faces to be used for creating cap of thin feature. Similarly, you can use the **Unique Thickness** button to specify different thickness for selected faces.
- After setting parameters, click on the **Preview** button to check the feature. Now, click on **Finish** button. Refer to Figure-90.

Figure-90. Thin Region created

- Press **ESC** to exit the tool.

RIB

The **Rib** tool is used to create support in the structures to increase their strength. You can find use of rib in various fixtures that are fastened to the wall or in the building columns to support objects. The procedure to create rib feature is given next.

- Click on the **Rib** tool from the **Thin Wall** drop-down in the **Solids** panel of **Home** tab in the **Ribbon**. The **Rib** command box will be displayed; refer to Figure-91. You will be asked to select a planar face or plane to be used as reference base for creating rib. Note that selected plane/face will be at the mid of rib feature.

Figure-91. Rib command box

- Select the sketch if there is an existing one for rib or select the plane/face and create an open sketch; refer to Figure-92.

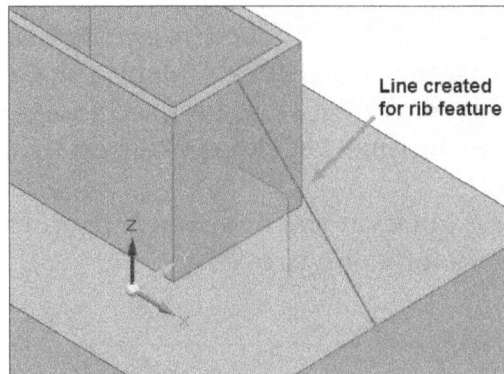

Figure-92. Sketch created for rib feature

- Close the sketch by clicking on the **Close Sketch** tool from the **Close** panel in the **Ribbon**. Preview of the rib feature will be displayed; refer to Figure-93.

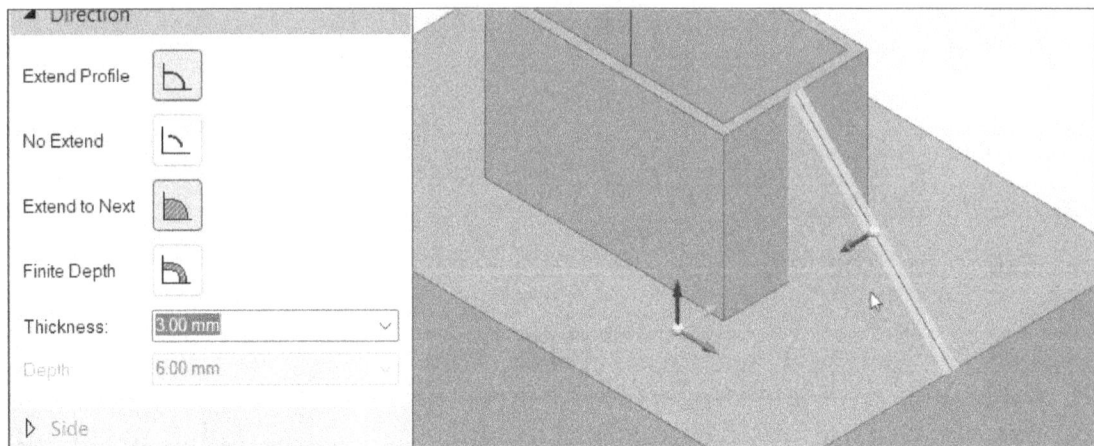

Figure-93. Preview of rib feature

- The direction of rib feature is defined by the position of cursor with respect to rib sketch. If you move cursor below the rib sketch then direction will be downward, if you move cursor to above the sketch then direction will be upward, similarly you can set left or right direction for rib feature.
- Select the **Extend Profile** button from the command box if you want to extend the rib sketch profile automatically so that it intersects with nearby faces to form closed profile. Select the **No Extend** button if you do not want to extend profile automatically.

- Select the **Extend to Next** button from the command box to automatically create rib feature up to next part surface in-line. If you want to specify the depth of rib feature then select the **Finite Depth** button from the command box and specify the value in the **Depth** edit box.
- After specifying desired parameters, click at desired location to create the feature. Preview of feature will be displayed; refer to Figure-94.

Figure-94. Rib feature created

- Click on the **Side** button from the command box if you want to change the side of rib feature with respect to sketch plane. You can place the rib feature on the left, right, or center of sketch plane.
- After setting desired parameters, click on the **Finish** button from the command box to create the feature. Press **ESC** to exit the tool.

WEB NETWORK

The **Web Network** tool is used to create web structure on entity using intersecting lines/curves. The procedure to use this tool is discuss next.

- Click on the **Web Network** tool from the **Thin Wall** drop-down. The **Web Network** command box will be displayed; refer to Figure-95 and you will be asked to select sketch entities or create a sketch on desired plane.

Figure-95. Web Network command box

- Select desired plane to create sketch or select the **Select from Sketch** option from the **Create-From** drop-down and select desired sketch entities; refer to Figure-96.

Figure-96. Sketch entities selected for web network feature

- After selecting sketch entities, click on the **OK** button from the command box. The **Web Network** command box will be displayed with preview of feature; refer to Figure-97 and you will be asked to define the direction of feature.

Figure-97. Web Network customization command box

- Specify thickness and direction of web network. The feature will be created; refer to Figure-98. Note that the direction should be towards close region so that it can add material. The other options in the command box are same as we discussed for rib feature.
- Click on **Finish** button from the command box to complete feature creation. Press **ESC** to exit the tool.

Figure-98. Web Network created

LIP

The **Lip** tool is used to create groove or lip on the surface of entity. Using this tool, you can add or remove material from the solid to form lip at the end face. The procedure of this tool is discussed next.

* Click on **Lip** tool from the **Thin Wall** drop-down in the **Solids** panel of **Home** tab in the **Ribbon**. The **Lip** command box will be displayed; refer to Figure-99 and you will be asked to select edge(s) of the model for creating lip feature.

Figure-99. Lip command box

* Select desired edge(s) of model and click on the **OK** button from the command box. The options to specify width and height of lip feature will be displayed in the command box; refer to Figure-100 and you will be asked to specify direction of feature.

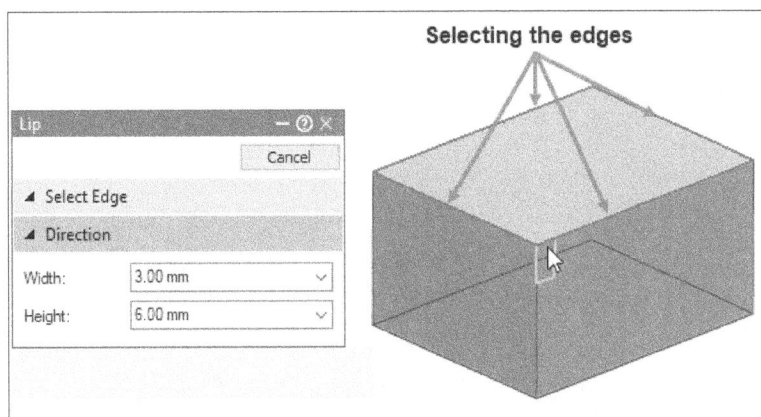

Figure-100. Lip customization command box

- Specify desired parameters in the command box and then click at desired side of selected edge to define direction of feature. Note that based on selected direction, the material will be added or removed from the model.
- After specifying direction, the preview of feature will be displayed; refer to Figure-101. Click on the **Finish** button from the command box to create the feature. Press **ESC** to exit the tool.

Figure-101. Lip created

VENT

The **Vent** tool is used create vent features. This feature is like web network but difference is that it removes material from solid face. To create vent, we need to create a sketch on the surface of solid. It must be closed sketch with intersecting curves. The procedure to use this tool is discuss next.

- Click on the **Vent** tool from the **Thin Wall** drop-down in the **Ribbon**. The **Vent Options** dialog box will be displayed with **Vent** command box; refer to Figure-102 and Figure-103.

Figure-102. Vent Command box

Figure-103. Vent Options dialog box

- Set desired parameters for ribs (horizontal) and spars (vertical) in respective edit boxes in the dialog box. Select the **Draft angle** check box and specify the angle if you want to create taper walls of vents. Select the **Round & fillet radius** check box if you want to apply fillets and rounds to the members of vent. Set desired values for selected parameters and click on the **OK** button from the dialog box. You will be asked to select the sketch boundary for vent.

- Select the closed loop existing sketch chain to define boundary of vent; refer to Figure-104 and click on the **OK** button from the command box. You will be asked to select sketch entities for ribs.

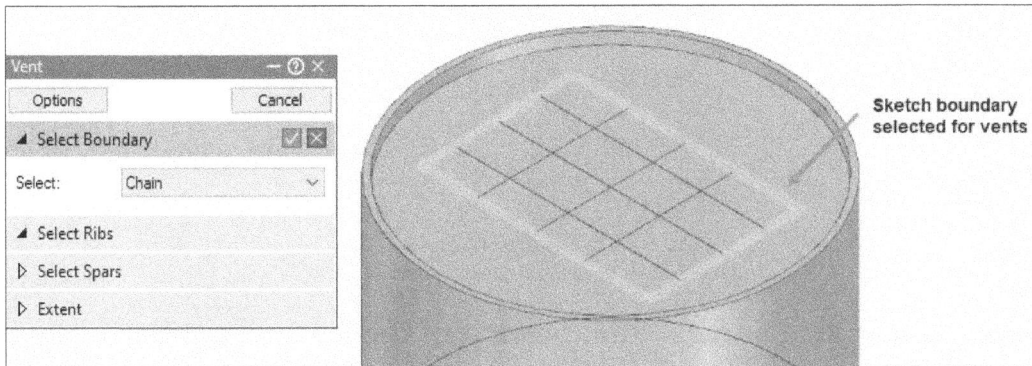

Figure-104. Selecting sketch boundary of vents

- Select the sketch entities for rib section of vent and press **ENTER**. You will be asked to select entities for spars.
- Select desired sketch entities for spars; refer to Figure-105 and press **ENTER**. You will be asked to define direction of cut.

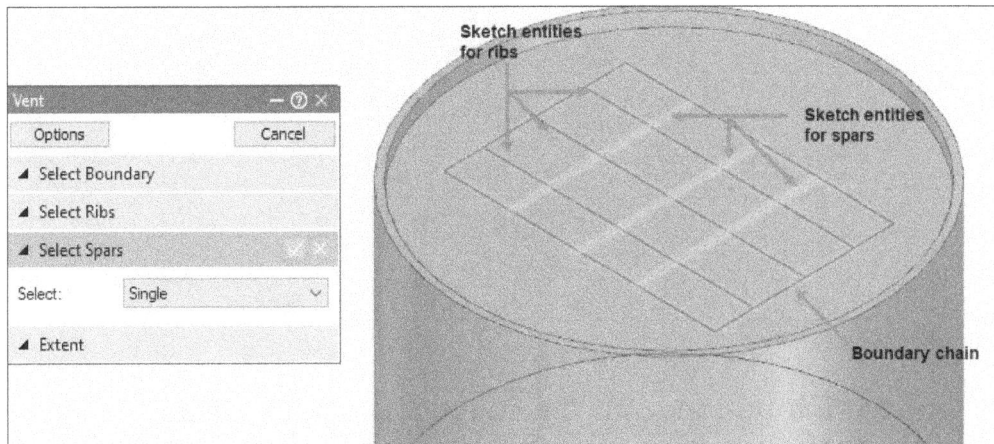

Figure-105. Sketch selection for vent feature

- Set the direction upward/downward as needed. Note that the direction must be towards the solid region on which vent should be created because vents remove material. Note that if you have created vent on solid filled model then it will automatically become hollow below the vents.
- If you want to manually specify the depth of vent then click on the **Extent** button from the command box. The command box will be displayed as shown in Figure-106. The options in this command box is same as discussed earlier for other features.

Figure-106. Vent customization command box

- After setting direction and extents, preview of the feature will be displayed; refer to Figure-107.

Figure-107. Vent created

- Click on the **Finish** button from the command box to create the feature. Press **ESC** to exit the tool.

MOUNTING BOSS

The **Mounting Boss** is used create simple cylindrical boss features with rib supports. You can add draft angle, round, and so on to the feature as discussed for vents. Mounting boss features are used to connect bodies with fasteners. The procedure to use this tool is given next.

- Click on the **Mounting Boss** tool from the **Thin Wall** drop-down in the **Solids** panel of **Home** tab in the **Ribbon**. The **Mounting Boss** command box will be displayed; refer to Figure-108, asking you to select plane/face where you want to place the mounting boss.

Figure-108. Mounting Boss command box

- Set desired option in the **Create-From** drop-down of the command box and select the plane on which you want to mount top of the feature; refer to Figure-109. Note that the feature height will be decided by distance of selected plane from the surface of existing solid. The base sketch of mounting boss feature will get attached to cursor and you will be asked to specify the positions of mounting boss features.

Figure-109. Selecting the plane to mount the feature

- Click at desired locations to specify the positions of features; refer to Figure-110. Note that you can also use the sketch dimensioning tools to locate sketch features.

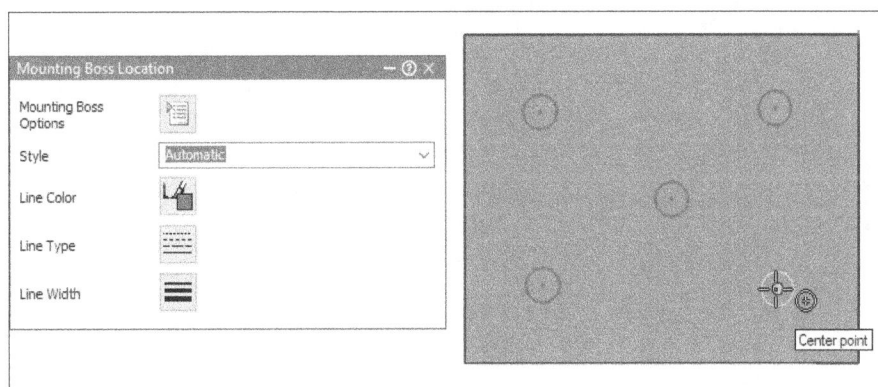

Figure-110. Specifying the position of feature

- After specifying positions, close the sketching environment by using the **Close Sketch** button from the **Ribbon**. You will be asked to specify direction of feature.
- Click in the direction towards face of solid. Preview of feature will be displayed; refer to Figure-111.

Figure-111. Preview of mounting boss feature

- Click on the **Options** button from the command box. The **Mounting Boss Options** dialog box will be displayed; refer to Figure-112.

Figure-112. Mounting Boss Options dialog box

- Specify desired parameters in the dialog box and click on the **OK** button from the dialog box. The mounting boss features will be modified accordingly; refer to Figure-113.

Figure-113. Mounting boss features after modification

- Click on the **Finish** button from the command box to create the feature. Press **ESC** to exit the tool.

The **Emboss** tool and **Etch** tool in **Thin Wall** drop-down will be discussed in chapter related to Sheet Metal Design in this book.

PRACTICAL 1

Create the model (isometric view) as shown in Figure-114. The views of the model with dimensions are given in Figure-115.

Figure-114. Practical Model 1

Figure-115. Views for practical 1

Before we start working on the practical, it is important to understand two terms; first angle projection and third angle projection. These are the standards of placing views in the engineering drawing. The views placed in the above figure are using third angle projection. In first angle projection, the top view of model is placed below the front view and right side view is placed at left of the front view. You will learn more about projection in chapter related to drafting.

Starting New Model and Creating Extrude Feature

- Double-click on the **Solid Edge** icon from desktop if you have not started Solid Edge yet.
- Click on the **New** tool from the **Menu bar** or press **CTRL+N**. The **New** dialog box will be displayed. Select **iso metric part.par** template from the dialog box and click on the **OK** button.
- You can create this part in **Synchronous** as well as **Ordered** environment. Here, we will create this part in **Synchronous** environment.
- Select the **Circle by Center Point** tool from the **Circle** drop-down in the **Draw** panel of **Home** tab in the **Ribbon** and select the **Top** plane from the **Base Reference Planes** node in the **Path Finder**.
- Draw a circle of diameter **200** at the centre of coordinate system.
- Again, draw a circle of diameter **50** on the center of coordinate system. Select the **Top** face of **ViewCube** to make the sketch plane parallel to screen.
- Click on the **Rectangle by Center** tool from the **Draw** panel in the **Home** tab of **Ribbon** and draw the rectangle as shown in Figure-116.
- Click on the **Trim** tool from the **Draw** panel in the **Home** tab of **Ribbon** and trim extra portion of the sketch; refer to Figure-117.

Figure-116. Sketch after creating rectangle

Figure-117. Sketch after trimming

- Click on the **Smart Dimension** tool from the **Dimension** panel in the **Home** tab of **Ribbon** and create the dimensions as shown in Figure-118.

Figure-118. Sketch of model

- Now, click on the **Extrude** tool from the **Solids** panel in the **Home** tab of **Ribbon**, and select the region between circle and inner section; refer to Figure-119. Right-click in the empty area. Preview of extrude feature will be displayed; refer to Figure-120.

Figure-119. Selecting region

Figure-120. Preview of extrude feature

- Specify the height of extrude feature as **50** in dynamic input box and press **ENTER**. The **Extrude** feature will be created; refer to Figure-121.

Figure-121. Extrude feature created

- Click on the **Thin Wall** tool from the **Solids** panel in the **Home** tab of **Ribbon**. You will be asked to select the faces to be removed and dynamic input box will be displayed to specify thickness of feature.
- Select the top face of feature and enter the thickness of feature as **10** in dynamic input box. The model will be created; refer to Figure-122.

Figure-122. Model created for Practical 1

PRACTICAL 2

Create the model (isometric view) as shown in Figure-123. The dimensions of the model are given in Figure-124.

Figure-123. Model for Practical 2

Figure-124. Practical 2 drawing views

Creating First Extrude Feature

We will create this model in the ordered sketching environment. So, we will first switch to Ordered environment. The procedure to switch environment and create first extrude feature is given next.

*   Start a part file in Solid Edge as discussed in previous practical.
*   Right-click in the empty area of drawing window and select the **Transition to Ordered** option from the shortcut menu. The Ordered environment will be activated.
*   Click on the **Sketch** tool from the **Sketch** panel in the **Home** tab of **Ribbon** and select the **Top** plane from the **Base Reference Planes** node in the **Path Finder**. Draw the sketch as shown in the Figure-125.

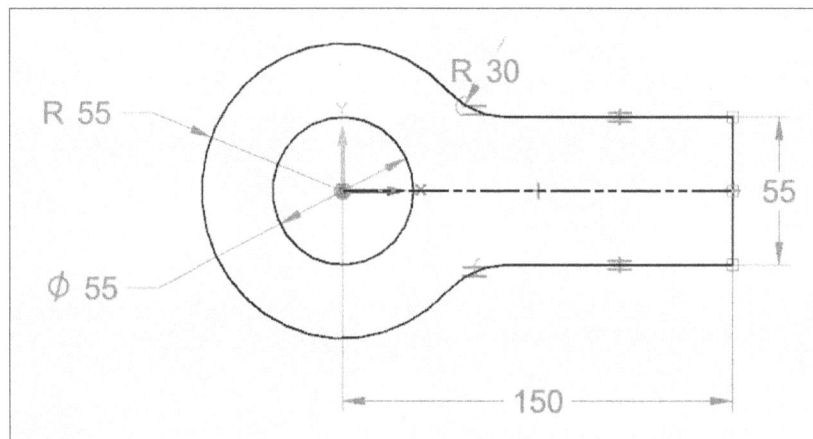

Figure-125. Sketch created at top plane

*   Close the sketch environment by clicking on the **Close Sketch** tool from the **Close** panel in the **Ribbon**.
*   Click on the **Extrude** tool from **Solids** panel in the **Ribbon**. The **Extrude** command box will be displayed.

- Select the **Select from Sketch** option from the **Create From Options** drop-down of the command box and select the complete sketch section created earlier. After selecting sketch, press **ENTER**. You will be asked to specify the depth of extrude feature (you might need to move cursor to display depth options).
- Specify the depth value as **120** in the **Distance** edit box of the command box and press **ENTER**. You will be asked to specify the side on which feature will be created with respect to sketch plane.
- Select the **Symmetric Extent** button from the command box to keep sketch plane at center and symmetrically create the feature. The preview of the feature will be displayed. Click on the **Finish** button to create the feature; refer to Figure-126 and press **ESC** to exit the tool.

Figure-126. Extrude feature created

Creating loft feature

- Click on the **Parallel** tool from the **More Planes** drop-down in the **Planes** panel of the **Ribbon** and select the side face of model as shown in Figure-127.
- Specify the offset distance as **180** and create the plane on the right side of selected face as shown in Figure-127. Press **ESC** to exit the tool.

Figure-127. Plane to be created

- Select this plane and create the sketch as shown in Figure-128.

Figure-128. Sketch to be created

- Click on the **Close Sketch** to exit from sketching and finish the sketch.
- Click on the **Loft** tool from the **Add** drop-down in the **Solids** panel of the **Ribbon**. The **Lofted Protrusion** command box will be displayed.
- Select the **Chain** option from the **Select** drop-down in command box and select the sketch created on new plane. Now, select the **Face** option from the **Select** drop-down and select the face as shown in Figure-129.

Figure-129. Selection for loft feature

- Click on the **Preview** button from the command box. The preview of loft feature will be displayed. Click on the **Finish** button to create the feature; refer to Figure-130. Press **ESC** to exit the tool.

Figure-130. Loft feature created

Creating Extrude Feature

*   Click on the **Extrude** tool from the **Solids** panel in the **Ribbon** and select the **Coincident Plane** option from the **Create From Options** drop-down of the command box.
*   Select the flat face of the loft feature recently created and draw a circle of **70** diameter at origin as shown in Figure-131.

Figure-131. Circle created for extrude feature

*   Click on the **Close Sketch** tool from the **Close** panel in the **Ribbon** to exit sketch environment. Preview of extrude feature will be displayed and you will be asked to specify height of feature. Enter the height as **220** in the **Distance** edit box of command box and click outside of feature to create feature. Preview of the feature will be displayed; refer to Figure-132.

Figure-132. Preview of model after extrusion

*   Click on the **Finish** button from the command box to create the feature. Press **ESC** to exit the tool.

Creating Extrude Cut feature

*   Click on the **Extrude** tool from the **Solids** panel in the **Home** tab of **Ribbon** and select the **Cut** button from **Operation** section in the dialog box. Select the left flat face of part. The sketching environment will be displayed.

- Draw a rectangle of **140 x 70** from the center of coordinate system; refer to Figure-133.

Figure-133. Extrude cut sketch

- Click on the **Close Sketch** tool from the **Close** panel in **Ribbon** and select **Symmetric Extent** button from the **Extent** rollout of dialog box displayed. Preview of cut feature will be displayed as you move cursor left/right in drawing area.
- Specify the distance value as **180** in the **Distance** edit box of the command box and press **ENTER**. Preview of cut feature will be displayed.
- Click on the **Finish** button to create the feature; refer to Figure-134. Press **ESC** to exit the tool.

Figure-134. Model after cut

Creating Round and Chamfers

- Click on the **Round** tool and specify the radius as **5** in the **Radius** edit box of command box.
- Select all the edges on which you want to create the round; refer to Figure-135.

Figure-135. Preview of round feature

- Click on the **OK** button from the command box to preview the feature. Click on the **Preview** button and then click on the **Finish** button from command box to create the feature. Press **ESC** to exit the tool.
- Similarly, click on the **Chamfer** tool from the **Round** drop-down in the **Solids** panel of the **Ribbon** and apply chamfer of **10 x 45** degree at the end face of model. Hide all the sketches and planes from the PathFinder to display a neat model; refer to Figure-136.

Figure-136. Final Model for Practical 2

- Save the file at desired location.

PRACTICE

Create 3D models for drawings shown in next figures.

Figure-137. Practice 1

Figure-138. Practice 2

Figure-139. Practice 3

Figure-140. Practice 4

SELF ASSESSMENT

1. In a 3D sketch, which of the following coordinates are used for defining the positions of entities?
A) X and Y
B) Y and Z
C) X, Y, and Z
D) X only

2. Which tool should be used to create a new 3D sketch in the Synchronous environment?
A) 3D Sketch Tool from the Sketch panel
B) New 3D Sketch Tool from the 3D Sketching tab
C) Extrude Tool
D) Project to Sketch Tool

3. In the Ordered environment, how do you create a 3D sketch after selecting the 3D Sketch tool?
A) Sketch directly without selecting any orientation
B) Use the View Cube to adjust the sketch's orientation
C) Select a plane or axis from OrientXpress
D) Change the dimensions manually

4. What is the purpose of the Routing Path tool in 3D sketches?
A) To create a sketch along a specified axis
B) To create guidelines for piping, electrical wiring, and other systems
C) To extrude sketch features
D) To modify the 3D model dimensions

5. Which tool in the Solids panel is used to create a solid volume by adding height to a selected sketch?
A) Extrude
B) Route
C) Sketch
D) Dimension

6. What is the difference between the Non-symmetric Extent and Symmetric Extent options in the Extrude tool?
A) Non-symmetric extends in both directions equally, while Symmetric only extends one direction
B) Non-symmetric allows different heights for both sides, while Symmetric extends equally in both directions
C) Non-symmetric is for sketching, while Symmetric is for modeling
D) There is no difference

7. What happens when you select the "Extrude - Through All" option in the Extrude tool?
A) The sketch will be extruded only on the selected face
B) The extrude feature will go through all previously created solids
C) The sketch is extruded to the next face
D) The sketch will be extruded to the selected face

8. What is the primary purpose of the "Project To Sketch" tool?
A) To create a new sketch on a selected face
B) To project edges, faces, or other entities into a new sketch
C) To create a 3D model
D) To add material to an existing feature

9. What happens if you select the "Copy elements (Non-associative)" radio button in the Tear-Off Sketch Options?
A) The copied elements will be linked to the original elements
B) The copied elements will be not linked with the original elements after copying
C) The copied elements will be moved to a new location
D) The copied elements will be locked

10. What is the function of the "Extrude Cut" tool?
A) To add material to the model
B) To remove material and create holes or cuts
C) To extrude the sketch in both directions
D) To modify existing cuts

11. Which option in the Extrude Cut command box is used to make a cut through the next face of the model?
A) Through All
B) Through Next
C) From/To Extent
D) Finite Extent

12. The "Coincident Plane" tool is used to:
A) Create a plane at an angle to a reference face
B) Create a plane coincident with a selected face or reference plane
C) Move an existing plane to a new location
D) Create a parallel plane

13. Which tool is used to create a coordinate system at a specific point with user-defined orientation?
A) Coordinate System tool
B) Extrude tool
C) Perpendicular Plane tool
D) Angled Plane tool

14. In the context of the 3D sketching, which of the following is used to change the orientation of a sketch while drawing it?
A) Change Sketch Orientation button
B) View Cube
C) Align tool
D) Reposition tool

15. What is the main use of the "More Planes" drop-down in the Planes panel?
A) To create various types of reference planes
B) To modify an existing plane's orientation
C) To delete unnecessary planes
D) To create new sketches

16. What is the difference between the Revolve and Revolve Cut operations?
A) Revolve adds material, while Revolve Cut removes material
B) Revolve Cut adds material, while Revolve removes material
C) Revolve and Revolve Cut both add material
D) Revolve and Revolve Cut both remove material

17. What is the purpose of the Hole tool in Solid Edge?
A) To create holes that comply with real machining tools
B) To create external threads on cylindrical surfaces
C) To add fillets at sharp edges of solid models
D) To create slots in the model

18. What is the first step to create a hole using the Hole tool in Solid Edge?
A) Select the desired face or plane to create the hole
B) Set the diameter and depth of the hole
C) Select the hole type from the Hole Options dialog box
D) Specify the radius for the hole

19. What type of hole can be created using the Hole Options dialog box in Solid Edge?
A) Simple Hole
B) Threaded Hole
C) Counterbore Hole
D) All of the above

20. Which button in the Hole Options dialog box is used to create a threaded hole?
A) Simple Hole
B) Threaded
C) Counterbore
D) Chamfer

21. How can a counterbore hole be created in Solid Edge?
A) By selecting the Counterbore button in the Hole Options dialog box
B) By selecting the Threaded button in the Hole Options dialog box
C) By using the Thin Wall tool
D) By selecting the Thread tool

22. What is the purpose of the Thread tool in Solid Edge?
A) To create holes in the model
B) To create threads on existing holes or cylindrical surfaces
C) To create slots in the model
D) To remove material from a solid model

23. How do you create a slot using the Slot tool in Solid Edge?
A) Select the plane and create a sketch for the slot
B) Define the slot width and depth in the command box
C) Select continuous tangent sketch elements for the slot
D) Both A and B

24. What is the purpose of the Recognize Holes tool in Solid Edge?
A) To create holes automatically
B) To recognize circular cut features as holes
C) To apply fillets to edges
D) To create thread patterns on holes

25. Which of the following tools is used to apply fillets at sharp edges or faces of a solid model?
A) Round
B) Chamfer
C) Draft
D) Rib

26. How is a constant radius round created in Solid Edge?
A) Select the Variable radius option in the Round Options dialog box
B) Select the Constant radius option and set the radius value
C) Select the Chamfer option in the Round dialog box
D) Use the Thin Wall tool to create rounded edges

27. Which option is used to define the shape of fillets in the Round Options dialog box?
A) Shape drop-down
B) Chamfer option
C) Angle option
D) Surface Continuity option

28. What is the purpose of the Chamfer tool in Solid Edge?
A) To apply fillets to sharp edges
B) To bevel the sharp edges of a solid model
C) To add material to the model
D) To apply a draft to faces

29. How is a chamfer created using the Angle and setback option in Solid Edge?
A) By selecting the face containing the edges to be chamfered
B) By specifying the desired radius for the chamfer
C) By selecting the type of slot to be created
D) By applying a variable radius

30. What is the Draft tool used for in Solid Edge?
A) To apply a taper to the faces of a solid model
B) To bevel sharp edges of the model
C) To create support ribs in structures
D) To recognize circular holes in the model

31. What is the Thin Wall tool used for in Solid Edge?
A) To create a hollow part by removing material from one or more faces
B) To apply a taper to faces
C) To create thread features in a model
D) To recognize circular cut features as holes

32. What is the Rib tool used for in Solid Edge?
A) To create a hollow structure by removing material
B) To create support in structures to increase their strength
C) To create a slot in the model
D) To create threads inside a hole

33. What is the purpose of the Web Network tool in Solid Edge?
A) To create cylindrical structures
B) To create web structures using intersecting lines/curves
C) To add material to a surface
D) To remove material from a solid

34. How do you create a web structure using the Web Network tool?
A) Select the desired plane and create a sketch, or select a sketch from existing entities
B) Use the Draft tool to define the shape
C) Select the desired solid face and apply a round
D) Use the Mounting Boss tool to define locations

35. What must you do after selecting the sketch entities in the Web Network tool?
A) Click on the Finish button to complete the feature
B) Define the thickness and direction of the feature
C) Apply a draft angle to the feature
D) Specify the depth of the feature

36. In the Web Network tool, which direction should the feature be created?
A) Towards the solid region to add material
B) Away from the solid region to remove material
C) Towards the closest region to add material
D) Any direction

37. What is the primary function of the Lip tool in Solid Edge?
A) To create slots on the surface
B) To add material to the edges of a model to form a lip
C) To remove material from a solid
D) To create fillets on sharp edges

38. How do you specify the direction for creating a lip feature?
A) By selecting a sketch entity
B) By defining the width and height of the lip
C) By clicking on the desired side of the selected edge
D) By selecting the thickness of the lip

39. What does the Vent tool in Solid Edge do?
A) Adds material to the solid face
B) Removes material from the solid face to create a vent feature
C) Creates a rib structure for the model
D) Creates support ribs for mounting bosses

40. What is required to create a vent feature using the Vent tool?
A) An open sketch with intersecting lines
B) A closed sketch with intersecting curves
C) A solid face with a draft angle
D) A plane with a circular boundary

41. How do you specify the direction of a vent feature in Solid Edge?
A) Select upward or downward direction based on the desired cut
B) Define the material thickness
C) Choose a horizontal or vertical direction
D) Specify the length and width of the vent

42. What happens when you create a vent on a solid-filled model?
A) It adds material to the model
B) It automatically becomes hollow below the vents
C) It removes material from the top surface
D) It adds a draft angle to the vent

43. Which tool in Solid Edge is used to create simple cylindrical boss features with rib supports?
A) Web Network
B) Mounting Boss
C) Lip
D) Rib

44. What is the purpose of the Mounting Boss feature in Solid Edge?
A) To add fillets to the edges of a model
B) To create cylindrical boss features with rib supports for fasteners
C) To remove material from a solid body
D) To create a hollow vent structure

45. How do you specify the positions of mounting boss features in Solid Edge?
A) Select the edge where the boss will be placed
B) Click on the desired locations in the sketch or use dimensioning tools
C) Select a plane on which the boss will be placed
D) Use the web network tool to define the boss locations

46. What options are available in the Mounting Boss Options dialog box in Solid Edge?
A) Specify material thickness and color
B) Define draft angle and round options for the boss
C) Set rib thickness and depth
D) Define the sketch boundary for ribs

47. Which tool is used to create ribs and spars for vent features in Solid Edge?
A) Vent tool
B) Lip tool
C) Mounting Boss tool
D) Web Network tool

FOR STUDENT NOTES

Chapter 4

Advanced Solid Modeling

Topics Covered

The major topics covered in this chapter are:

- *Advanced Solid Modeling tools*
- *Creating Pattern*
- *Modifying tools*
- *Practical 1*
- *Practical 2*
- *Practice*

INTRODUCTION

In previous chapter, you have learned about the basic solid modeling tools which are regularly needed in 3D modeling using any CAD software. In this chapter, you will learn about some advanced 3D features like Sweep, loft, helix, and so on. The tools to create these features are available in drop-downs of **Solids**, **Pattern**, and **Modify** panels of **Ribbon**. These tools are discussed next.

ADD DROP-DOWN

There are various tools in **Add** drop-down used to create 3D features like Sweep, loft, fillet weld, and so on; refer to Figure-1. These tools are discussed next.

Figure-1. Add drop down

CREATING SWEEP FEATURE

The **Sweep** tool is used to create a solid volume by moving a sketch along the selected path. In other words, if you move a sketch along a path then the volume that is covered by moving sketch boundary is called sweep feature. Note that to use this tool, you must have a sketch section and a path. The steps to create sweep feature are given next.

*   Click on the **Sweep** tool from the **Add** drop-down in the **Solids** panel of **Home** tab in the **Ribbon**. The **Swept Protrusion** command box and **Sweep Options** dialog box will be displayed; refer to Figure-2 and Figure-3.

Figure-2. Swept Protrusion command box

Figure-3. Sweep Options dialog box

You can create mainly two type of sweep features; Sweep using single path and single cross-section or Sweep using multiple paths and multiple cross-sections. The procedures to create both type of sweep sections are discussed next.

Sweep with Single Path and Single Cross-section

- Select the **Single path and cross section** radio button from the dialog box if you want to create sweep feature with single path and cross section.
- Select the **No merge** radio button from the **Face Merging** area of the dialog box if you do not want to merge faces while creating sweep feature. Select the **Full Merge** radio button to merge as many faces as possible while creating feature. Select the **Along path** radio button to merge faces along the path curve selected for sweep feature.
- Select the **Normal** radio button from the **Section Alignment** area of dialog box to keep section plane perpendicular to path curve. Select the **Parallel** radio button if you want the sections to be in parallel plane along the path while creating the feature.
- Select the **Tangent continuous** radio button or **Curvature continuous** radio button from the **Face Continuity** area of the dialog box to define how continuity will be followed for different sections of sweep feature.
- Select **Thin Wall** check box from the **Thin Wall** area to define the details for adding a thin wall to the feature.
- Select **Add End Caps** check box to include end caps on the thin wall.
- Select **Remove Inside Material** check box to delete the inner material of the thin wall.

- Click in the **Thickness** edit box and specify desired value to define the thin-wall thickness applied to the body.
- Select **Offset Outside** radio button to apply the wall thickness to the exterior of the solid.
- Select **Offset Inside** radio button to apply the wall thickness to the interior of the solid.
- Select **Symmetrical** radio button to evenly distribute half of the wall thickness on both the inside and outside of the solid.
- Click on the **OK** button from the dialog box to apply parameters.
- Select path from sketch along which you want to sweep the sketched profile. The path sketch can be an open or closed curve. After selecting path, click on the **OK** button from the **Swept Protrusion** command box. The options in command box will be modified and you will be asked to define cross section for sweep feature.
- Select the **Plane Normal To Curve** option from the drop-down and click on the curve created for path of sweep feature. A plane will get attached to cursor.
- Click at desired location on curve to place the plane. The sketching environment will displayed and you will be asked to draw section for sweep.
- Create a closed loop sketch for cross-section; refer to Figure-4.

Figure-4. Sketch for sweep cross-section

- After creating sketch, close the sketching environment by clicking on the **Close Sketch** tool from the **Ribbon**. You will be asked to specify starting point of the cross-section.
- Click at desired point of cross-section to define start point and then right-click in the empty area. The sweep feature will be created; refer to Figure-5.

Figure-5. Swept Protrusion created

Note that if the sweep feature is not created and an error message is displayed then you need to first check the options specified in the **Sweep Options** dialog box and then make sure cross-section profile is a closed loop sketch created using plane perpendicular to path curve.

*   Click on the **Finish** button from the command box to create the feature and press **ESC** to exit the tool.

Multi Section Sweep

The **Multiple paths and cross sections** radio button in **Sweep Options** dialog box is used to create sweep feature using two or more sketch paths and cross sections at a time. The procedure to create multi section sweep feature is discussed next.

*   Click on the **Sweep** tool from the **Add** drop-down in the **Solids** panel of the **Ribbon**. The **Sweep Options** dialog box will be displayed as discussed earlier.
*   Select the **Multiple paths and cross sections** radio button from the dialog box to create feature with multiple paths and cross sections.
*   Select the **Parametric** radio button from the **Section Alignment** area of the dialog box to create sweep feature up to the end points of path curves. Note that to accomplish this, the orientation of cross section will change along the path. Select the **Arc length** radio button from the dialog box if you want to vary the orientation of cross sections according to points on path curves. Nota that these points will have proportional arc length distance along the path curves.
*   The other options in the dialog box are same as discussed earlier. Click on the **OK** button from the dialog box. The command box will be displayed and you will be asked to select sketch curves for paths.
*   Select the first path curve, right-click in the empty area and then select the next path curve. Repeat the step until all the path curves are selected.
*   After selecting path curves, click on the **Next** button from the command box. You will be asked to select the cross section sketches.
*   One by one select the cross sections from their starting points. If you do not want the feature to be twisted then these starting points of sections should be in alignment; refer to Figure-6.

Figure-6. Multiple sections for sweep

- After selecting sketch entities, click on the **Preview** button. The feature will be displayed; refer to Figure-7.

Figure-7. Multiple Cross Sections Sweep Protrusion created

- Click on the **Finish** button from the command box to create the feature. Press **ESC** to exit the tool.

Twist Sweep

You can twist the section while sweeping along the path to create drill bit type of shape. Note that to use this option, you need to use single path and single cross section option. The procedure to create twist sweep feature is given next.

- Select the **Sweep** tool from the **Add** drop-down in the **Solids** panel of **Ribbon**. The **Sweep Options** dialog box will be displayed.
- Select the **Single path and cross section** radio button from the dialog box and click on the **OK** button from the dialog box. You will be asked to select sketch curve for path.
- Select the curve earlier created for path and right-click in the empty area. You will be asked to select sketch for cross section. Select or create sketch as discussed earlier. You will be asked to specify start point of cross section.
- Click at desired point of cross section sketch and then right-click in the empty area. The preview of feature will be displayed.
- Now, click on the **Sweep Options** button from the command box. The **Sweep Options** dialog box will be displayed as shown in Figure-8.

Figure-8. Sweep Options dialog box with Twist options

- Select the **Scale along path** check box if you want to increase/decrease the size of cross section while moving from one end of path curve to another. After selecting check box specify desired scale factor values in the **Start scale** and **End scale** edit boxes.

- Select the **Number of turns** radio button from the **Twist** area if you want to specify how many turns will be applied on cross section along the path to create twisted sweep feature. Select the **Turns per length** radio button if you want to specify number of turns of cross section in specified length along the path curve. Select the **Angle** radio button if you want to specify start angle and end angle value of cross section. For example, if you want to give 10 turns along the full length of path curve then specify 0 in **Start angle** and 3600 in **End angle** edit boxes. After selecting the radio button, specify respective values in edit boxes of **Twist** area.

- After setting desired parameters, click on the **OK** button from the dialog box. Preview of twisted sweep feature will be displayed; refer to Figure-9.

Figure-9. Twist Sweep created

- Click on the **Finish** button from command box and press **ESC** to exit the tool.

SOLID SWEEP

This tool is same as **Sweep** discussed earlier but for cross section we will use solid body in place of sketch for creating the feature. The procedure to create solid sweep feature is given next.

- Click on the **Solid Sweep** tool from the **Add** drop-down in the **Solids** panel of the **Home** tab of the **Ribbon**. You will be asked to select sketch and **Solid Sweep Protrusion** command box will be displayed; refer to Figure-10.

Figure-10. Solid Sweep Protrusion command box

- Select the sketch curve for path and right-click in the empty area. You will be asked to select the body to be used for sweep.
- Select desired solid body and press **ENTER**. You will be asked to select an edge, cylindrical face, or axis to be used as axis for revolving the solid body.
- Select desired axis/edge and press **ENTER**. You will be asked to define the axis for direction of sweep.
- Select desired entity; refer to Figure-11. After selecting the entities, press **ENTER**. The preview of feature will be displayed; refer to Figure-12.

Figure-11. Entities selected for solid sweep

Figure-12. Preview of solid sweep feature

*   Click on the **Finish** button from command box and press **ESC** to exit the tool.

LOFT

The **Loft** tool is used to create solid by using a series of sections. These sections define the resulting shape of feature. The procedure to create loft feature is discussed next.

*   Click on the **Loft** tool from the **Add** drop-down in the **Solids** panel of **Home** tab in the **Ribbon**. The **Lofted Protrusion** command box will be displayed; refer to Figure-13 and you will be asked to select the sketch entities.

Figure-13. Lofted Protrusion command box

*   One by one select the section sketches in sequence from their start point and right-click in empty area. The preview of feature will be displayed; refer to Figure-14.

Figure-14. Preview of loft feature

- If you want to use a guide curve then click on the **Guide Curve** button from the command box and select desired curve. Right-click in the empty area to apply the settings.
- Click on the **Finish** button from the command box and press **ESC** to exit the tool.

HELIX

The **Helix** tool is used to create helical feature or spring like structures. The procedure to use this tool is discuss next.

- Click on the **Helix** tool from the **Add** drop-down in the **Solids** panel of **Home** tab in the **Ribbon**. The **Helix** command box will be displayed; refer to Figure-15 and You will be asked to select existing sketch or create sketch after selecting plane.

Figure-15. Helix command box

- For creating helix feature, there must be an axis of revolution and a closed loop cross section in the sketch; refer to Figure-16. Select desired plane and create the sketch.

Figure-16. Sketch for helix feature

- After creating sketch, click on the **Close Sketch** tool from the **Close** panel in the **Ribbon**. You will be asked to specify start point of the feature.
- Select desired point on the axis to specify start point. You will be asked to specify helix parameters and the command box will be displayed as shown in Figure-17.

Figure-17. Helix customization command box

- Specify desired parameters and press **ENTER**. You will be asked to specify start point of helix feature.
- Click at the start point of axis. You will be asked to specify the parameters.
- Click on the **Parameters** rollout and set desired parameters. Preview of helix will be displayed; refer to Figure-18.

Figure-18. Preview of helix

- If you want to specify the extent of feature then click on the **Extent** rollout from the command box and set desired parameters.
- After setting the parameters, click on the **Finish** button from the command box to create feature and press **ESC** to exit the tool.

Before using the **Normal** tool, we need a sketch projected on the face of solid part. The tool to project sketch is a surfacing tool but we will discuss it here.

PROJECTING SKETCH CURVES ON SURFACE/FACE

The **Project** tool in **Surfacing** tab is used to project curves of selected sketch onto a face/surface. The procedure to do so is given next.

- Click on the **Project** tool from the **Curves** panel in the **Surfacing** tab of the **Ribbon**. The **Project** command box will be displayed; refer to Figure-19 and you will be asked to select the curves to be projected.

Figure-19. Project command box

- Select desired curves and right-click in the empty area. You will be asked to select the face on which curves are to be projected.
- Select desired face and right-click in empty area. You will be asked to specify the direction in which the sketch curve will be projected.
- Click above or below the curve plane to define direction. The curves will be projected; refer to Figure-20.

Figure-20. Projecting curves on face

- Click on the **Finish** button from the command box to create the feature and press **ESC** to exit the tool.

NORMAL

The **Normal** tool is used to create protrusion normal to a face using closed loop sketch section projected on the face. The procedure to use this tool is discuss next.

- Click on the **Normal** tool from the **Add** drop-down in the **Solids** panel of the **Home** tab in the **Ribbon**. The **Normal Protrusion** command box will be displayed; refer to Figure-21 and you will be asked to select curves projected on a face.

Figure-21. Normal Protrusion command box

- Select desired curves and right-click in the empty area. The command box will be displayed as shown in Figure-22 and you will be asked to specify which side of section will be used for protrusion.

Figure-22. Height edit box for protrusion height

- Specify desired value in the **Height** edit box of the command box and then click inside or outside of the section to define the side to be used for protrusion. The protrusion will be created; refer to Figure-23.

Figure-23. Normal Protrusion created

• Click on the **Finish** button from the command box and press **ESC** to exit the tool.

THICKEN

The **Thicken** tool is used to thicken selected surface or modify thickness of existing solid. Using this tool you can add material to model without modifying actual sketch. The steps to use this tool are discuss next.

• Click on the **Thicken** tool from the **Add** drop-down in the **Solids** panel of the **Home** tab in the **Ribbon**. The **Thicken** command box will be displayed; refer to Figure-24. Note that this tool will be active only when there is a solid or surface body already created in the modeling area.

Figure-24. Thicken command box

• Select desired option from the **Select** drop-down to define scope of selection.
• Select desired body/faces/surfaces to be thicken and right-click in the empty area.
• Specify distance value up to which you want to thicken selected objects in the **Distance** edit box of command box and then click on desired side of surface/ face/body to define direction of thickening. The preview of thicken feature will be displayed; refer to Figure-25.

Figure-25. Thicken created

• Click on the **Finish** button from the command box and press **ESC** to exit the tool.

CREATING FILLET WELD

The **Fillet Weld** tool is used to create weld beam at intersection of two faces. The weld is applied during fabrication for permanent joint. The procedure to use this tool is given next.

• Click on the **Fillet Weld** tool from the **Add** drop-down in the **Solids** panel of **Home** tab in the **Ribbon**. The **Fillet Weld** command box and **Fillet Weld Options** dialog box will be displayed; refer to Figure-26 and Figure-27.

Figure-26. Fillet Weld command box

Figure-27. Fillet Weld Options dialog box

- Specify desired parameters in the dialog box to define weld parameters. Note that tips for various options are displayed when you hover cursor on the options in the dialog box.
- After setting desired parameters in the dialog box, click on the **OK** button from the dialog box. The **Fillet Weld** command box will be displayed and you will be asked to select faces for base set.
- Select desired face(s) and right-click in empty area. You will be asked to select faces of target set.
- Select desired faces and right-click in empty area. Click on the **Preview** button from the command box. The preview of weld will be displayed; refer to Figure-28.

Figure-28. Fillet Weld preview

- Click on the **Finish** button from the command box to create the feature and press **ESC** to exit the tool.

CREATING GROOVE WELD

The **Groove Weld** tool is used to create a weld bead joining two parallel or inclined intersecting faces. The procedure to use this tool is given next.

- Click on the **Groove Weld** tool from the **Add** drop-down in the **Solids** panel of **Home** tab in the **Ribbon**. The **Groove Weld Options** dialog box will be displayed similar to **Fillet Weld Options** dialog box discussed earlier.
- Specify desired parameters in the dialog box and click on the **OK** button. The **Groove Weld** command box will be displayed; refer to Figure-29 and you will be asked to select faces for base set.

Figure-29. Groove Weld command box

- Select desired face(s) and right-click in the empty area. You will be asked to select faces for target set.
- Select desired face(s) and right-click in the empty area. You will be asked to select the top boundary of weld bead.
- Select the top edge of target/base face and right-click. You will be asked to define the bottom boundary of weld bead.
- Select desired edge; refer to Figure-30 and press **ENTER**. Click on the **Preview** button from the command box. Preview of weld bead will be displayed; refer to Figure-31.

Figure-30. Selection for Groove weld

Figure-31. Preview of groove weld bead

- Click on the **Finish** button from the command box to create the feature and press **ESC** to exit the tool.

CREATING STITCH WELD

The **Stitch Weld** tool is used to add cutouts to a weld bead (Available in Ordered Environment). Note that there must be weld beads in the model for using this tool. The procedure to use this tool is given next.

- Click on the **Stitch Weld** tool from the **Add** drop-down in the **Solids** panel of **Home** tab in the **Ribbon**. The **Stitch Weld Options** dialog box will be displayed; refer to Figure-32.

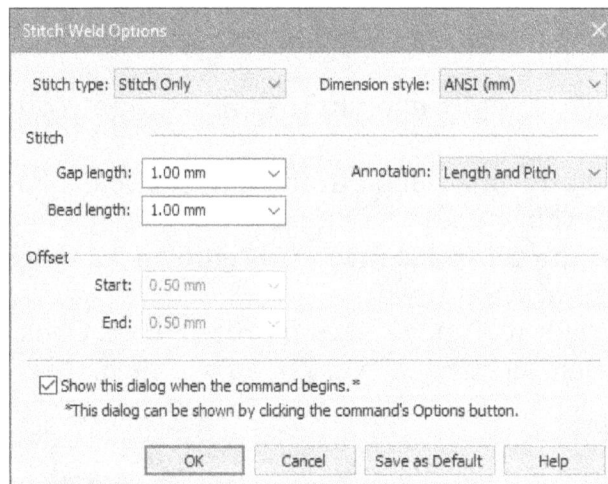

Figure-32. Stitch Weld Options dialog box

- Select desired option from the **Stitch type** drop-down to define whether you want to create stitches in weld bead, offset the weld bead along with stitch, or want to offset the weld bead.
- Select desired option from the **Dimension style** drop-down to define the dimensioning style.
- Set desired values for gap between two consecutive bead lengths and length of bead in respective edit boxes of the **Stitch** area in the dialog box. Select desired option from the **Annotation** drop-down to define how stitch weld annotation will be created.

- If you have selected option for offset then specify desired values of start and end offset in respective edit box. After setting parameters, click on the **OK** button from the dialog box. The **Stitch Weld** command box will be displayed; refer to Figure-33 and you will be asked to select the edge of weld bead.

Figure-33. Stitch Weld Command box

- Select desired edge and then right-click in empty area. The preview of stitch weld will be displayed; refer to Figure-34.

Figure-34. Preview of stitch weld

- Click on the **Finish** button from the command box to create feature and press **ESC** to exit the tool.

CREATING LABEL WELD

The **Label Weld** tool is used to create user defined weld label for selected edges (Available in Ordered Environment). Using this tool, you can apply the weld annotation to any edge for fabrication. The procedure to use this tool is given next.

- Click on the **Label Weld** tool from the **Add** drop-down in the **Solids** panel of **Home** tab in the **Ribbon**. The **Label Weld Options** dialog box will be displayed similar to **Fillet Weld Options** dialog box as discussed earlier.
- Set desired parameters in the dialog box and click on the **OK** button. You will be asked to select an edge or edge chain.
- Select desired edges where you want to apply weld label and click on the **OK** button from the command box. The preview of weld will be displayed with dark edges.
- Click on the **Finish** button to create the feature and press **ESC** to exit the tool.

CUT FEATURES

There are five tools available in the **Cut** drop-down to create different type of cut features; refer to Figure-35. These tools are discussed next.

Figure-35. Cut drop down

Sweep Cutout

The **Swept Cutout** tool is used to remove material from solid model by using swept feature. The **Swept** tool works in the same way as discussed for **Sweep** tool. The procedure to use this tool is given next.

*   Click on the **Swept Cutout** tool from the **Cut** drop-down in the **Solids** panel of **Home** tab in the **Ribbon**. The **Sweep Options** dialog box will be displayed as discussed earlier.
*   Specify the parameters as discussed for **Sweep** tool and click on the **OK** button from the dialog box. The **Swept Cutout** command box will be displayed similar to **Swept Protrusion** command box; refer to Figure-36.

Figure-36. Swept Cutout command box

*   Select path and cross-section sketches which intersects with the solid body; refer to Figure-37. The preview of swept cutout feature will be displayed; refer to Figure-38.

Figure-37. Swept cutout sketch created

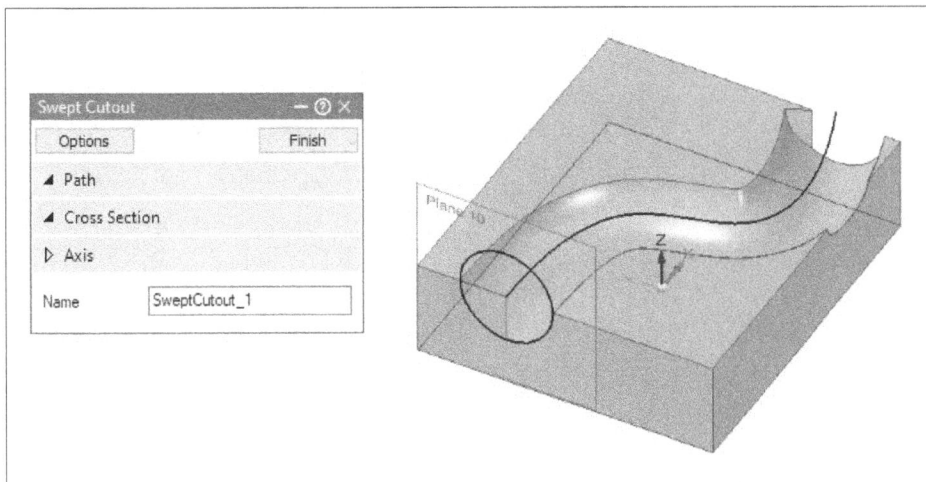

Figure-38. Swept cutout feature created

- Click on the **Finish** button from the command box to create feature and press **ESC** to exit the tool.

Solid Sweep Cut

The **Solid Sweep Cutout** tool is used to remove material from a solid model by using another solid body and a path. If you have multiple bodies in the design then you need to define the body on which operation will be performed. To do so, select the body from which material will be removed from the **Path Finder** and right-click on it. The right-click shortcut menu will be displayed; refer to Figure-39. Select the **Activate Body** option from the shortcut menu to activate the base body. Now, click on the **Solid Sweep Cutout** tool from the **Cut** drop-down in the **Solids** panel of **Home** tab in the **Ribbon**. The **Solid Sweep Cutout** command box will be displayed; refer to Figure-40. Rest of the steps are same as we used for Solid Sweep. Figure-41 shows an example of solid sweep cutout feature.

Figure-39. Activate body

Figure-40. Solid Sweep Cutout command box

Figure-41. Solid Sweep Cutout created

Loft Cutout

The **Loft Cutout** tool is used to remove material by using multiple sections. The working steps are same as we discussed earlier for **Loft** tool. Refer to Figure-42, Figure-43, and Figure-44.

Figure-42. Loft Cutout sketch sections

Figure-43. Lofted Cutout command box

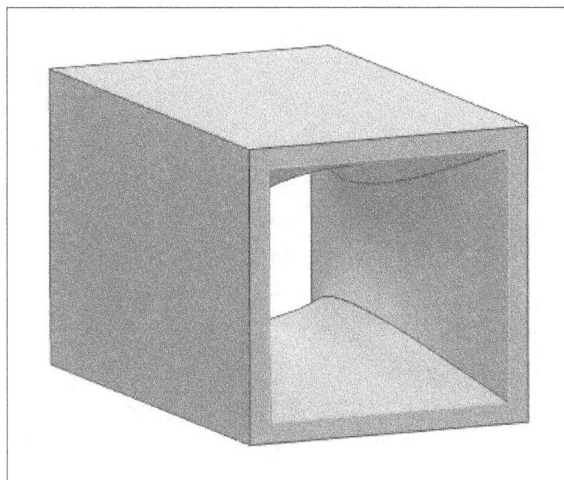

Figure-44. Loft Cutout created

Helix Cut

The **Helix Cutout** tool is used to create helix cut on solid model while removing material. This tool is available in the **Cut** drop-down. The procedure to create helix cut is similar to creating helix feature. Refer to Figure-45 and Figure-46.

Figure-45. Helical Cutout command box

Figure-46. Helical Cutout created

Normal cut

The **Normal Cutout** tool is similar to **Normal** tool discussed earlier in **Add** drop-down. Using the **Normal Cutout** tool, you can remove material from the solid by using sketch projected on solid face; refer to Figure-47.

Figure-47. Normal Cutout created

ADD BODY FEATURES

There are eight tools available in the **Add body** drop-down to create different type of cut features; refer to Figure-48. These tools are discussed next.

Figure-48. Add body drop-down

ADD BODY

The **Add Body** tool is used to add another body in the model separate from existing bodies. The procedure to use this tool is discuss next.

• Click on **Add Body** tool from the **Solid** group. The **Add Body** dialog box will be displayed; refer to Figure-49.

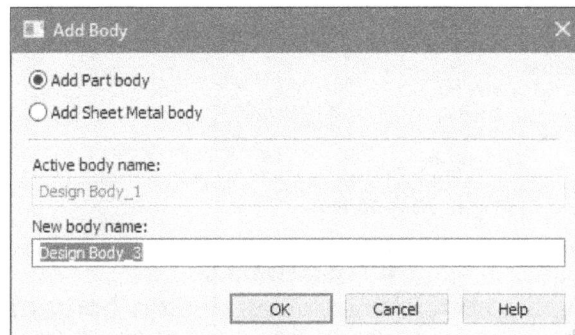

Figure-49. Add Body dialog box

• Select the **Add Part body** radio button to add a new part body. Select the **Add Sheet Metal body** radio button to create a new sheet metal body. You will learn about sheet metal model later in the book.
• Select the **Add Part body** radio button and specify desired names of existing body and new body in the respective edit boxes.
• Click on the **OK** button from the dialog box. New body will be activated.
• Create desired solid using the solid modeling tools discussed earlier; refer to Figure-50.

Figure-50. New Body added

Note that by default new body is activated after creation. So, if you now create any object then it will be created in new body. To activate any previous body, select it from the **Design Bodies** node in the **Path Finder**, right-click on it, and then select the **Activate Body** option; refer to Figure-51.

Figure-51. Activate Body option

Enclosure

The **Enclosure** tool is used to enclose existing design body into box or cylinder. The procedure is discussed next.

• Click on the **Enclosure** tool from the **Add Body** drop-down in the **Solids** panel of **Home** tab in the **Ribbon**. The **Enclosure** command box will be displayed; refer to Figure-52.

Figure-52. Enclosure command box

- Select desired enclosure shape from the **Shape** drop-down in the command box.
- Select desired option from the **Select** drop-down in the command box and select desired object for which enclosure is to be created.
- After selecting object, right-click in the empty area. You will be asked to select a plane/face for defining normal direction of enclosure.
- Select desired face/plane to give direction of enclosure and then specify the value in **Offset** edit box of command box to increase side of enclosure; refer to Figure-53.

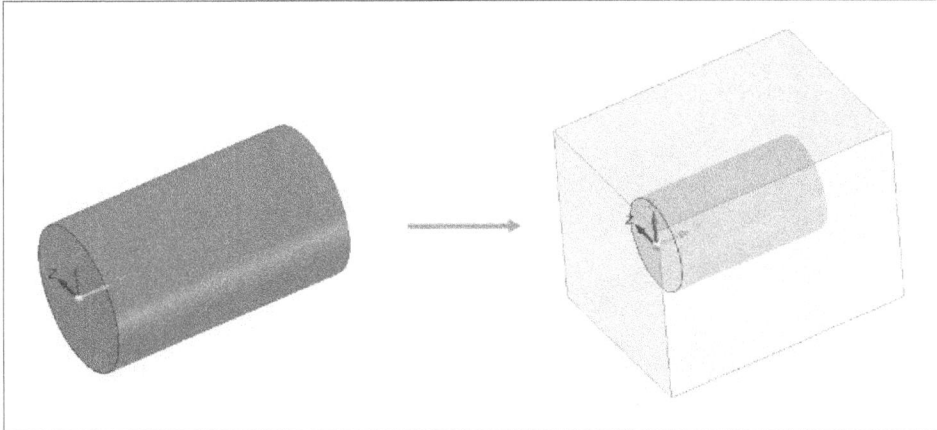

Figure-53. Enclosure created

- Click on the **Finish** button from the command box to create the feature and press **ESC** to exit the tool.

Multi-body Publish

The **Multi-body Publish** tool is used to publish multiple design bodies in the active document to separate documents and optionally an assembly document. The procedure to use this tool is discussed next.

- Click on the **Multi-body Publish** tool from the **Add Body** drop-down in the **Solids** panel of **Home** tab in the **Ribbon**. The **Multi-body Publish** command box will be displayed; refer to Figure-54. You must save the file before opening this tool.

Figure-54. Multi-body Publish dialog box

- The **Body** column of the dialog box displays the design bodies found in the multi-body file.
- The **Status** column displays current status for each published file.
- The **Filename** column displays the location or url, and filename of the published files.

- Select the **Create Assembly** check box to create an assembly after publishing the design bodies.
- Click on the **Set Path** button. The **Select Folder** dialog box will be displayed to specify the path of file to be published; refer to Figure-55.

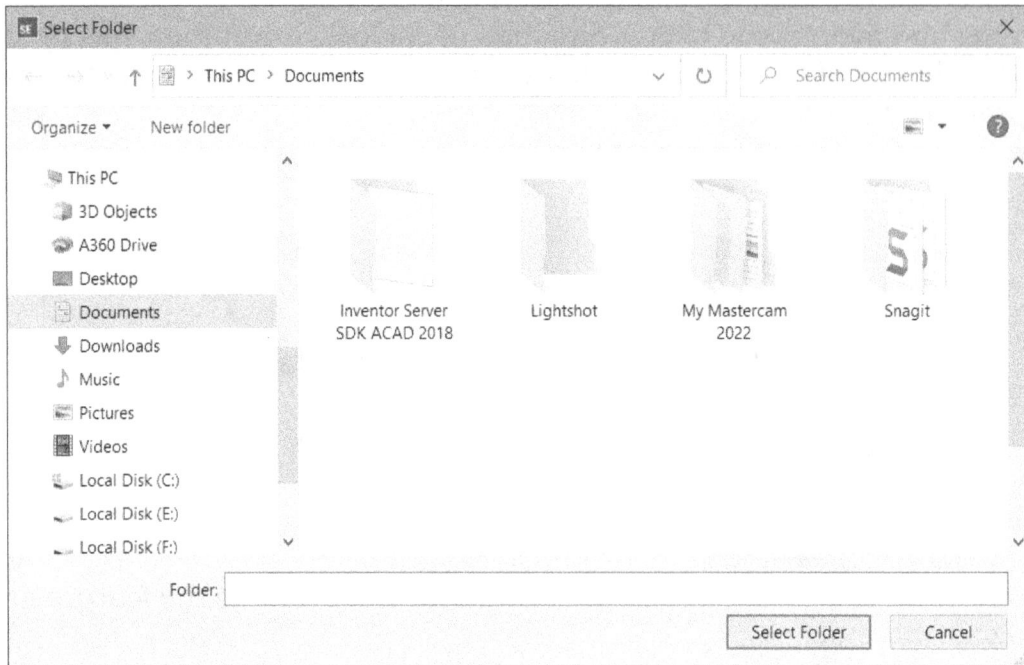

Figure-55. Select folder dialog box

- Click on **Save Files** button to publish and save the files and click on **Close** button to close the dialog box.

Union

The **Union** tool is used to combine multiple bodies into single body. The procedure to use this tool is given next.

- Click on the **Union** tool from the **Add Body** drop-down of **Solids** panel in the **Ribbon**. The **Union** command box will be displayed; refer to Figure-56.

Figure-56. Union command box

- Select first body and then select the second body.
- Click on the **OK** button from the command box. The two bodies will be converted into single entity which can be checked in **Path Finder**; refer to Figure-57.

Figure-57. Union created

Subtract

The **Subtract** tool is used to remove material from main body using a tool body. Note that there must be at least two intersecting bodies for this tool to work. The procedure to use this tool is discuss next.

*   Click on the **Subtract** tool from the **Add Body** drop-down in the **Solids** panel of **Home** tab in the **Ribbon**. The **Subtract** command box will be displayed; refer to Figure-58.

Figure-58. Subtract command box

*   Click on the **Options** button from the command box to modify output of operation. The **Subtract** dialog box will be displayed; refer to Figure-59.

Figure-59. Subtract dialog box

- Set desired parameters in the dialog box and click on the **OK** button. You will be asked to select the target body.
- Select entity from which you want to subtract another body and right-click in the empty area. You will be asked to select the tool body.
- Select the second object to be used as tool for subtraction. Preview of the subtraction feature will be displayed; refer to Figure-60.

Figure-60. Subtraction feature created

- Click on the **Finish** button to create the feature and press **ESC** to exit the tool.

Intersect

The **Intersect** tool is used to create common volume by using intersection of two solid bodies. The procedure of using this tool is discussed next.

- Click on the **Intersect** tool from **Add Body** drop-down. The **Intersect** command box will be displayed; refer to Figure-61 and you will be asked to select the target body.

Figure-61. Intersect command box

- Select the first entity which intersects with another body. You will be asked to select the tool body.
- Select the second entity. Preview of the feature will be displayed; refer to Figure-62.

Figure-62. Intersection feature created

- Click on the **OK** button from the command box to create the feature and press **ESC** to exit the tool.

Split

The **Split** tool is used to break a solid body into two individual bodies. The procedure to use this tool is discussed next.

- Click on the **Split** tool from the **Add Body** drop-down in the **Solids** panel of **Home** tab in the **Ribbon**. The **Split** command box will be displayed; refer to Figure-63.

Figure-63. Split command box

- Select entity which you want to split. You will be asked to select cutting tool (plane, face, body).
- Select desired plane, face, or body as a split tool and click on the **OK** button from the command box. Preview of split feature will be displayed; refer to Figure-64.

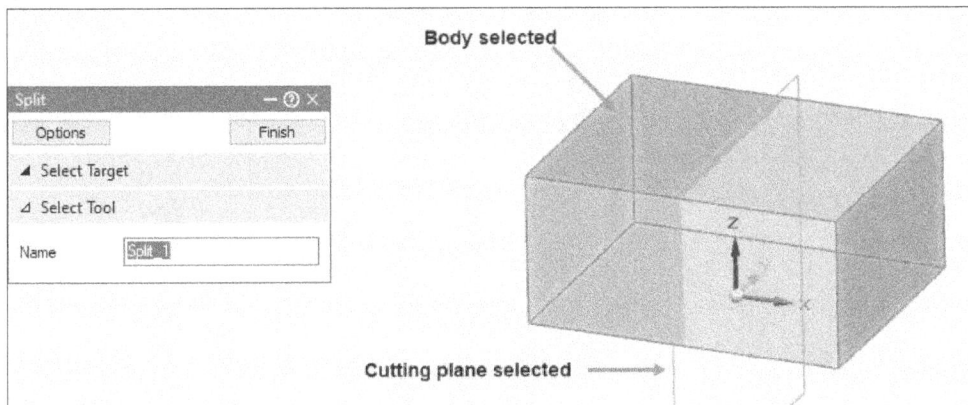

Figure-64. Split of solid body created

- Click on the **Finish** button from command box and press **ESC** to exit the tool.

Scale Body

The **Scale Body** tool is used to enlarge or diminish body by using specified scale factor. The procedure of using this tool is discussed next.

- Click on the **Scale Body** tool from **Add Body** drop-down in the **Solids** panel of **Home** tab in the **Ribbon**. The **Scale Body** command box will be displayed; refer to Figure-65.

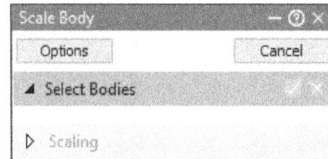

Figure-65. Scale Body command box

- Select desired body to be scaled up/down and right-click in the empty area. The command box will be modified and you will be asked to specify the scale value in the **Scale** edit box of command box.
- Click on the **Preview** button to check preview of scaling; refer to Figure-66. Click on the **Finish** button to create the feature and press **ESC** key to exit the tool.

Figure-66. Scale Body created

PATTERN

We have already discussed **Pattern** tools in sketching, now we are going to discuss about creation of solid pattern. The tools to create pattern are available in **Pattern** drop-down of **Pattern** panel in the **Ribbon**; refer to Figure-67. Various tools in this drop-down are discussed next.

Figure-67. Pattern drop down

CREATING PATTERN

The **Pattern** tool in **Pattern** drop-down is used to create multiple instances of selected body in rectangular as well as circular style. The procedure to use this tool is given next.

Rectangular Pattern

In rectangular pattern style, the object instances are created in two perpendicular linear directions on a plane. This tool is similar to the sketch rectangular pattern. The procedure to create rectangular pattern is given next.

*   Click on the **Pattern** tool from the **Pattern** drop-down in **Pattern** panel of the **Ribbon**. The **Pattern** command box will be displayed; refer to Figure-68.

Figure-68. Pattern command box

*   Select the **Rectangular** option from the **Pattern Type** drop-down at the top in the dialog box to create rectangular pattern.
*   Select all the features that you want to pattern and right-click in the empty area. You will be asked to select a planar face or plane for defining direction references of pattern.
*   Create a rectangular shape at desired face/plane to define X axis and Y axis directions for the rectangular pattern; refer to Figure-69. The options in **Pattern** dialog box will be displayed as shown in Figure-70.

Figure-69. Defining rectangular pattern directions

Figure-70. Rectangular Pattern command bar

- Click on the **Stagger Options** button from the command box to define how instances will be placed in the rectangular boundary. The **Stagger Options** dialog box will be displayed; refer to Figure-71.

Figure-71. Stagger Options dialog box

- Select the **Row** option from the drop-down to offset instances alternatively in rows or select the **Column** option to offset instances alternatively in columns. The preview of staggering will be displayed in the dialog box.

- Specify desired value of stagger distance in the dialog box using desired radio button. If you do not want pattern instances to go outside the boundary then clear the **Include last column** check box. Click on the **OK** button from dialog box to apply the changes.

- Specify desired parameters in the input boxes to define occurrences in X and Y directions. You can also specify the width, height, and inclination angle for boundary rectangle in the command box or by using handles in the graphics area. Preview of the pattern will be displayed; refer to Figure-72.

Figure-72. Preview of Rectangular Pattern

- By default, the **Pattern - Fit** button is selected in the **Type** section of the command box so all the instances of pattern will try to fit in the rectangle created by specified width and height. Note that in this case, gap between instances will be increasing/decreasing to fit instances based on specified number of instances. Select the **Pattern - Fill** button from the command box if you want to completely fill the rectangular region created by specified width and height. In this case, number of instances will be increasing/decreasing to fill the rectangle depending on specified gap between instances. Select the **Pattern - Fixed** button from **Type** section if you want to specify the number of instances and gap between two consecutive instances manually. In this case, size of rectangular boundary will automatically change based on specified parameters.

- If you do not want to create an instance then click on the **Suppress Occurence** toggle button from the **Pattern Parameters** section of the command box and select the green dots for instances that you do not want to generate in pattern.

- By default, the **Pattern - Fast** button from the **Occurrence Type** area from the **Body Selection** section of the dialog box so the orientation of all the instances is fixed. Select the **Pattern - Smart** button from the dialog box if you want the orientation of parts to be changed for accommodating maximum instances in the same boundaries.

Figure-73. Suppressing instance

- Click on the **Finish** button to create the pattern and press **ESC** key to exit the tool.

Circular Pattern

You can use the **Pattern** tool to create circular style pattern in the same way as discussed for rectangular pattern. The procedure to create circular pattern is given next.

- Click on the **Pattern** tool from the **Pattern** drop-down of **Pattern** panel in the **Home** tab of **Ribbon**. The **Pattern** command box will be displayed as discussed earlier.
- Select the **Circular** option from the **Pattern Type** drop-down in the command box.
- Select all the features that you want to pattern and click on the **OK** button from the command box. You will be asked to select a sketching plane/face.
- By default, the **Pattern - Quick Pattern** button is selected in the **Placement Type** section, so you are asked to specify location of center point and construction circle on which pattern will be created.
- Click at desired location on face of the feature and create circle. Pattern will be generated. Now, drag the center point at desired location to define diameter for construction circle of pattern; refer to Figure-74.
- If you want to use specified sketch circle for pattern then click on the **Pattern - By Sketch** button from the **Placement Type** section of the dialog box and select desired face/plane. You will be redirected to sketching environment.

Figure-74. Quick circular pattern

- Draw a circle and specify the direction in which next instances will be created. Preview of pattern will be displayed in the form of sketch dots; refer to Figure-75. Specify desired parameters in the command box as discussed earlier.

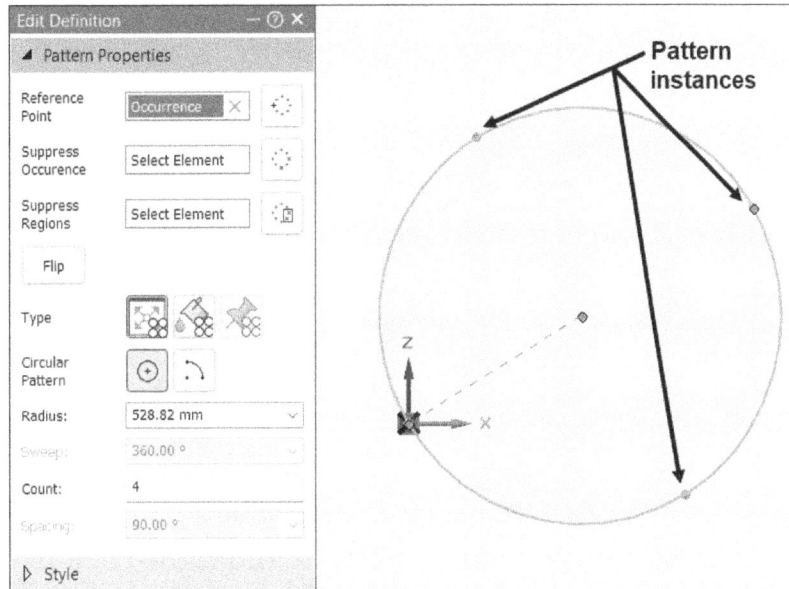

Figure-75. Preview of pattern in the form of sketch dots

- Click on the **Close Sketch** tool from the **Close** panel in the **Home** tab of **Ribbon**. The preview of circular pattern will be displayed; refer to Figure-76.

Figure-76. Circular Pattern created

- Click on the **Finish** button from the command box to create the feature and press **ESC** to exit the tool.

Note that in Synchronous environment, the **Rectangular** and **Circular** tools are separately available in the **Pattern** drop-down of **Pattern** panel to create the two types of patterns.

Along Curve Pattern

The **Along Curve** tool is used to create multiple instances of a solid features along the selected path. To create along curve pattern, there should be a curve and solid body available in the modeling area. The procedure to create along curve pattern is given next.

- Click on the **Along Curve** tool from the **Pattern** drop-down in the **Pattern** panel of the **Home** tab in **Ribbon**. The **Along Curve** command box will be displayed; refer to Figure-77.

Figure-77. Along Curve command box

- Select all the features that you want to pattern and right-click in empty area. You will be asked to select an edge chain or curve.
- Select curve path along which pattern is to be created and right-click in empty area. You will be asked to specify anchor point (start point) for pattern.
- Select desired point on the path curve. If you want to offset the anchor point then specify the respective value in the **Offset** edit box of command box.
- Click on the **OK** button from the command box. Preview of pattern feature will be displayed with modified command box; refer to Figure-78.

Figure-78. Preview of curve pattern

- Specify desired values of number of instances in **Count** edit box and space between two consecutive instances of pattern in **Spacing** edit box of command box.
- Note that you can also use a second curve path if you want to create pattern along two curves by selecting the **Path Curve Step** button from the command box.
- After setting desired parameters, click on **Next** button from command box.
- If you want to suppress any instance in pattern then click on the **Suppress Occurrence** button from command box and select the green dots for instances that you want to suppress from the preview.
- Click on the **Preview** button to check the preview; refer to Figure-79 and click on the **Finish** button to create the feature.

Figure-79. Along Curve Pattern created

Pattern by Table

The **Pattern by Table** tool is used to create multiple instances of a features as per the coordinates specified in the table. Note that you will need a Microsoft Excel Sheet file for specifying coordinates of pattern instances. The procedure to create pattern by table is given next.

• Click on the **Pattern by Table** tool from the **Pattern** drop-down in the **Pattern** panel of **Ribbon**. The **Pattern by Table** command box will be displayed; refer to Figure-80.

Figure-80. Pattern By Table command box

• Select the features which you want to pattern and right-click. You will be asked to select a coordinate system for reference.
• Select desired coordinate system from the modeling area. The **Instance Table** dialog box will be displayed; refer to Figure-81. Click on the **Browse** button from the top area in the dialog box. The **Select Excel File** dialog box will be displayed; refer to Figure-82.

Figure-81. Instance Table dialog box

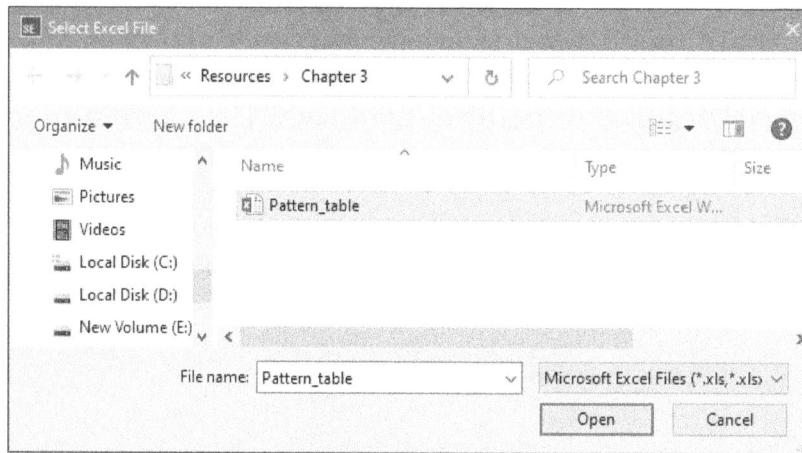

Figure-82. Select Excel File dialog box

- Select desired file in which you have specified coordinates for pattern instances. Preview of pattern will be displayed; refer to Figure-83.

Figure-83. Preview of Pattern by table

- If you want to edit the table then click on the **Edit** button from the dialog box. The sheet will open in Microsoft Excel if installed.
- Click on the **Close** button from the dialog box to exit and click on the **Finish** button from the command box to create the pattern feature.

Fill Pattern

The **Fill Pattern** tool is used to generate a pattern of a selected feature that automatically fills the entire specified area. The procedure to use this tool is discussed next.

- Click on the **Fill Pattern** tool from the **Pattern** drop-down in the **Pattern** panel of **Home** tab in the **Ribbon**. The **Fill Pattern** command box will be displayed; refer to Figure-84.

Figure-84. Fill pattern command box

- Select desired feature by which you want to fill the pattern and click on the ☑ button in the command box. You will be asked to define the pattern geometry.
- Select the face on which you want to pattern the geometry, create closed loop sketch to be filled by pattern feature, and click on the ☑ button in the command box. The preview of the fill pattern will be displayed; refer to Figure-85.

Figure-85. Preview of fill pattern

- Other parameters in the dialog box have been discussed earlier.
- After specifying desired parameters, right-click in the graphics window and then click on the **Finish** button from the dialog box. The fill pattern will be created; refer to Figure-86.

Figure-86. Fill pattern created

Duplicate Pattern

The **Duplicate** tool is used to create duplicate pattern from one body to another. There must be a reference coordinate on which duplicate pattern is to be created. The procedure of duplicate pattern is discussed next.

- Click on the **Duplicate** tool from the **Pattern** drop-down in the **Pattern** panel of **Home** tab in the **Ribbon**. The **Duplicate** command box will be displayed; refer to Figure-87.

Figure-87. Duplicate command box

- Select all pattern bodies you want to duplicate and press **ENTER**. You will be asked to specify coordinate system to be used as reference for copying the bodies.
- Select coordinate system of bodies as parent references and then select coordinate system with respect to which you want to place the duplicate bodies. The preview of duplicate pattern will be displayed in wire frame style. Click on the **Preview** button to check solid preview of feature; refer to Figure-88.

Figure-88. Preview of Duplicate pattern

- Click on the **Finish** button from command box to create the feature.

MIRROR

Earlier, we have discussed about working of **Mirror** tool for sketch elements. Now we will discuss the procedure to mirror solid part or features. The tools to perform mirror operation are available in the **Mirror** drop-down of **Pattern** panel in the **Home** tab of **Ribbon**; refer to Figure-89.

Figure-89. Mirror drop down

Mirror Copy Feature

The **Mirror Copy Feature** tool is used to create mirror copy of selected features of a solid body. The procedure to create mirror copy feature is given next.

- Click on the **Mirror Copy Feature** tool from the **Mirror** drop-down in the **Mirror** panel in the **Home** tab of **Ribbon**. The **Mirror Copy Feature** command box will be displayed; refer to Figure-90.

Figure-90. Mirror Copy Feature command box

- Select the features which you want to mirror copy and press **ENTER**. You will be asked to select a planar face/plane to be used as mirror.
- Select desired plane about which you want to mirror selected feature. The preview of mirror feature will be displayed; refer to Figure-91.

Figure-91. Preview of mirror copy feature

- Click on the **Finish** button to create the feature.

Mirror Copy Part

The **Mirror Copy Part** tool is used to mirror part or solid body. Note that all the features of selected part/body will be mirror copied. The procedure to use this tool is similar to **Mirror Copy Feature** tool.

Save As Mirrored Part

The **Save As Mirrored Part** tool is used to create a new part file that contains a mirrored copy of the original part. The procedure to use this tool is discussed next.

- Click on the **Save As Mirrored Part** tool from the **Mirror** drop-down in the **Mirror** panel of **Home** tab in the **Ribbon**. The **Save As Mirrored part** command box will be displayed; refer to Figure-92.

Figure-92. Save As Mirrored Part command box

- Select **Link to Original** check box to link mirror copy with original so that any changes in original will be reflected in mirror copy as well.
- Select desired plane type from the **Create From** drop-down to create the profile. The **Save As Mirrored Part** dialog box will be displayed; refer to Figure-93.

Figure-93. Save As Mirrored Part dialog box

- Select desired location in the dialog box and specify desired name of file in the **File name** edit box.
- Click on the **Save** button from the dialog box. The file will be saved.

MOVE FACES

The tools to perform moving operation are available in the **Move Faces** drop-down of **Modify** panel in the **Home** tab of **Ribbon**; refer to Figure-94.

Figure-94. Move Faces drop down

Moving faces

The **Move Faces** tool is used to move face at a finite extend. The procedure to use this tool is discussed next.

- Click on the **Move Faces** tool from **Move Faces** drop-down in the **Modify** panel of **Home** tab in the **Ribbon**. The **Move Faces** command box will be displayed; refer to Figure-95.

Figure-95. Move Faces command box

- Click on the **Connected Faces** button and select desired option from the drop-down. Select **Extend/Trim** option to trim and extend the adjacent faces. Select **Tip** option to trim, extend, or change the current orientation of connected faces. Select **Lift** option to lift the select set and add faces to the model.
- Click on the **Precedence** button and select desired option from the drop-down. On selecting the **Select Set Priority** option, the selected and other moving faces have priority over non-moving faces. Also, on selecting the **Model Priority** option, the non-moving faces have priority over moving faces.
- Select desired option from the **Select** drop-down to filter selection of objects and then select the face(s) of model to be moved. After selecting faces, press **ENTER**. You will asked to select a direction reference along which the select faces will move.
- Select desired edge/axis to define direction. You will be asked to specify start point for moving faces.
- Select desired reference point from the model and move it along direction earlier selected.
- Click at desired location to create the feature. Preview of feature will be displayed; refer to Figure-96.

Figure-96. Move Faces created

- Click on the **Finish** button from the command box to create feature and press **ESC** to exit the tool.
- The preview of moving faces by selecting other options in the command box are displayed next; refer to Figure-97 and Figure-98.

Figure-97. Connected Faces options preview

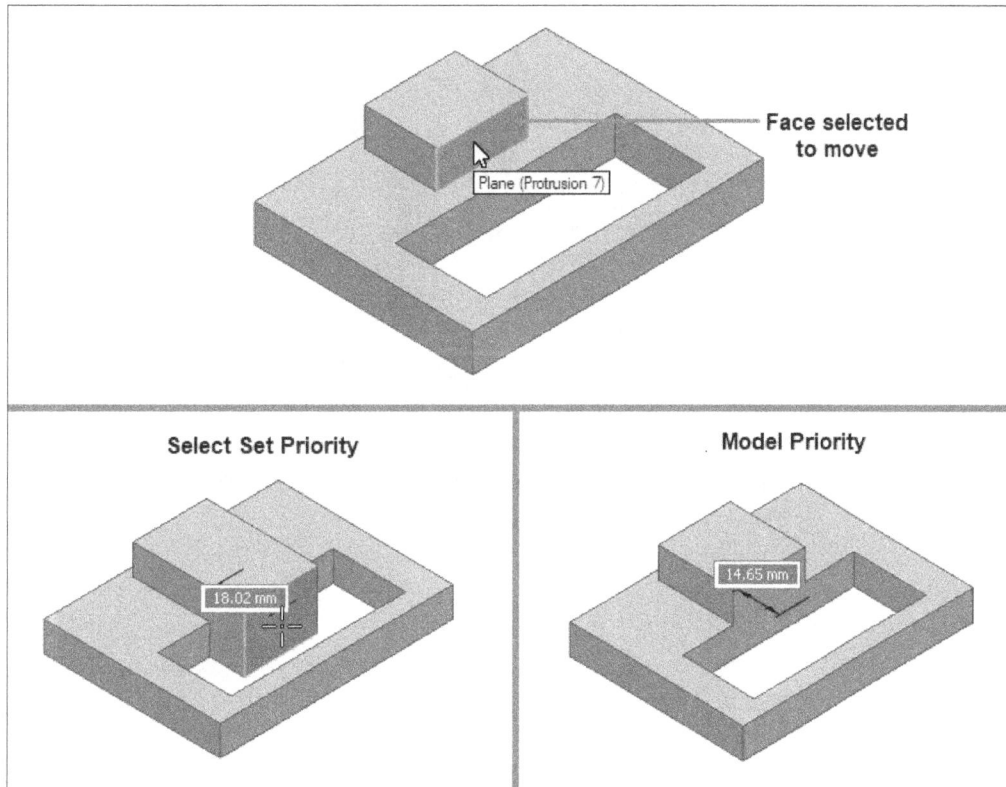

Figure-98. Precedence options preview

Rotate faces

The **Rotate Face** tool is used to rotate face, feature, and body along revolution of axis. The procedure is given next.

- Click on the **Rotate Faces** tool from the **Move Faces** drop-down in the **Modify** panel of the **Ribbon**. The **Rotate Faces** command box will be displayed; refer to Figure-99.

Figure-99. Rotate Faces command box

- The options displayed in the drop-down by clicking on the **Connected Faces** and **Precedence** buttons is same as discussed earlier.
- Choose selection type from the **Select** drop-down in the command box and select desired objects. After selecting objects, press **ENTER**. You will be asked to select an axis step about which selected objects will be rotated.
- Select desired option from **Movement Type** drop-down in **Select Axis** rollout; refer to Figure-100.

Figure-100. Movement Type drop down

- Select desired option from the drop-down and specify the parameters displayed. Previews of rotation of feature will be displayed; refer to Figure-101.

Figure-101. Previews of rotation

- Click on the **Finish** button from the command box to create the feature and press **ESC** to exit the tool.

Offset faces

The **Offset Faces** tool is used to offset faces to a finite extent. The procedure to use this tool is given next.

- Click on the **Offset Faces** tool from the **Move Faces** drop-down. The **Offset Faces** command box will be displayed; refer to Figure-102.

Figure-102. Offset Faces command box

- Select desired option from the **Select** drop-down and select objects to be offset Press **ENTER** or right-click, you will be asked to specify the offset distance.
- Specify offset distance value in command box and click to specify direction of offset. Preview of offset will be displayed; refer to Figure-103.

Figure-103. Offset Faces created

- Click on the **Finish** button to create the feature and press **ESC** to exit the tool.

DELETE

The **Delete** tools are used to delete some unwanted elements of solid model. These tools are available in **Modify** panel; refer to Figure-104. Various tools in this drop-down are discussed next.

Figure-104. Delete drop down

Deleting Faces

The **Faces** tool is used to delete faces from solid body. The procedure to use this tool is discuss next.

- Click on the **Faces** tool from the **Delete** drop-down in the **Modify** panel of **Home** tab in **Ribbon** in **Ordered** environment. The **Faces** command box will be displayed; refer to Figure-105.

Figure-105. Faces command box

- Select desired option from the **Select** drop-down and select the entities from model to be deleted.
- After selecting faces, click on the **OK** button from the command box. Preview of delete feature will be displayed; refer to Figure-106.

Figure-106. Delete Faces created

- Click on the **Finish** button from command box. The selected faces will be deleted. Press **ESC** to exit the tool.

Deleting Regions

The **Regions** tool is used to delete particular regions from solid surface. The procedure to use this tool is discuss next.

- Click on the **Regions** tool from the **Delete** drop-down. The **Regions** command box will be displayed; refer to Figure-107.

Figure-107. Regions command box

- Select desired option from the **Select** drop-down in the command box and select edges bounding the faces to delete.
- After selecting edges, press **ENTER**. You will be asked to select faces to be deleted.
- Select desired face of part to define side to be deleted and click on the **Preview** button. Preview of the feature will be displayed; refer to Figure-108.

Figure-108. Delete Regions created

- Click on the **Finish** button to create the feature.

Deleting Holes

The **Holes** tool is used to delete holes from solid model. After activating this tool from the **Delete** drop-down in **Modify** panel of **Ribbon**, select holes from solid model. Note that you can set the selection filter in **Holes** command box as discussed earlier; refer to Figure-109. Click on the **OK** button from the command box. Preview of deletion will be displayed; refer to Figure-110. Click on the **Finish** button to create the feature.

Figure-109. Holes command box

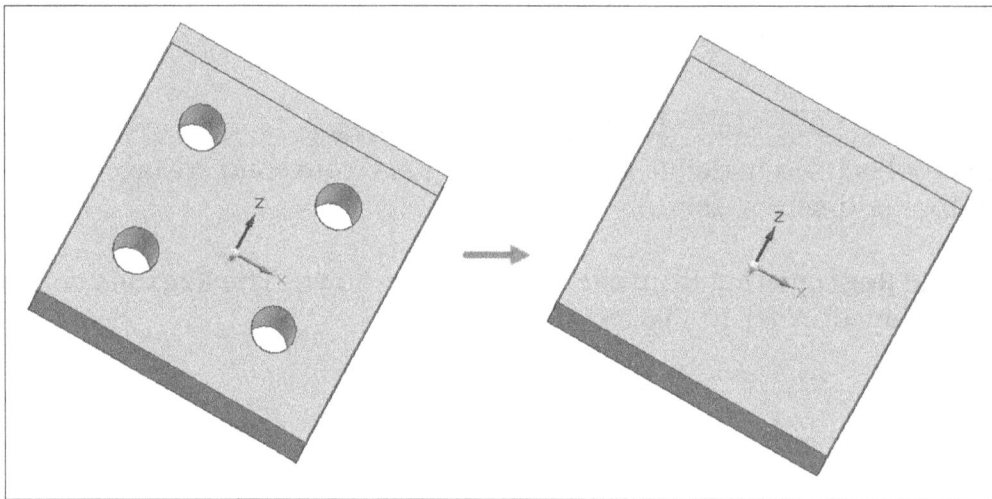

Figure-110. Delete Holes created

Deleting Rounds

The **Rounds** tool is used to delete rounds/fillets applied on solid model. The procedure is similar to deleting holes. The command box and preview of deletion are shown in Figure-111 and Figure-112.

Figure-111. Rounds command box

Figure-112. Delete Rounds created

Re-sizing Holes

The **Resize Holes** tool is used to change size of selected holes. The procedure to use this tool is given next.

*   Click on the **Resize Holes** tool from the **Modify** panel in the **Home** tab of **Ribbon**. The **Resize Holes** command box will be displayed; refer to Figure-113.

Figure-113. Resize Holes command box

*   Select the hole(s) from solid model and press **ENTER**. You will be asked to specify hole size in the command box.
*   Select desired option from the **Standard** drop-down in command box and enter the new size in edit box.
*   Click on the **Preview** button to check preview of new holes; refer to Figure-114.
*   Click on the **Finish** button to create the feature and press **ESC** to exit the tool.

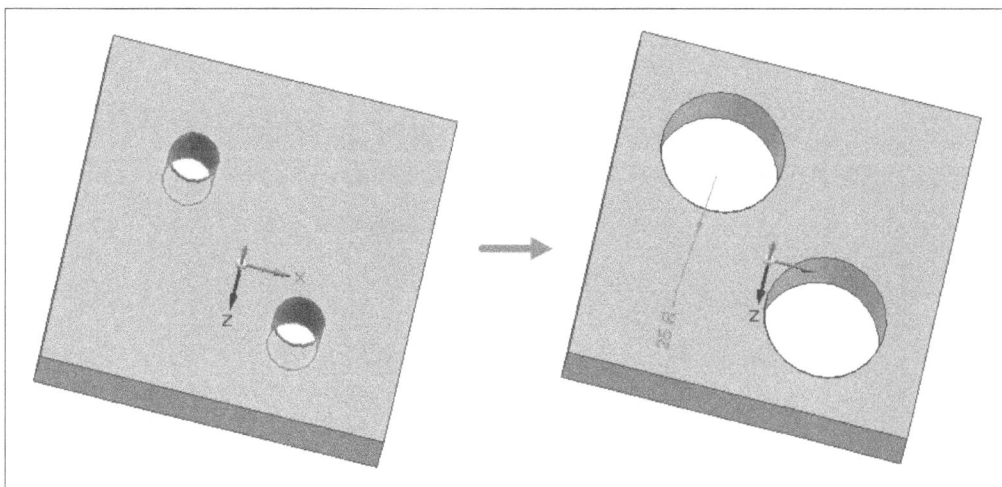

Figure-114. Resize Holes created

Resize Rounds

The **Resize Rounds** tool is used to change size of round/fillet applied on the model. The procedure to use this tool is discuss next.

- Click on the **Resize Rounds** tool from the **Modify** panel in the **Home** tab of **Ribbon**. The **Resize Rounds** command box will be displayed; refer to Figure-115.

Figure-115. Resize Rounds command box

- Select the round(s) to be modified and press **ENTER**. You will be asked to specify the size of round.
- Specify new round size value in the **New Radius** edit box of command box and press **ENTER**. The preview of modified size will be displayed; refer to Figure-116.

Figure-116. Resize Rounds created

- Click on the **Finish** button to create the feature.

PRACTICAL - 1

Create the model (isometric view) as shown in Figure-117. The dimensions are given in Figure-118.

Figure-117. Model for Practical 1

Figure-118. Practical 1 drawing view

Creating Sketches for Hook

- Switch to Ordered modeling style as discussed earlier.
- Start a new sketch on **Front** plane.
- Click on the **Circle by Center Point** tool and create the circle of diameter **120** taking coordinate system as center.
- Draw a straight line starting from top quadrant point of circle and having length of approximately **40** (No need to be accurate).

- Click on the **Arc by 3 Points** tool from **Arc** drop-down in the **Draw** panel of **Home** tab in the **Ribbon** and create a tangent arc at the bottom of the circle as shown in Figure-119.
- Click on the **Fillet** tool from **Draw** panel in **Ribbon** and apply fillet between the vertical line and the circle of radius **20**. Refer to Figure-119.

Figure-119. Sketch after creating three points arc

- Click on the **Trim** tool from the **Draw** panel in the **Ribbon** and trim the portion between the straight line and the arc; refer to Figure-120.

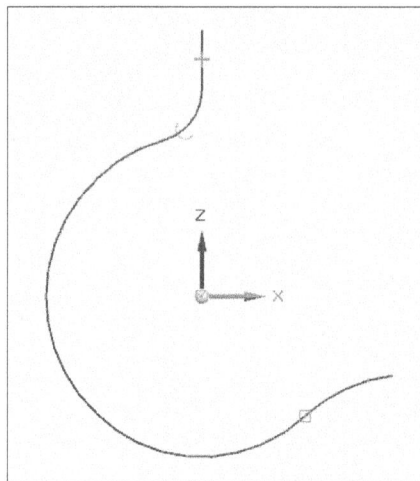

Figure-120. Sketch after trimming circle

- Click on the **Smart Dimension** tool from the **Dimension** panel in the **Home** tab of **Ribbon** and apply the dimensions as shown in Figure-121. Click on the **Close Sketch** tool from **Ribbon** and exit the sketching environment. We have created the sketch of the path. Now, we will create the section sketch.
- Click on the **Normal to curve** tool from the **More Planes** drop-down in the **Planes** panel of the **Ribbon** and select the curve near straight vertical line. A plane will be displayed attached to cursor.
- Click near the start point of the curve to create the plane; refer to Figure-122.

Figure-121. Sketch after applying dimensions

Figure-122. Creating plane

- Click on the **Sketch** tool from the **Sketch** panel in the **Home** tab of **Ribbon** and select the newly created plane. The sketching environment will be displayed.
- Create a circle of diameter **25** using the coordinate system origin as center of circle; refer to Figure-123. Close the sketch and finish sketch creation.

Figure-123. Circle created on the normal to plane

Creating Sweep Feature

Now, we have all the sketches to create the sweep feature. The procedure to create the sweep features using these sketches is given next.

- Click on the **Sweep** tool from the **Add** drop-down in the **Solids** panel of the **Ribbon**. The **Sweep Options** dialog box will be displayed.
- Select the **Single path and cross section** radio button from the **Default Sweep Type** area of the dialog box. Select the **No merge, Normal,** and **Tangent continuous** radio buttons from the dialog box. Click on the **OK** button. You will be asked to select the curve for path.
- Select the open curve and right-click in the empty area. You will be asked to select sketch section for profile.
- Select the **Select from Sketch/Part Edges** option from the **Create From** drop-down in the command box and select the circle created earlier. Preview of the sweep feature will be displayed; refer to Figure-124.

Figure-124. Sweep feature created

- Click on the **Finish** button from the command box and press **ESC** to exit the tool.

Applying Round at End

As we all know that hooks do not end with sharp edges. So, we need to apply round at the end. The steps to do so are given next.

- Click on the **Round** tool from the **Solid** panel in the **Home** tab of the **Ribbon**. The **Round** command box will be displayed.
- Click on the **Options** button from the command box and select the **Round - Blend** radio button from the dialog box displayed. Click on the **OK** button from the dialog box to apply settings.
- Select the **Bevel** option from the **Shape** drop-down in the command box. The options in command box will be displayed as shown in Figure-125.

Figure-125. Round command box with Bevel options

- Select the front face and side faces at the bottom end of hook. Preview of round will be displayed; refer to Figure-126.

Figure-126. Face selection for round

- Specify the value as **10** in **Setback** edit box, **1** in the **Value** edit box of command box, and click on the **OK** button from the command box. Click on the **Preview** button and then **Finish** button from the command box to create the round feature. Press **ESC** to exit the tool.
- Hide the extra sketches and planes. The final model will be displayed; refer to Figure-127.

Figure-127. Final model created

PRACTICE 1

Create the model as shown in Figure-128. The dimensions are given in Figure-129.

Figure-128. Practice 1

Ø63
4X R13
5X R72
2R
3°
R35
Ø24
Ø35
43
26
A
Ø52
(55)
2X R83
4X Ø10
12
150
30
250
A

2X R78
2.5°
7
88
70
42
5
28
24
(320)

SECTION A-A

Figure-129. Dimensions of the Practice 1 model

PRACTICE 2

Create the model using the drawings shown in Figure-130.

Figure-130. Rope Pulley

PRACTICE 3

Create the model as shown in Figure-131. Dimensions are given in Figure-132. Assume the missing dimensions.

Figure-131. Practice 3 model

Figure-132. Practice 3 drawing

PRACTICE 4

Create the model by using the dimensions given in Figure-133.

Figure-133. Practice 4 drawing

PRACTICE 5

Create the model by using the dimensions given in Figure-134.

Figure-134. Practice 5 drawing

PRACTICE 6

Create the model by using the dimensions given in Figure-135.

Figure-135. Practice 6 drawing

PRACTICE 7

Create a ring nut with value of **D** as **5,6,8,** and **10** using equation and design table. Dimensions are given in Figure-136.

Figure-136. Ring Nut

SELF ASSESSMENT

1. What is the primary function of the Sweep tool in the "Add" drop-down menu?
a) To create a sketch from a 3D path
b) To move a sketch along a selected path to create a 3D feature
c) To rotate a sketch along an axis
d) To create a solid volume by mirroring a path

2. What must be present to use the Sweep tool in creating a sweep feature?
a) A solid body
b) A cross-section and a path
c) A guide curve
d) A helical shape

3. In the Sweep with Single Path and Single Cross-section, what happens when you select the "Full Merge" option?
a) It merges as many faces as possible
b) It prevents merging of faces
c) It merges faces along the path curve
d) It creates a gap between faces

4. What option in Sweep Options allows the cross-section to remain perpendicular to the path curve?
a) Tangent continuous
b) Normal
c) Parametric
d) Curvature continuous

5. What is the difference between Sweep and Solid Sweep?
a) Sweep uses a sketch profile; Solid Sweep uses a solid body
b) Sweep creates a helical shape; Solid Sweep does not
c) Solid Sweep can only be used for open paths
d) Solid Sweep does not require a path

6. What is required to create a Loft feature?
a) A series of sections defining the shape
b) A single path and cross-section
c) A helix path
d) A twist parameter

7. Which tool is used to create a spring-like structure?
a) Loft Tool
b) Sweep Tool
c) Helix Tool
d) Thicken Tool

8. What is the main function of the Project tool in the Surfacing tab?
a) To extrude sketches
b) To project curves onto a face or surface
c) To define protrusions
d) To add material thickness

9. Which tool is used to create a protrusion normal to a face from a projected sketch?
a) Thicken Tool
b) Sweep Tool
c) Normal Tool
d) Helix Tool

10. What does the Thicken tool do?
a) Creates a new body by lofting multiple sections
b) Adds material to a model without modifying the sketch
c) Defines the height of a protrusion
d) Removes material from a model

11. What is the primary use of the Fillet Weld tool?
a) To add material to a body
b) To create a weld beam at the intersection of two faces
c) To project a sketch onto a surface
d) To add a hole to a face

12. What is the Groove Weld tool used for?
a) Creating a welded joint between two intersecting faces
b) Joining multiple bodies using a sweep
c) Adding labels to welds
d) Thinning a solid body

13. The Stitch Weld tool adds what to a weld bead?
a) Cutouts to the weld bead
b) Extra weld material
c) Weld labels
d) A solid body

14. Which tool is used to create a user-defined weld label for selected edges?
a) Label Weld Tool
b) Stitch Weld Tool
c) Groove Weld Tool
d) Fillet Weld Tool

15. What is the function of the Sweep Cutout tool?
a) To add material to a model
b) To remove material from a solid model using a swept feature
c) To create a solid sweep
d) To define a new path for a feature

16. How does the Solid Sweep Cut tool remove material from a solid model?
a) By using a sketch profile
b) By using another solid body and a path
c) By applying a guide curve
d) By adding material thickness

17. What is the primary use of the Loft Cutout tool?
a) To remove material using multiple sections
b) To create a 3D shape by joining two sketches
c) To add material using multiple profiles
d) To create a circular cutout

18. Where can the Helix Cutout tool be found?
a) In the Insert menu
b) In the Cut drop-down
c) In the Edit menu
d) In the View tab

19. What is the purpose of the Normal Cutout tool?
a) To create a hole in a part
b) To add material along a normal direction
c) To remove material by using a sketch projected onto a solid face
d) To merge two bodies into one

20. What does the Add Body tool do in a solid model?
a) Combines two bodies into one
b) Adds another body separate from the existing ones
c) Creates a mirrored version of an object
d) Subtracts material from an existing body

21. What type of body can be added using the Add Body tool?
a) Only solid bodies
b) Only sheet metal bodies
c) Part bodies and sheet metal bodies
d) Only hollow bodies

22. How does the Enclosure tool work?
a) Creates a hollow space within a body
b) Encloses an existing body into a box or cylinder
c) Adds a material enclosure around an object
d) Separates a body into multiple smaller bodies

23. What is the function of the Multi-body Publish tool?
a) To combine multiple bodies into one
b) To publish multiple bodies to separate documents
c) To remove material from multiple bodies
d) To create a new body in an existing part

24. What is the purpose of the Union tool?
a) To remove material from one body
b) To split a body into two
c) To combine multiple bodies into one
d) To mirror a body

25. How is the Subtract tool used?
a) To add material to a body
b) To remove material using a tool body
c) To combine two bodies
d) To scale a body

26. What is the purpose of the Intersect tool?
a) To create a mirror image of a body
b) To remove material from an intersecting body
c) To create the common volume of two bodies
d) To scale a body by a specific factor

27. What does the Split tool do in solid modeling?
a) Combines two bodies
b) Breaks a solid body into two separate bodies
c) Removes material from a body
d) Creates a copy of a body

28. How does the Scale Body tool work?
a) It adds material to a body
b) It breaks a body into smaller parts
c) It enlarges or diminishes a body by a specified scale factor
d) It mirrors a body

29. What does the Rectangular Pattern tool do?
a) Creates multiple instances of a body in a circular pattern
b) Creates a single instance of a body at a new location
c) Creates multiple instances of a body in two perpendicular directions
d) Copies a body along a curve path

30. What does the Circular Pattern tool create?
a) Instances of a body arranged along a curve
b) A single instance of a body at a new location
c) Multiple instances of a body arranged in a circular pattern
d) Multiple instances of a body in rectangular directions

31. How does the Along Curve Pattern tool work?
a) It arranges instances along a specified curve path
b) It mirrors instances along a curve
c) It creates a circular pattern along a defined axis
d) It scales a body along a curve

32. What is the Pattern by Table tool used for?
a) To create multiple instances of a body based on specified coordinates in a table
b) To duplicate an instance of a body in a table format
c) To mirror bodies using table coordinates
d) To scale bodies based on table values

33. What is the purpose of the Duplicate Pattern tool?
a) To create multiple copies of a body at different locations
b) To create duplicate instances of a body along a specified curve
c) To mirror bodies in a new location
d) To copy bodies using a reference coordinate system

34. How does the Mirror Copy Feature tool function?
a) It creates a mirrored copy of a solid body
b) It creates a copy of a body using a reference plane
c) It creates a duplicate of the selected feature
d) It adds a mirrored part to a new location

35. What does the Save As Mirrored Part tool do?
a) It creates a new part file containing a mirrored copy of the original part
b) It mirrors the part in the same file
c) It creates a reference to a mirrored part
d) It saves a mirrored copy in the same location as the original part

36. Where can the Move Faces tool be found in the Ribbon?
a) View tab
b) Home tab in the Modify panel
c) Insert tab
d) Modify tab in the View panel

37. Which option in the Move Faces tool allows you to trim and extend adjacent faces?
a) Tip
b) Lift
c) Extend/Trim
d) Rotate

38. What does the Precedence button in the Move Faces tool control?
a) The selection priority of faces to move
b) The direction of face movement
c) The type of surface to move
d) The visibility of the selected faces

39. In the Offset Faces tool, how do you define the direction of the offset?
a) By selecting a direction reference
b) By specifying the offset distance value
c) By selecting the desired edge/axis
d) By pressing ENTER after face selection

40. What is the purpose of the Delete Faces tool?
a) To delete an entire solid body
b) To delete specific faces from a solid model
c) To remove unwanted edges
d) To move faces within the solid model

41. How do you delete a region from the solid surface using the Delete Regions tool?
a) Select faces directly to delete
b) Select edges bounding the region to delete
c) Use the Move tool to remove regions
d) Specify the offset distance for deletion

42. What is the function of the Holes tool in the Delete drop-down?
a) To resize holes in a model
b) To delete faces with holes
c) To delete holes from a solid model
d) To delete entire regions with holes

43. How can you resize a hole in the model using the Resize Holes tool?
a) Select a hole and input the new size
b) Specify the hole size after selecting faces
c) Use the Resize tool to move the hole
d) Select edges to define hole size

44. How is the Resize Rounds tool used in modifying a model?
a) It allows you to modify the round/fillet size by entering a new radius
b) It deletes the rounds applied to the model
c) It rotates the rounds to adjust their shape
d) It adjusts the offset distance of the round

FOR STUDENT NOTES

Chapter 5

Selection Tools, 3D Printing, and Views

Topics Covered

The major topics covered in this chapter are:

- *Selection Tools*
- *Introduction to 3D Printing*
- *3D Printing Tools*
- *Introduction to Views*
- *View Tools*

INTRODUCTION

There are various features of objects in the model like face, vertex, body, part, and so on. Sometimes, default selection does not allow you to select desired feature of model. The tools in the **Select** group are used to define criteria for selecting objects in the model; refer to Figure-1. The **Select** group is available in multiple tabs of **Ribbon** like **Home** tab, **Features** tab, **PMI** tab, **Simulation** tab, and so on. Various tools in this group are discussed next.

Figure-1. Select group

SELECT DROP-DOWN

The tools in the **Select** drop-down are used to specify which feature of model takes priority while selecting objects; refer to Figure-2. Note that some of the options in this drop-down will be available in the Assembly environment. The options in this drop-down are given next.

Figure-2. Select drop down

- Select the **Face Priority** option from the **Select** drop-down if you want to select faces from the model.
- Select the **Part Priority** option from the **Select** drop-down in the **Select** group to select parts from the assembly model.
- Select the **Normal Select Mode** option from the **Select** drop-down to use standard selection mode. In this selection mode, you can select multiple objects while holding the **SHIFT** key and remove objects from selection while holding the **CTRL** key during selection.
- After selecting an object from the model, select the **Add/Remove Mode** option from the **Select** drop-down if you want to add or remove objects from the selection.
- Similarly, you can use the **Add Mode** and **Remove Mode** options from the **Select** drop-down to add and remove objects from selection.
- Select the **Selection Manager Mode** option from the drop-down to use selection manager for selecting object. After selecting this option, select desired face of synchronous model. The **Selection Manager** menu will be displayed; refer to Figure-3. Various options of this menu are discussed next.

Figure-3. Selection Manager menu

- Click on the **Deselect** option to deselect latest selected face.
- Click on the **Clear Selection** option from the menu to remove all the faces from selection.
- Click on the **3D Box Select** option from the menu to draw a 3D box. All the objects falling inside 3D box will be selected.
- Hover the cursor on **Connected** option and select desired option from the cascading menu. Select the **Connected** option to select all the faces connected to latest selected face. Select the **Interior Faces** option from the cascading menu to select all the connected interior faces. Select the **Exterior Faces** option from the cascading menu to select all the connected exterior faces.
- Select the **Related Items** option to select the objects which are linked to selected object.
- Select the **Sets** option from the menu to select all the features of object whose face is selected. The **Quick Pick** dialog box will be displayed; refer to Figure-4. Select desired feature to be selected from the dialog box.

Figure-4. Quick Pick dialog box

- Select desired option from the **Recognize** cascading menu in the **Selection Manager** menu to select respective features identified in selection.
- Select desired option from the **Parallel** cascading menu to select faces parallel to selected face; refer to Figure-5. On selecting an option, you will be asked to draw a 3D box for selected parallel faces. Draw desired box and all the faces parallel to selected face inside the 3D box will be selected.

Figure-5. Parallel cascading menu

- Similarly, you can use the **Perpendicular**, **Coplanar**, **Tangent Faces**, and **Tangent Chain** options from the **Selection Manager** menu.
- Select desired option from the **Symmetric About** cascading menu and then select desired face. All the faces symmetric about selected plane will be also get selected.
- Deselect the **Use Box Selection** option to use single object selection.
- Select the **Deselect Items** option from the menu to deselect the items.
- Press **ESC** to exit the **Selection Manager** menu.

SELECTION FILTER

The **Selection Filter** tool is used to apply filter on selection like you can select only dimensions, annotations, model geometry, and so on. On clicking the **Selection Filter** tool from the **Select** group of the **Home** tab in the **Ribbon**. The **Selection Filter** menu will be displayed; refer to Figure-6. Select desired check boxes from the menu to define what type of objects can be selected.

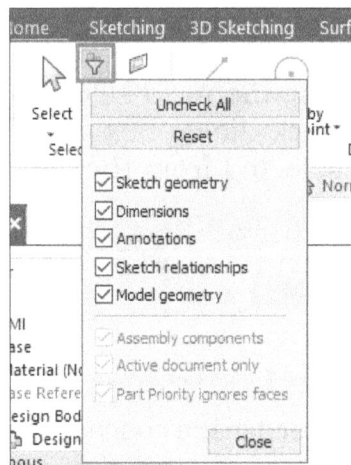

Figure-6. Selection Filter menu

OVERLAPPING SELECTION BOX

Select the **Overlapping** toggle button from the **Directional Fence** drop-down of **Select** group in the **Home** tab of the **Ribbon** to use a selection box which can be used to select objects which intersect with selection box and fall inside it; refer to Figure-7. Press **ESC** to exit the selection.

Figure-7. Overlapping selection box

RECTANGULAR AND POLYGONAL/LASSO SELECTION FENCE

By default, the **Rectangular Fence** option is selected in the **Rectangular Fence** drop-down of **Select** panel in the **Home** tab of **Ribbon**, so when you create box selection for selecting object, rectangular shape is generated for the selection box. If you want to use a polygonal shape for selecting objects then select the **Polygon/Lasso Fence** option from the drop-down or press **ALT** key when drawing first line of selection box to create polygonal selection and draw the polygon; refer to Figure-8. After creating desired polygonal fence, right-click to complete the selection fence and select the objects.

Figure-8. Polygonal Fence selection

3D PRINTING

3D Printing also known as Additive Manufacturing is not a new concept as it was developed in 1981 but since then 3D Printing technology is continuously evolving. In early stages, the 3D printers were able to create only prototypes of objects using the polymers. But now a days, 3D printers are able to produce final products using metals, plastics, and biological materials. 3D printers are being used for making artificial organs, architectural art pieces, complex design objects etc. Although, 3D Printing technique was created for manufacturing industry but now, it has found more applications in medical field.

In Solid Edge, there is a very simple and robust mechanism for 3D printing. The procedure of 3D Printing itself is not difficult but it is important to prepare your part well for 3D printing. We will first discuss the part preparation for 3D Printing and then we will use Solid Edge tools for performing 3D print.

PART PREPARATION FOR 3D PRINTING

Part preparation is very important step for 3D Printing. If your part is not stable in semi molten state then it is less suitable for 3D printing. Stability of model is directly dependent on the material you are using for 3D Printing. We will learn more about part preparation but before that it is important to understand different type of processes available in 3D Printing.

3D Printing Processes

Not all 3D printers use the same technology. There are several ways to print and all those available are additive, differing mainly in the way layers are build to create the final object.

Some methods use melting or softening material to produce the layers. Selective laser sintering (SLS) and fused deposition modeling (FDM) are the most common technologies using this way of 3D printing. Another method is when we talk about curing a photo-reactive resin with a UV laser or another similar power source one layer at a time. The most common technology using this method is called stereo lithography (SLA).

In 2010, the American Society for Testing and Materials (ASTM) group "ASTM F42 – Additive Manufacturing", developed a set of standards that classify the Additive Manufacturing processes into 7 categories according to Standard Terminology for Additive Manufacturing Technologies. These seven processes are:

1. Vat Photopolymerisation
2. Material Jetting
3. Binder Jetting
4. Material Extrusion
5. Powder Bed Fusion
6. Sheet Lamination
7. Directed Energy Deposition

Brief introduction to these processes is given next.

Vat Photopolymerisation

A 3D printer based on the Vat Photopolymerisation method has a container filled with photopolymer resin which is then hardened with a UV light source; refer to Figure-9.

Figure-9. 3D Printing via vat-photopolymerisation

The most commonly used technology in this processes is Stereolithography (SLA). This technology employs a vat of liquid ultraviolet curable photopolymer resin and an ultraviolet laser to build the object's layers one at a time. For each layer, the laser beam traces a cross-section of the part pattern on the surface of the liquid resin. Exposure to the ultraviolet laser light cures and solidifies the pattern traced on the resin and joins it to the layer below.

After the pattern has been traced, the SLA's elevator platform descends by a distance equal to the thickness of a single layer, typically 0.05 mm to 0.15 mm (0.002" to 0.006"). Then, a resin-filled blade sweeps across the cross section of the part, re-coating it with fresh material. On this new liquid surface, the subsequent layer pattern is traced, joining the previous layer. The complete three dimensional object is formed by this project. Stereolithography requires the use of supporting structures which serve to attach the part to the elevator platform and to hold the object because it floats in the basin filled with liquid resin. These are removed manually after the object is finished.

Material Jetting

In this process, material is applied in droplets through a small diameter nozzle, similar to the way a common inkjet paper printer works, but it is applied layer-by-layer to a build platform making a 3D object and then hardened by UV light; refer to Figure-10.

Figure-10. 3D Printing via Material-Jetting

Binder Jetting

With binder jetting, two materials are used: powder base material and a liquid binder. In the build chamber, powder is spread in equal layers and binder is applied through jet nozzles that "glue" the powder particles in the shape of a programmed 3D object; refer to Figure-11. The finished object "glued together" by binder remains in the container with the powder base material. After the print is finished, the remaining powder is cleaned off and used for 3D printing the next object.

Figure-11. 3D Printing via binder-jetting

Material Extrusion

The most commonly used technology in this process is Fused deposition modeling (FDM). The FDM technology works using a plastic filament or metal wire which is unwound from a coil and supplying material to an extrusion nozzle which can turn the flow on and off. The nozzle is heated to melt the material and can be moved in both horizontal and vertical directions by a numerically controlled mechanism, directly controlled by a computer-aided manufacturing (CAM) software package; refer to Figure-12. The object is produced by extruding melted material to form layers as the material hardens immediately after extrusion from the nozzle. This technology is most widely used with two plastic filament material types: ABS (Acrylonitrile Butadiene Styrene) and PLA (Polylactic acid) but many other materials are available ranging in properties from wood filed, conductive, flexible etc.

Figure-12. 3D Printing via Fused Deposition Modeling

In the above figure:
1 – nozzle ejecting molten material (plastic),
2 – deposited material (modelled part),
3 – controlled movable table.

Powder Bed Fusion

The most commonly used technology in this processes is Selective laser sintering (SLS). This technology uses a high power laser to fuse small particles of plastic, metal, ceramic or glass powders into a mass that has desired three dimensional shape. The laser selectively fuses the powdered material by scanning the cross-sections (or layers) generated by the 3D modeling program on the surface of a powder bed; refer to Figure-13. After each cross-section is scanned, the powder bed is lowered by one layer thickness. Then a new layer of material is applied on top and the process is repeated until the object is completed.

Figure-13. 3D Printing via Selective Laser Sintering

All untouched powder remains as it is and becomes a support structure for the object. Therefore there is no need for any support structure which is an advantage over SLS and SLA. All unused powder can be used for the next print.

Sheet Lamination

Sheet lamination involves material in sheets which is bound together with external force. Sheets can be metal, paper or a form of polymer. Metal sheets are welded together by ultrasonic welding in layers and then CNC milled into a proper shape; refer to Figure-14. Paper sheets can be used also, but they are glued by adhesive glue and cut in shape by precise blades.

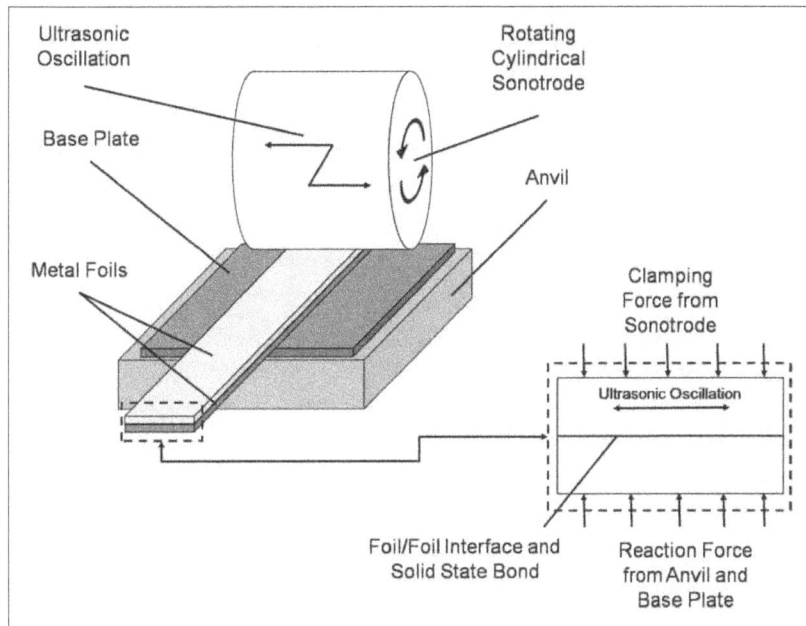

Figure-14. 3D Printing via Ultrasonic Sheet Lamination

Directed Energy Deposition

This process is mostly used in the high-tech metal industry and in rapid manufacturing applications. The 3D printing apparatus is usually attached to a multi-axis robotic arm and consists of a nozzle that deposits metal powder or wire on a surface and an energy source (laser, electron beam or plasma arc) that melts it, forming a solid object; refer to Figure-15.

Figure-15. 3D Printing via Direct Energy Deposition

Now, you know about different 3D Printing techniques so it is clear from different processes that the main work of 3D printer is to solidify material at your will, through different techniques. Now, we will learn about the points to be taken care of while preparing model for 3D printing.

Part Preparation for 3D Printing

Various important points to remember while preparing part for 3D printing are given next.

*   You should avoid holes in the areas where model is not supported. Holes can cause material to flow out in various 3D printing processes.
*   Use the Solid modeling tools. It does not mean that you cannot use surface modeling tools but you should avoid surfaces in your model. Thicken your surfaces after performing modeling operations.
*   Make sure that you have not left any unwanted piece inside the model enclosure after performing boolean operations.
*   Shell your model after creating it. You can 3D print solid model but if a hollow box can do your work then why to waste material on solid cube. Less material means cost efficient.
*   If you want to write text on your model then check the specification of your printer for possible font range.
*   The color and textures you apply on model in SolidEdge will not be exported for 3D printing so do not waste your time on them.
*   3D printing is closer to mesh modeling than solid modeling. So, check your mesh model by exporting solid model in .stl file.
*   If your file has quite a bit of text and multiple emboss/engrave features, try exporting the file as a vector. Vector files are more appropriate for extremely complex files. Try .iges or .step.
*   Orientation is particularly important when it comes to 3D printing in order to determine the interior and exterior of an object. Orient your part the same way as your want it in 3D Printing.
*   Don't make a multi-body model for 3D printing. Your printer may die thinking what to do when two separate bodies overlap each other!!
*   SolidEdge do not has capability to edit mesh model so you should prepare your model carefully to get desired mesh model.

3D PRINT PART PREPARATION IN SOLID EDGE

The tools to prepare part model for 3D printing are available in the **3D Print** tab of the **Ribbon**; refer to Figure-16. Various tools related to 3D printing in this tab are discussed next.

Figure-16. 3D Print tab

Converting Cosmic Threads to Physical

The **Physical Thread** tool is used to convert cosmic threads into physical threads. The procedure to use this tool is given next.

*   Click on the **Physical Thread** tool from the **Prepare** group of **3D Print** tab in the **Ribbon**. The **Physical Thread** command box will be displayed; refer to Figure-17.

Figure-17. Physical Thread command box

- Click on the **Select All Threads** button from the command box to select all the cosmic threads in the model or one by one click on the threads to be included.
- After selecting cosmic threads, click on the **OK** button from the command box. The threads will be converted to physical threads and a message box will be displayed.
- Click on the **OK** button from the message box and press **ESC** to exit the tool.

Deleting Internal Voids

The **Delete Voids** tool is used to delete internal voids in the model. The procedure to use this tool is given next.

- Click on the **Delete Voids** tool from the **Prepare** group of **3D Print** tab in the **Ribbon**. The **Delete Voids** command box will be displayed; refer to Figure-18.

Figure-18. Delete Voids command box

- Click on the **Options** button from the command box to define colors of bodies with voids and without voids in **Voids Settings** dialog box; refer to Figure-19.

Figure-19. Voids Settings dialog box

- Click on the **Check the voids** button from the command box to check whether the part has internal voids or not; refer to Figure-20, we have created an intentional void.

Figure-20. Internal void

- Click on the **Delete** button in graphics area to delete desired void. After deleting the voids, press **ESC** to exit the tool.

Reorienting Model for 3D Printing

The **Reorient** tool in **3D Print** tab is used to change the orientation of model as per the table of 3D Printer. The procedure to use this tool is given next.

- Click on the **Reorient** tool from the **Prepare** group of the **3D Print** tab in the **Ribbon**. The **Reorient** command box will be displayed with a preview box showing orientation of model in 3D Printing; refer to Figure-21.

Figure-21. Reorient command box

- Click on the **Options** button from the command box to define size of the 3D printing volume. The **Reorient Settings** dialog box will be displayed; refer to Figure-22. Specify desired values of width, depth, and height of 3D printing volume as per your 3D printer specifications and select the resolution option from the **Display resolution** drop-down. Note that display resolution does not reflect the printing resolution. After setting desired parameters, click on the **OK** button.

Figure-22. Reorient Settings dialog box

- Click on the **Settle** button from the command box to place nearest flat face of model on the bed. Note that it might not be the most suitable orientation of model for 3D Printing.
- Click on the **Auto Orient** button from the command box to orient model to next best fit for 3D printing using same base face.
- Click on the **Scale to Fit** button from the command box to automatically increase/ decrease the size of model so that it fits in the printing volume.
- Click on the **Reset Orientation** button from the command box to resets the model to its original orientation.
- Click on the **Undo** button from the command box to revert model to the last action.
- Click on the **Redo** button from the command box to apply the previously undone action.
- By default, the **Move** option is selected in the **Reorient** drop-down of the command box. Specify desired values in **X**, **Y**, and **Z** edit boxes to move the model in respective directions.
- Select the **Rotate** option from the **Reorient** drop-down of command box to rotate the model. Specify desired angle values in the **X**, **Y**, and **Z** edit boxes.
- After making desired changes, press **ESC** to exit the tool.

Defining Print Material

The **Print Material** tool is used to define the material to be used in printer for 3D printing. The procedure to use this tool is given next.

- Click on the **Print Material** tool from the **Prepare** panel in the **3D Print** tab of the **Ribbon**. The **Print Material** command box will be displayed; refer to Figure-23.

Figure-23. Print Material command box

- Select desired option from the **Material** drop-down in the command box to define material being used for 3D printing the model.
- Click on the **Color** button from the command box to define color for model to be printed. The **Color** dialog box will be displayed. Select desired color from the dialog box and click on the **OK** button.
- Click on the **OK** button from the command box to apply new material.

Wall Thickness Validation

The **Wall Thickness** tool in **3D Print** tab is used to check if all the features in the model have valid thickness or not. The procedure to use this tool is given next.

- Click on the **Wall Thickness** tool from the **Validate** group of **3D Print** tab in the **Ribbon**. The **Wall Thickness** command box will be displayed; refer to Figure-24.

Figure-24. Wall Thickness command box

- Click on the **Options** button from the command box. The **Wall Thickness Settings** dialog box will be displayed to specify colors for which are less thick and more thick than specified minimum thickness value.
- Specify desired value of minimum thickness your 3D Printer can produce in the **Minimum Thickness** edit box of the command box.
- Click on the **Toggle Visibility** button after running analysis to check features which have less thickness than required. Note that sometimes you may get false problematic areas so you need to carefully check the results.
- Click on the **Perform Analysis** button from the command box to check whether all the features in the model have minimum thickness.
- Press **ESC** to exit the tool.

Checking Overhangs

The **Overhang** tool is used to check whether there are overhangs in the model at angle more than specified values. The procedure to use this tool is given next.

- Click on the **Overhang** tool from the **Validate** group of the **3D Print** tab in the **Ribbon**. The **Overhang** command box will be displayed; refer to Figure-25.

Figure-25. Overhang command box

- Specify desired value of overhang angle in the **Maximum Angle** edit box.
- Click on the **Toggle Visibility** button from the command box. Only the problematic areas will be displayed.
- Click on the **Perform Analysis** button. The problematic area of model will be displayed.
- Press **ESC** to exit the tool.

Print Validations

The **Print Validations** tool is used to check whether model passes all the 3D print checks. The procedure to use this tool is given next.

- Click on the **Print Validations** tool from the **Validate** group in the **3D Print** tab of the **Ribbon**. The **Print Validations** dialog box will be displayed; refer to Figure-26.

Figure-26. Print Validations dialog box

- Select desired options from the **Model Options** area to define material and quality of 3D print.
- Select check boxes from the **Checks** area to define the tests to be performed on model for its validity for 3D printing.
- Set the other parameters as desired and click on the **Compute** button. The results of checks will be displayed in the **Results** area. If there is any problem then modify the model accordingly.
- Click on the **OK** button to exit the dialog box.

Exporting Model to STL File

The **Export** tool is used to generate an STL file using the 3D model prepared. Click on the **Export** tool from the **Export** group of **3D Print** tab in the **Ribbon**. The **3D Printing Export** dialog box will be displayed; refer to Figure-27. Specify desired name of file and click on the **Save** button.

Figure-27. 3D Printing Export dialog box

The other options for 3D Printing has been discussed earlier in Chapter 1.

VIEW TAB

The tools and options in the **View** tab are used to create and manage various views of the model. This tab is available in various environments like Part modeling, assembly, sheet metal, and so on. The tools in this tab are discussed next.

SHOWING PANES

The options in the **Panes** drop-down are used to display/hide various panes in the interface; refer to Figure-28. To display this drop-down, click on the **Panes** button from the **Show** group of the **View** tab in the **Ribbon**. Select desired options from the drop-down to display respective interface elements.

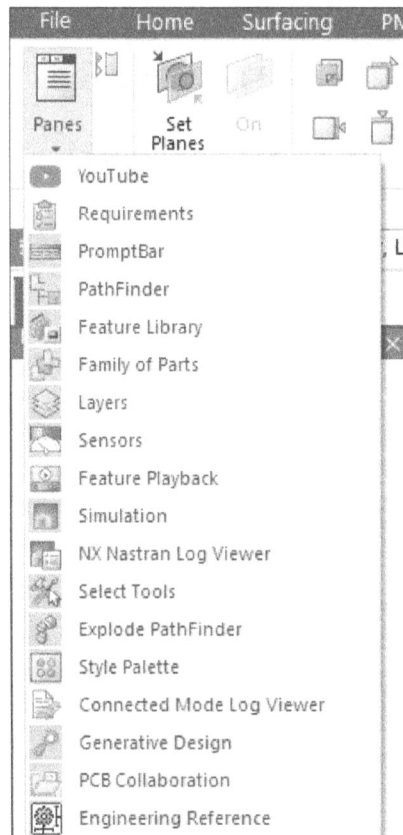

Figure-28. Panes drop down

CONSTRUCTION DISPLAY OPTIONS

The **Construction Display** button in **Show** group is used to show/hide construction elements. The procedure to use this button is given next.

- Click on the **Construction Display** button from the **Show** group in the **View** tab of the **Ribbon**. The **Show All/Hide All** dialog box will be displayed; refer to Figure-29.

Figure-29. Show All Hide All dialog box

- Select the check boxes under **Show All** or **Hide All** columns for various construction objects to be displayed or hidden.
- After setting desired parameters, click on the **OK** button from the dialog box.

SETTING PLANES

The **Set Planes** tool is used to create reference planes for clipping the model. The procedure to use this tool is given next.

- Click on the **Set Planes** tool from the **Clip** group in the **Views** tab of the **Ribbon**. The **Set Planes** command box will be displayed; refer to Figure-30.

Figure-30. Set Planes command box

- Select desired plane or face to be used as reference for creating clipping planes. Note that the planes will be created parallel to selected face/plane.
- Click at desired location to specify starting clipping plane. You will be asked to specify end clipping plane.
- Click at desired location to specify end clipping plane. The model between created clipping planes will be displayed when **Clipping On** button is active in the **Clip** group of **View** tab in the **Ribbon**; refer to Figure-31.
- To modify the planes after creating them, select **Dynamic Clipping** button from the command box and specify desired value in **Distance** edit box to set the distance to the offset the clipping plane.
- Click on the **Finish** button from the command box to create clipped model.

Figure-31. Clipping

VIEWS

The buttons in the **Saved Views** box are used to set the orientation of parts and assemblies as per saved orientation styles; refer to Figure-32.

Figure-32. Saved Views

Front View

The **Front View** button in the **Saved Views** box is used to change the view to front view. The hot key used for it is **CTRL+F**.

Back View

The **Back View** button in the **Saved Views** box is used to change the view to back view. The hot key used for it is **CTRL+K**.

Left View

The **Left View** button in the **Saved Views** box is used to change the view to left view. The hot key used for it is **CTRL+L**.

Right View

The **Right View** button in the **Saved Views** box is used to change the view to right view. The hot key used for it is **CTRL+R**.

Top View

The **Top View** button in the **Saved Views** box is used to change the view to top view. The hot key used for it is **CTRL+T**.

Bottom View

The **Bottom View** button in the **Saved Views** box is used to change the view to bottom view. The hot key used for it is **CTRL+B**.

Dimetric View

The **Dimetric View** button in the **Saved Views** box is used to change the view to dimetric view. In Dimetric view, two of the angles between the projection of axes are equal. The hot key used for it is **CTRL+J**.

ISO View

The **ISO View** is used to change the view to ISO View. The ISO view is the method for visually representing three dimensional objects in two dimensions in technical and engineering drawing. It is an axonometric projection in which the three coordinate axes appear equally foreshortened and the angle between any two of them is 120 degree. The hot key used for it is **CTRL+I**.

Trimetric View

The **Trimetric View** button in the **Saved Views** box is used to change the view to trimetric view. In Trimetric view, the projection of the angles between the axes are unequal. Thus, separate scales are needed to generate trimetric projection of an object. The hot key used for it is **CTR1+M**.

Saving Current View

The **Save Current View** option in the **Saved Views** box is used to save current view of model in the **Saved Views** box. On clicking this button, the **New Named View** dialog box will be displayed; refer to Figure-33. Specify desired name in the **New view name** edit box and description of your view in the **Description** edit box of the dialog box. Click on the **OK** button from the dialog box to create the view. The new view will be displayed above **Save Current View** option from the **Saved Views** box.

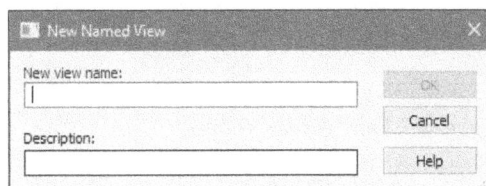

Figure-33. New Named View dialog box

View Manager

The **View Manager** option in the **Saved Views** box is used to manager standard and custom views. Click on the **View Manager** option from the **Saved Views** box in the **Views** group of **View** tab in the **Ribbon**. The **Named Views** dialog box will be displayed; refer to Figure-34. Select desired view from the list and click on the **Apply** button to apply selected view. Select desired view and click on the **Redefine to Current** button from the dialog box if you want to change the selected view based on how model is oriented in graphics area. Select desired view and click on the **Delete** button. Click on the **Close** button from the dialog box to exit.

Figure-34. Named Views dialog box

ORIENT

We have already discussed most of the tools in previous chapters. Here, we are going to discuss the view tools which have not been discussed earlier.

Look at Face

The **Look at Face** tool is used to make the selected face parallel to the screen. To use this tool, click on the **Look at Face** tool from the **Orient** group of the **View** tab in the **Ribbon** and select the face that you want to make parallel to screen.

Previous View

The **Previous View** tool is used to restore previous view. On selecting this tool, the part or assembly will be displayed in previous view.

Spin About

The **Spin About** tool is used to rotate the model about a selected face. Click on this tool and select the face along which you want to rotate the model; refer to Figure-35.

Figure-35. Spin About feature

Refresh View

The **Refresh View** tool is used to redraw the active view. To use this tool, click on the **Refresh View** tool from the **Orient** group of the **View** tab in the **Ribbon**.

VIEW STYLES

The buttons in the **View Styles** box are used to change the view style of the model. Click on the **View Styles** drop-down from **Style** group in the **View** tab of the **Ribbon**. The **View Styles** box will be displayed; refer to Figure-36.

Figure-36. View Styles box

Wire Frame

The **Wire Frame** tool is used to apply wire frame display mode to the model. On selecting this tool the model changes to wire frame; refer to Figure-37.

Figure-37. Wire Frame model

Visible and Hidden Edges

The **Visible and Hidden Edges** tool is used to display visible and hidden edges of the model; refer to Figure-38.

Figure-38. Visible and Hidden Edges model

Visible Edges

The **Visible Edges** tool is used to display visible edges only in the model; refer to Figure-39.

Figure-39. Visible Edges model

Shaded

The **Shaded** tool is used to display model in shaded display mode. On selecting this tool, the model display will change to shaded display; refer to Figure-40.

Figure-40. Shaded model

Shaded With Visible Edges

The **Shaded with Visible Edges** tool is used to display shaded model with dark visible edges; refer to Figure-41.

Figure-41. Shaded with visible edges model

FLOOR REFLECTION

The **Floor Reflection** tool is used to show reflection of model on the floor. Click on the **Floor Reflection** tool from the **Style** group of the **View** tab in the **Ribbon** to display floor reflection; refer to Figure-42.

Figure-42. Floor Reflection displayed

FLOOR SHADOW

The **Floor Shadow** tool is used to display shadow of model on the floor; refer to Figure-43.

Figure-43. Floor Shadow displayed

HIGH-QUALITY RENDERING

The **High-Quality Rendering** button is used to improve the appearance of low face-count models and enables display of Bump Maps. This setting is an overrides to the active 3D view style.

EDGE COLOR

Select the **Single Edge Color** button from the **Single-Color Edges** drop-down of **Style** group in the **View** tab of the **Ribbon** to display all the edges in one color. You can set desired color for edges by using **Select Edge Color** tool from **Single-Color Edges** drop-down.

SHARPEN

The **Sharpen** button in the **Style** drop-down is used to sharpen edges of the model. You can also select sharpening level by selecting desired option from the **Sharpness Set** cascading menu of **Sharpen** drop-down; refer to Figure-44.

Figure-44. Sharpness Set cascading menu

PERSPECTIVE VIEW

The **Perspective** button in the **Style** group of **View** tab in the **Ribbon** is used to display model in perspective view.

SHOWING FACET EDGES

The **Show Facet Edges** toggle button is used to display facet edges of the mesh body; refer to Figure-45.

Figure-45. Mesh edges displayed

COLOR MANAGER

The **Color Manager** tool is used to change color style for various objects in the model. The procedure to use this tool is given next.

• Click on the **Color Manager** tool from the **Style** group in the **View** tab of the **Ribbon**. The **Color Manager** dialog box will be displayed; refer to Figure-46.

Figure-46. Color Manager dialog box

- Select the **Use Solid Edge Options color settings** radio button to use the color scheme specified in **Solid Edge Options** dialog box.
- If you want to use individual part styles for various objects then select the **Use individual part styles** radio button. Select desired color/material styles from the drop-downs for various objects in the **Base Styles** section of dialog box.
- After setting desired parameters, click on the **OK** button.

VIEW OVERRIDES

The **View Overrides** tool is used to apply overrides to the active 3D view style. The procedure to use this tool is given next.

- Click on the **View Overrides** tool from the **Style** group in the **View** tab of **Ribbon**. The **View Overrides** dialog box will be displayed; refer to Figure-47.

Figure-47. View Overrides dialog box

Rendering Tab

- By default, the **Rendering** tab is selected in the dialog box. Set desired options in this tab to define smoothness and style in the rendering.
- Select desired option from the **Perspective** drop-down to set focus level of perspective view in the dialog box.
- Select desired option from the **Antialias level** drop-down to specify edge quality of model.
- Select desired option from the **Render mode** drop-down to specify rendering mode.
- Select desired option from the **Hidden lines** drop-down to display hidden lines in the rendering of model.
- Select desired check boxes from the dialog box to display various elements of rendering in the dialog box.

Lights Tab

- Select the **Lights** tab in the dialog box to set the location, intensity, and color of light; refer to Figure-48.

Figure-48. Lights tab in View Overrides dialog box

- Set desired color and intensity using the sliders.
- Select desired button for light from the dialog box to define direction and density of light in rendering.

Background Tab

- Select the **Background** tab from the dialog box to apply color or image in the background of model; refer to Figure-49.

Figure-49. Background tab in View Overrides dialog box

- Select the **Solid** radio button and then select desired color from the drop-down next to the radio button to apply a solid color in background of model.
- Select the **Gradient** radio button to use gradient background using two colors selected in the drop-downs. You can also select direction of gradient in the drop-down next to the radio button.
- Select the **Image** radio button to use desired image as background of the model and click on the **Browse** button. The **Open a File** dialog box will be displayed; refer to Figure-50. Select desired image file and click on the **Open** button to select it.

Figure-50. Open a File dialog box

- Select the **Mirror up/down** radio button to mirror the image background from up to down. Similarly, select the **Mirror left/right** radio button mirror image from left to right.

Environment Tab

- Select **Panoramic image** option from the **Type** drop-down to specify a panoramic image reflected by the model. Select **Reflection box** option to specify a six sided environment in which the model resides. Select **Sphere map** option to specify a hemispherical shaped image reflected by the model.
- Select **Use for background** check box and select desired option from the drop-down below it. Select **Static** option to define fixed background image that does not change with the camera. Select **Immersive** option to define dynamic background that changes with the view showing the model and environment as the user reorients the view.
- Click on **Browse** button. The **Open a File** dialog box will be displayed; refer to Figure-51.

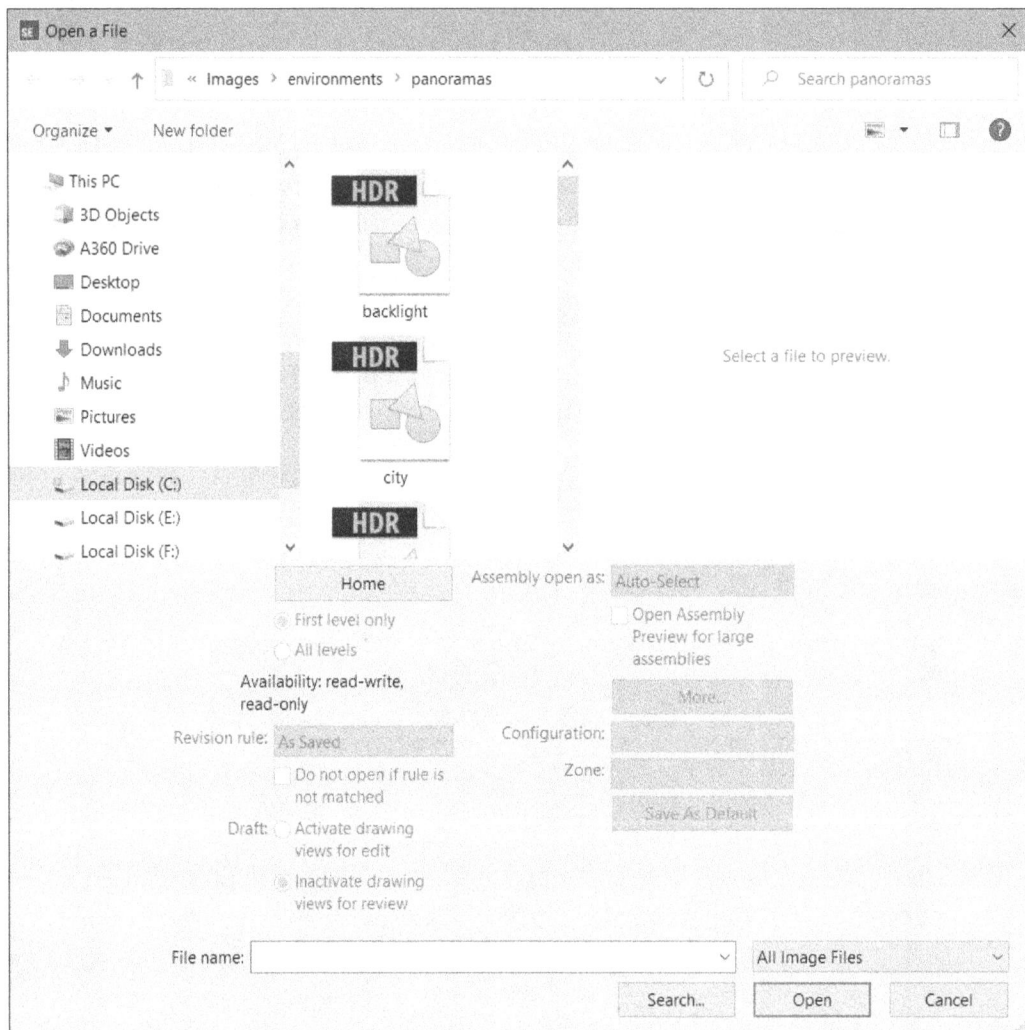

Figure-51. Open a file dialog box

- Select desired image from the dialog box and click on the **Open** button. The name and path of the image will be displayed in the **File name** box.
- Click on the **Orient** button. The **Orient Reflection Box** dialog box will be displayed; refer to Figure-52.

Figure-52. Orient Reflection Box dialog box

- The **Direction** area of the dialog box changes the direction from which the image is viewed. The **Roll** area rolls the viewing angle of the image. The **Tilt** area tilts the viewing angle of the image. The **Zoom** area changes the viewing area of the image. You can dynamically drag the image dialog box to set desired orientation. After specifying desired parameters, click on **OK** button from the dialog box.
- Click on **Save As** button. The **New Style from Overrides** dialog box will be displayed; refer to Figure-53.

Figure-53. New style overrides dialog box

- Specify desired style name in the **New style name** edit box and click on **OK** button from the dialog box.
- After specifying desired parameters in **Environment** tab, click on **OK** button from **View Overrides** dialog box.

APPLYING APPEARANCES TO PART

The **Part Painter** tool is used to apply appearances to model like gold paint, brass paint, and so on. This tool is available for Part Design. The procedure to use this tool is given next.

- Click on the **Part Painter** tool from the **Style** group in the **View** tab of the **Ribbon**. The message box to ask whether you want to display **Part Painter** dialog box or **Style Palette** will be displayed; refer to Figure-54.
- Select the **Yes** button if you want to use the **Style Palette** for applying appearances to model. The **Style Palette** will be displayed at the right in the application window; refer to Figure-55. Select the **No** button from the message box if you want to use **Part Painter** command box; refer to Figure-56. Note that Part Painter can use only appearances saved in current document not the library.

Figure-54. Message box for Style Palette

Figure-55. Style Palette

Figure-56. Part Painter command box

Using Part Painter

- Select desired color/appearance from the **Style** drop-down.
- Click on the **Select** drop-down to define the selection filter for selecting object. By default, **Any** is selected in the drop-down so you can select any of the object type from the model.
- Select desired option from the **Currently Applied Styles** drop-down to define whether you want to replace the previous display style or you want to just apply the appearance and rest of the style will remain same.
- After setting desired options, click on the model to apply selected style.
- Click on the **Close** button from the command box to exit the tool.

Using Style Palette

- Select desired material category from the **Material Table** section of the **Style Palette**. The list of material textures/appearances will be displayed.
- Select desired material appearance from the palette and drag it on the model.

SELF ASSESSMENT

Q1. What is the purpose of the Face Priority option in the Select drop-down menu?

A) To select parts from the assembly model
B) To select faces from the model
C) To use standard selection mode
D) To add or remove objects from selection

Q2. Which of the following processes is NOT a method of 3D printing?

A) Vat Photopolymerisation
B) Sheet Lamination
C) Layer Compression
D) Material Extrusion

Q3. What is the main function of the Wall Thickness tool in the 3D Print tab?

A) To check for overhangs
B) To validate the model for printability
C) To check if all features have valid thickness
D) To define the printing material

Q4. Which technology uses a high power laser to fuse small particles of materials in 3D printing?

A) Fused Deposition Modeling (FDM)
B) Material Jetting
C) Selective Laser Sintering (SLS)
D) Stereolithography (SLA)

Q5. What must be done before using the Delete Voids tool in Solid Edge?

A) Check the voids in the model
B) Specify the material for printing
C) Select the model to be printed
D) Reorient the model

Q6. When using the 3D Printing feature in Solid Edge, what is the importance of part stability?

A) It affects the model's color and texture
B) It determines the model's complexity
C) It ensures the model does not collapse during printing
D) It reduces the overall material cost

Q7. Which button is used to change the view to the Front View?

A) CTRL+K
B) CTRL+F
C) CTRL+T
D) CTRL+B

Q8. What feature does the High-Quality Rendering button provide?

A) It saves the current model
B) It improves the appearance of low face-count models
C) It changes the color of edges
D) It rotates the model

Q9. What can you do using the Color Manager tool?

A) Change the background of the model
B) Adjust the model's light settings
C) Change color styles for various objects in the model
D) Set the rendering mode

Q10. What does the Sharpen button do in the Style drop-down?

A) It applies a color to the model
B) It sharpens the edges of the model
C) It creates a new view
D) It changes the view perspective

Q11. What is the hotkey for switching to the Right View?

A) CTRL+T
B) CTRL+B
C) CTRL+R
D) CTRL+J

Q12. Which view is represented by the hotkey CTRL+T?

A) Bottom View
B) Top View
C) Dimetric View
D) ISO View

Q13. What is the function of the View Manager?

A) To create new views only.
B) To manage standard and custom views.
C) To delete all views permanently.
D) To reset all views to default settings.

Q14. Which view style displays the model as a series of lines and vertices?

A) Shaded
B) Wire Frame
C) Perspective
D) Shaded with Visible Edges

Q15. What is the purpose of the Shaded with Visible Edges tool?

A) To show the model in wireframe mode.
B) To display the model in shaded mode with outlines.
C) To hide all edges of the model.
D) To create a floor reflection of the model.

Q16. Which tool allows you to apply color styles to various objects in the model?

A) View Styles
B) Color Manager
C) Part Painter
D) View Overrides

Q17. What is the function of the Spin About tool?

A) To rotate the model freely.
B) To rotate the model about a selected face.
C) To zoom in on the selected face.
D) To change the viewing angle of the model.

Q18. What is the primary function of the Orient tool?

A) To save the current model.
B) To change the orientation of the model view.
C) To adjust color settings.
D) To display hidden edges.

Q19. Which hotkey is used to change to Trimetric View?

A) CTRL+I
B) CTRL+M
C) CTRL+J
D) CTRL+B

Q20. Which option would you select to modify the current view in the View Manager?

A) Apply
B) Save
C) Redefine to Current
D) Delete

Q21. In the Perspective View, what is the effect of changing the focus level?

A) It alters the color of the model.
B) It affects the depth perception of the model.
C) It changes the lighting conditions.
D) It modifies the rendering quality.

Q22. What is the purpose of the Show Facet Edges toggle button?

A) To hide all edges.
B) To display facet edges of the mesh body.
C) To apply a solid color to edges.
D) To convert the model to wireframe.

Q23. Which button is used to apply a solid color to all edges in the model?

A) Single Edge Color
B) Color Manager
C) Edge Color Tool
D) Visible Edges Tool

Q24. What is the purpose of the Physical Thread tool?

A) To create new threads
B) To convert cosmic threads into physical threads
C) To delete internal voids
D) To validate wall thickness

Q25. What does the Overhang tool analyze?

A) The number of threads
B) The color of the model
C) The presence of overhangs at specified angles
D) The size of the printing volume

Q26. What is the purpose of the Export tool in 3D printing?

A) To check wall thickness
B) To generate an STL file from the 3D model
C) To analyze overhangs
D) To define printing material

Q27. What does the Set Planes tool do?

A) It changes the model's color
B) It creates reference planes for clipping the model
C) It converts cosmic threads
D) It validates print material

Q28. Which button in the Saved Views box is used to change the view to front view?
A) CTRL+K
B) CTRL+F
C) CTRL+L
D) CTRL+M

Q29. What is the purpose of the Physical Thread tool in the 3D Print tab?

A) To delete internal voids in a model
B) To convert cosmic threads into physical threads
C) To check wall thickness
D) To define print material

Q30. Which button do you click to check for internal voids in the model?

A) Select All Threads
B) Check the voids
C) Delete button
D) OK button

Q31. What does the Scale to Fit option do in the Reorient tool?

A) Resets the model to its original orientation
B) Deletes unwanted voids in the model
C) Adjusts the model's size to fit within the 3D printing volume
D) Changes the color of the model

Q32. What is the function of the Wall Thickness tool?

A) To check for overhangs in the model
B) To delete internal voids
C) To verify that all features have valid thickness
D) To define the material used for printing

Q33. Which of the following is NOT an option when using the Overhang tool?

A) Specify desired overhang angle
B) Toggle visibility of problematic areas
C) Define print material properties
D) Perform analysis on overhangs

Q34. How can a user save the current view of the model?

A) By clicking the Save Current View option
B) By using the Look at Face tool
C) By selecting the Refresh View tool
D) By adjusting the View Styles

Q35. Which command is used to apply color to the model being printed?

A) Wall Thickness
B) Print Material
C) View Overrides
D) Part Painter

Q36. What does the 'Toggle Visibility' button do in the Wall Thickness tool?

A) Displays all edges of the model
B) Hides or shows problematic features with insufficient thickness
C) Resets the model orientation
D) Applies new material to the model

Q37. What feature does the Look at Face tool provide?

A) It allows the user to save the current view.
B) It makes the selected face parallel to the screen.
C) It checks the wall thickness of the model.
D) It deletes internal voids.

Q38. What is the purpose of the Environment tab in the Rendering dialog box?

A) To set the lighting conditions
B) To define the color of the model
C) To apply background colors or images
D) To check for overhangs

Q39. How do you apply appearances like gold paint to a model?

A) By using the Print Material tool
B) By using the Part Painter tool
C) By exporting the model as STL
D) By checking for internal voids

Q40. What action does the Refresh View tool perform?

A) It applies a new color to the model
B) It redraws the active view
C) It saves the current view settings
D) It exits the tool

Chapter 6

Surfacing

Topics Covered

The major topics covered in this chapter are:

- *Introduction to Surfacing*
- *Creating Curves for Surfacing*
- *Creating Surfaces*
- *Modifying Surfaces*
- *Surface Visualization Settings*
- *Displaying Section Curvature*

SURFACING

Surfacing is a separate world in the field of CAD. The complicated shapes which are difficult for solid modeling are most of the time easy for surfacing. Basic tools of surfacing are very similar to the solid creation tools discussed earlier like extrude, revolve, swept, and so on. But, there are some other tools that allow to modify the part shape freely in 3D. In Solid Edge, the tools to create surface models are available in the **Surfacing** tab of the **Ribbon** in the Part environment; refer to Figure-1. By default, most of the surfacing tools are not active. These tools will be active automatically when related objects are available in the drawing area.

Figure-1. Surfacing tab in Ribbon synchronous

SUBDIVISION MODELING

The tools in the **Subdivision Modeling** tab are used to create shapes that are not easily modeled with traditional methods. Click on the **Subdivision Modeling** button from **Free-Form** panel in the **Surfacing** tab of the **Ribbon**. The tools are displayed in the **Ribbon**; refer to Figure-2. These tools are discussed next.

Figure-2. Subdivision Modeling tools

Box

The **Box** tool is used to create a box-shaped primitive feature with smooth or hard edges as the base for your subdivision feature. The procedure to use this tool is discussed next.

*   Click on the **Box** tool from **Shapes** panel in the **Home** tab of the **Ribbon**. The **Box** command box will be displayed with the box-shaped primitive feature; refer to Figure-3.

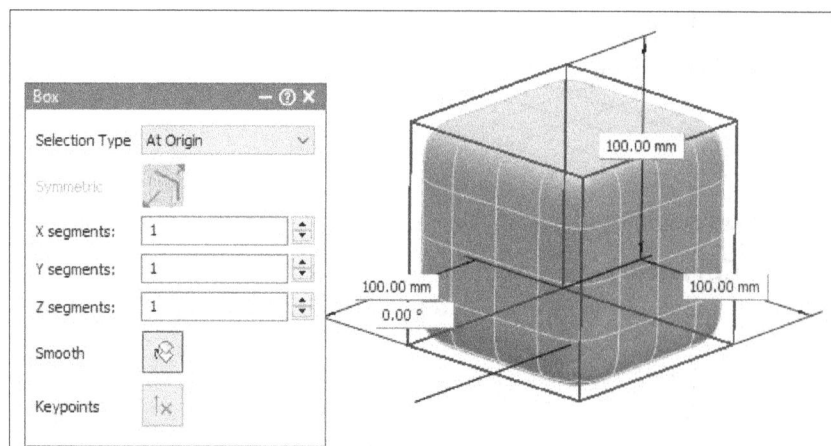

Figure-3. Box command box with primitive

- Select **At Origin** option from the **Selection Type** drop-down to create a box cage at the origin of the base coordinate system. Select **By Center** option to create a box cage by defining the center and the shape of the feature cross-section. Select **By 2 Points** option to create a box cage by defining the start corner and shape of the feature cross-section. Select **By 3 Points** option to create a box by defining the width and height of the feature cross-section by clicking 3 points.

- Select the **Symmetric** button from the command box to create an extrusion symmetrically about the selection.

- Specify the number of segments which you want to use along the x-axis, y-axis, and z-axis of the box cage in the **X-segments**, **Y-segments**, and **Z-segments** edit boxes of the command box, respectively.

- Click on the **Smooth** toggle button to create the shape with smooth edges.

- Select **Center and End Point** option from the **Key point** drop-down to sets the key point filter to locate the display center and end point. Select **All** option from the drop-down to sets the key point filter to locate the display all key point type. Select **End Point** option from the drop-down to sets the key point filter to locate the display end points. Select **Center Point** option from the drop-down to sets the key point filter to locate the display center point. Select **Mid Point** option from the drop-down to sets the key point filter to locate the display mid points. Select **Tangent** option from the drop-down to sets the key point filter to locate the display tangent points. Select **Silhouette** option from the drop-down to sets the key point filter to locate the display silhouette points. Select **Edit Point** option from the drop-down to sets the key point filter to locate the display points on a curve. Select **None** option from the drop-down to turn off locate and display of key points.

- After specifying desired parameters, click **RMB** in an empty area. The box shaped primitive feature will be created; refer to Figure-4.

Figure-4. Box shaped primitive feature created

Cylinder

The **Cylinder** tool is used to create a cylindrical-shaped primitive feature with smooth or hard edges as the base for your subdivision feature. The procedure to use this tool is discussed next.

- Click on the **Cylinder** tool from **Shapes** panel in the **Home** tab of the **Ribbon**. The **Cylinder** command box will be displayed with the cylinder-shaped primitive feature; refer to Figure-5.

Figure-5. Cylinder command box with feature

- Select **At Origin** option from the **Selection Type** drop-down to create a cylindrical cage at the origin of the base coordinate system axially aligned with the Z-axis. Select **By 2 Points** option to create a cylindrical cage by defining the center of the polygon and the midpoint of the side of the feature on the top face.
- Select the **Symmetric** button from the command box to create an extrusion symmetrically about the selection.
- Specify desired number of segments used for the side face of the cylindrical cage in the **Side Segments** edit box.
- Specify desired number of segments used for the base sides of the cylindrical cage in the **Base Segments** edit box.
- Click on the **Smooth** toggle button to create the shape with smooth edges.
- After specifying desired parameters, click **RMB** in an empty area. The cylindrical shaped primitive feature will be created; refer to Figure-6.

Figure-6. Cylindrical shaped primitive feature created

Sphere

The **Sphere** tool is used to create a spherical shape as the base for your subdivision feature. The procedure to use this tool is discussed next.

- Click on the **Sphere** tool from **Shapes** panel in the **Home** tab of the **Ribbon**. The **Sphere** command box will be displayed with the sphere-shaped primitive feature; refer to Figure-7.

Solid Edge 2026 Black Book

Figure-7. Sphere command box with feature

- Select **At Origin** option from the **Selection Type** drop-down to create a spherical cage at the origin of the base coordinate system. Select the **By 2 Points** option to create a spherical cage by defining the center of the sphere and size of the cage.
- Specify desired number of level used to determine the subdivision segments of the spherical cage in the **Level** edit box.
- After specifying desired parameters, click **RMB** in an empty area. The sphere shaped primitive feature will be created; refer to Figure-8.

Figure-8. Sphere shaped primitive feature created

Torus

The **Torus** tool is used to create a toroidal shape as the base for your subdivision feature. The procedure to use this tool is discussed next.

- Click on the **Torus** tool from **Shapes** panel in the **Home** tab of the **Ribbon**. The **Torus** command box will be displayed with the toroidal-shaped primitive feature; refer to Figure-9.

Figure-9. Torus command box with feature

- Select **At Origin** option from the **Selection Type** drop-down to create a torus cage at the origin of the base coordinate system axially aligned with the Z-axis. Select **By Axis and 2 Points** option to create a torus cage by defining the center, torus axis, the cage major diameter (including segment angle), and the cage minor diameter (including the segment angle).
- Specify desired number of segments used for the major diameter of the torus cage in the **Major Diameter Segments** edit box.
- Specify desired number of segments used for the minor diameter of the torus cage in the **Minor Diameter Segments** edit box.
- After specifying desired parameters, click **RMB** in an empty area. The toroidal shaped primitive feature will be created; refer to Figure-10.

Figure-10. Toroidal shaped primitive feature created

Start Symmetry

The **Start Symmetry** tool allows you to work on one side of the model to apply symmetric edits on both sides of the model. Any edits to the parent side are reflected to the symmetric side. The procedure to use this tool is discussed next.

- Click on the **Start Symmetry** tool from **Symmetry** panel in the **Home** tab of the **Ribbon**. You are asked to select a face or reference plane.
- Select desired face or reference plane as the mirror plane for cage symmetry. The symmetric feature will be created; refer to Figure-11.

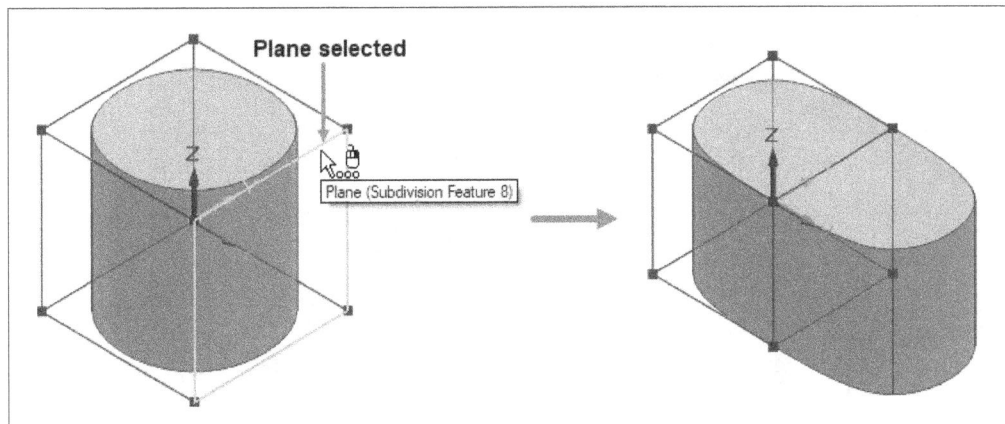

Figure-11. Symmetric feature created

Stop Symmetry

The **Stop Symmetry** tool returns you to the regular editing.

Scale

The **Scale** tool is used to make an entire model—or portions of a model—larger or smaller by precise amounts. The procedure to use this tool is discussed next.

- Click on the **Scale** tool from **Modify** panel in the **Home** tab of the **Ribbon**. The **Scale** command box will be displayed; refer to Figure-12. You are asked to select an edge or face.

Figure-12. Scale command box

- Select **3 Axes Scaling** option to scale the body along 3 axes. Select **Planar Scaling** option to scale the body along 2 axes. Select **Linear Scaling** option to scale the body along 1 axis. Select **Uniform** option to scale the body uniformly when 3 axes or planar scaling is used.
- Select desired edge or face from where you want to scale the model. A steering wheel will be attached to the model.
- Select the **Recenter** option to recenter the steering wheel to the geometric center of the select set.
- Click and drag the steering wheel axis to scale the model or specify desired value in the edit box displayed and press **Enter**. The body will be scaled; refer to Figure-13.

Figure-13. Body scaled

Blend

The **Blend** tool is used to sharpen or smooth the hard edges of a subdivision model. The procedure to use this tool is discussed next.

- Click on the **Blend** tool from **Modify** panel in the **Home** tab of the **Ribbon**. The **Blend** dialog box will be displayed; refer to Figure-14.

Figure-14. Blend dialog box

- Select desired edges or faces which you want to blend and drag the slider in the dialog box to specify desired value of blending between 0 and 3.
- After specifying desired value, click **RMB** in an empty area. The model will be blended; refer to Figure-15.

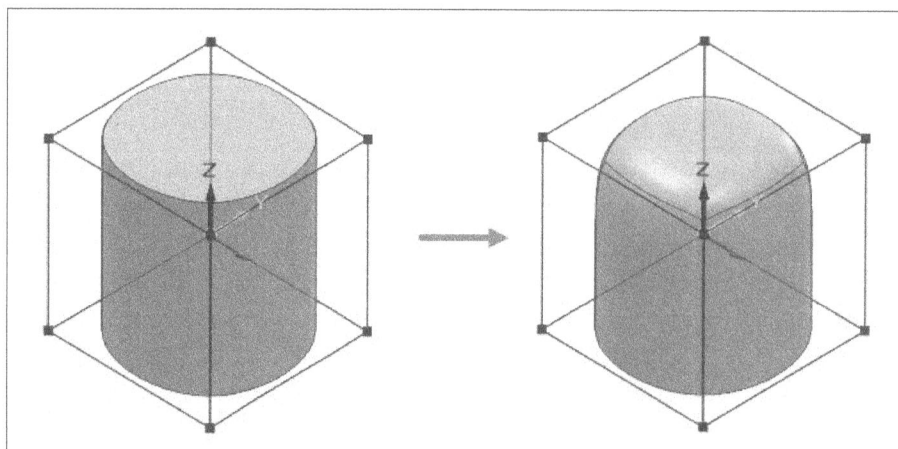

Figure-15. Model blended

- Click on the **Close** button to close the command box.

Split

The **Split** tool is used to add faces to your model by splitting one or more faces in the body cage. The procedure to use this tool is discussed next.

* Click on the **Split** tool from **Split** drop-down in the **Modify** panel of **Home** tab in the **Ribbon**. The **Split** command box will be displayed; refer to Figure-16.

Figure-16. Split command box

* Select **Single** option from the **Select** drop-down to split a single face and select **Chain** option from the drop-down to select multiple faces.
* Select desired face of the model to split and then select reference face to split the selected face.
* Specify desired number of split in the **Number** edit box and click on the **Accept** button from the command box. The faces of the model split; refer to Figure-17.

Figure-17. Faces splitted

* Click on the **Close** button to close the command box.

Split with Offset

The **Split with Offset** tool is used to split one or more faces in the body cage by offsetting them inward by a specified value. The procedure to use this tool is discussed next.

* Click on the **Split with Offset** tool from **Split** drop-down in the **Modify** panel of **Home** tab in the **Ribbon**. The **Split with Offset** command box will be displayed; refer to Figure-18.

Figure-18. Split with Offset command box

* Select desired face which you want to split and specify desired offset value in the **Offset** edit box displayed.
* After specifying desired value, click on the **Accept** button from the command box. The face will be split; refer to Figure-19.

Figure-19. Face split with offset

Fill

The **Fill** tool is used to fill missing faces in the subdivision model. The procedure to use this tool is discussed next.

- Click on the **Fill** tool from **Modify** panel in the **Home** tab of the **Ribbon**. The **Fill** command box will be displayed; refer to Figure-20.

Figure-20. Fill command box

- Select three or more edges of the cage which are required to create a face. The preview of filled face of the model will be displayed; refer to Figure-21.

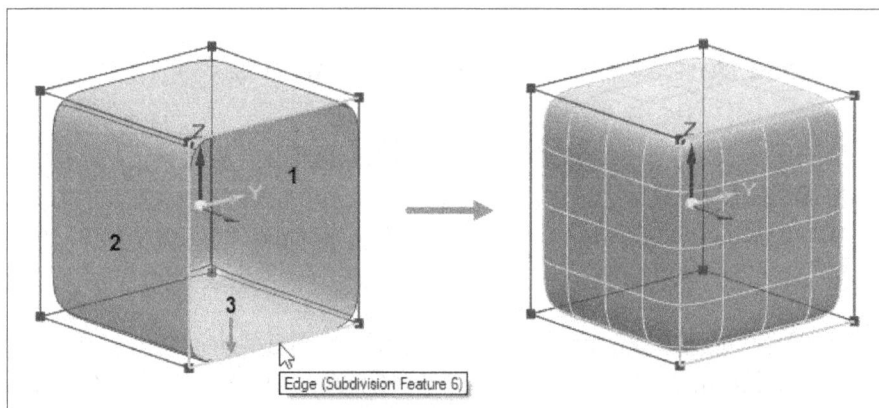

Figure-21. Preview of filled face

- Click on the **Accept** button from the command box to fill the face of the model.

Bridge

The **Bridge** tool is used to connect the edges of selected cage elements by bridging the space between them with a lofted structure. You can define a curve to control the shape of the bridge. The procedure to use this tool is discussed next.

- Click on the **Bridge** tool from **Modify** panel in the **Home** tab of the **Ribbon**. The **Bridge** command box will be displayed; refer to Figure-22.

Figure-22. Bridge command box

- Click on the **From Section** button and select the edges of the cage to define the start of the bridge and then select the **To Section** button and select the edges of another cage to define the end of the bridge; refer to Figure-23. The preview of bridge will be displayed; refer to Figure-24.

Figure-23. Defining the sections

Figure-24. Preview of bridge

- Click on the **Shape Curve** button and select the edge or curve used for the shape curve for the bridge, if required.
- Select desired option method from **Selection** drop-down.
- Specify desired number of cage segments for the resultant bridge in the **Segments** edit box.
- After specifying desired parameters, click on the **Accept** button from the command box to create the bridge.

Offset

The **Offset** tool is used to move or lift a cage face by offsetting it by a positive or negative specified value. The procedure to use this tool is discussed next.

- Click on the **Offset** tool from **Modify** panel in the **Home** tab of the **Ribbon**. The **Offset** command box will be displayed; refer to Figure-25.

Figure-25. Offset command box

- Click on the **Connected Faces** button to specify the faces connected to the selection during the modification. Click on the **Select Faces** button and select face(s) of the model which you want to offset.
- After selecting the face(s), click on the **Accept** button from the command box. The preview of offsetted face will be displayed; refer to Figure-26.

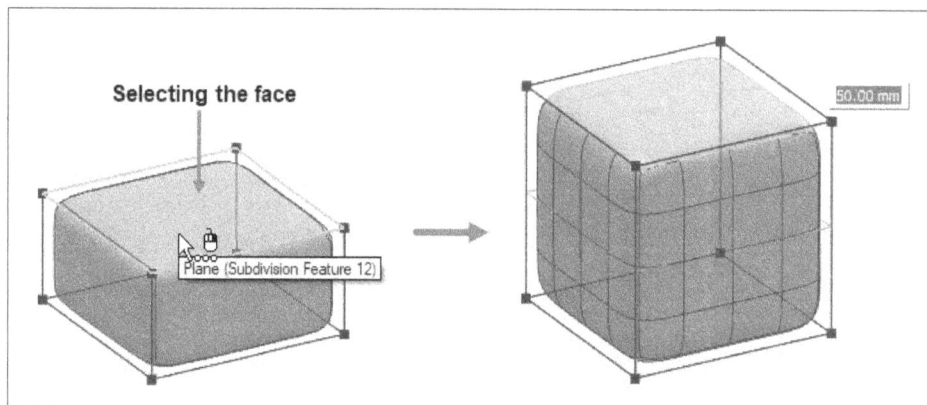

Figure-26. Face of the model offsetted

- Specify desired offset value in the edit box displayed and press **Enter**. The face of the model will be offsetted.
- Click on the **Close** button to close the command box.

Align to Curve

The **Align to Curve** tool is used to align a shape to a curve. The input curve can be a curve, sketch, or a set of user-defined gestures. The procedure to use this tool is discussed next.

- Click on the **Align to Curve** tool from **Modify** panel in the **Home** tab of the **Ribbon**. The **Align to Curve** command box will be displayed; refer to Figure-27.

Figure-27. Align to Curve command box

- Select the **Vertices** button of the model that you want to move to the sketch curve.
- Click on the **Alignment Direction** button and specify the alignment direction of the vertices.
- Click on the **Curves** button and select the curve which you want to align to the vertices.
- Select desired option from the **Selection** drop-down to set the element selection method.
- After specifying the parameters, click on the **Accept** button from the command box. The preview of alignment will be displayed; refer to Figure-28.

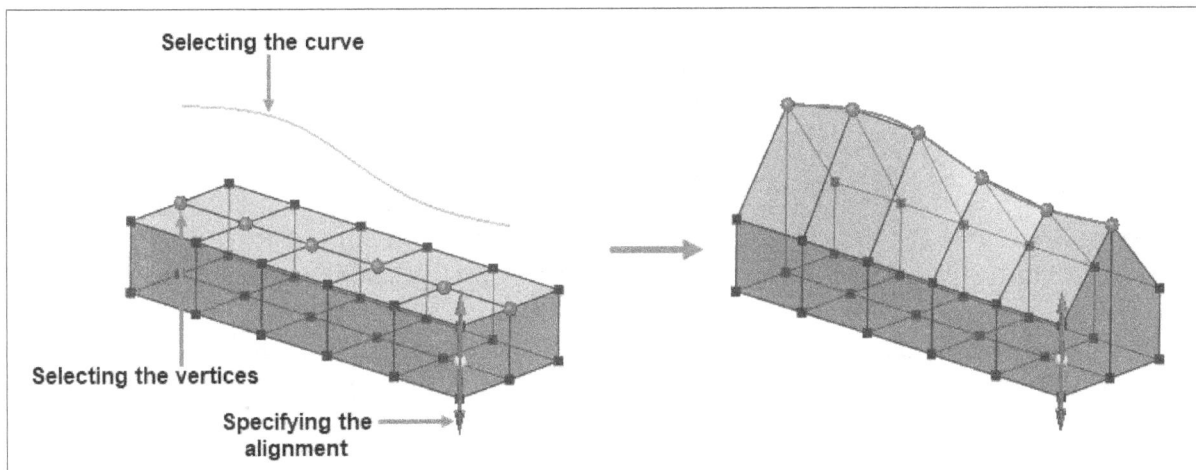

Figure-28. Preview of alignment

- Click on the **Back** or **Forward** button to move back or forward in the curve sequence, respectively.
- Click on the **Close** button to close the command box.

Cage Style

The **Cage Style** tool is used to apply user-defined color and size values to the edges and vertices of the cage. The procedure to use this tool is discussed next.

- Click on the **Cage Style** tool from **Style** panel in the **Home** tab of the **Ribbon**. The **Cage Style** dialog box will be displayed; refer to Figure-29.

Figure-29. Cage Style dialog box

- Select desired color and size of edges and vertices of cage from **Color** and **Size** drop-down of **Edges** and **Vertices** section of the dialog box, respectively.
- Click on the **Reset** button to reset the parameters.
- After specifying desired parameters, click on the **OK** button from the dialog box. The cage style will be applied; refer to Figure-30.

Figure-30. Cage style applied

- Click on the **Close Subdivision Modeling** button from the **Close** panel to exit the Subdivision Modeling environment.

KEY POINT CURVE

The **Keypoint Curve** tool is used to draw curve through selected key points or points which are created in sketch. The procedure to use this tool is given next.

- Click on the **Keypoint Curve** tool from the **Keypoint** drop-down in the **Curves** panel of **Surfacing** tab in the **Ribbon**. The **Keypoint Curve** command box will be displayed; refer to Figure-31.

Figure-31. Keypoint Curve command box

- Select desired key points for curve and right-click in the empty area. Select Preview button of the curve will be displayed; refer to Figure-32.

Figure-32. Keypoints curve

- If you want to modify the end condition of curve then click on the **End Conditions** button from the command box. The options for end conditions will be displayed in the command box; refer to Figure-33. Click on the **Open** button from the command box to sets the curve type to be open depending on specified keypoints. Click on the **Closed** button from the command box to close the curve by joining start and end points. Click **Show/Hide Tangency** button from the command box to set the tangency control option at the start and end point of the curve.

Figure-33. Buttons for modifying end condtions

- If you want to modify the length of curve then click on the **Curve Length** button from the command box and set the parameters as discussed earlier.
- Click on the **Preview** button from the command box and then click on the **Finish** button to create the curve.

CURVE BY TABLE

The **Curve by Table** tool is used to draw curve by using the coordinates specified in the table. The procedure is discuss next.

- Click on the **Curve by Table** tool from the **Keypoint** drop-down in the **Curves** panel of the **Surfacing** tab in the **Ribbon**. The **Insert Object** dialog box will be displayed as shown in Figure-34.

Figure-34. Insert Object dialog box

- Select the **Create from file** radio button if you want to use already existing Microsoft Excel file. If you want to create a new sheet then select the **Create new spreadsheet** radio button. After setting desired parameters, click on the **OK** button. The worksheet will open in Microsoft Excel. Make sure you have Microsoft Excel installed in your system.
- Specify desired values of X, Y, and Z coordinates of key points for curve; refer to Figure-35.

Figure-35. Create New table in Excel file

- After specifying parameters in the excel sheet, close the Microsoft Excel application. Preview of the curve will be displayed; refer to Figure-36.

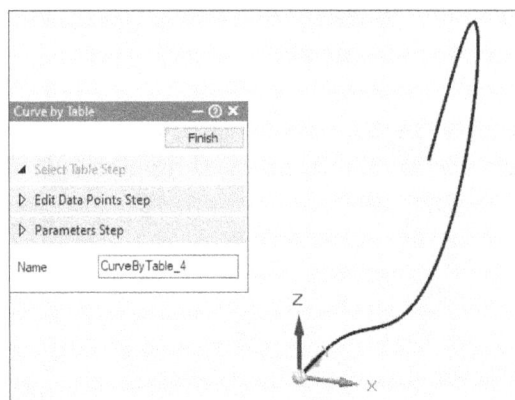

Figure-36. Preview of curve by table

- Click on the **Finish** button to create the feature.

INTERSECTION

The **Intersection** tool is used to create curve at the intersection of two surfaces or surface and plane. The procedure to use this tool is given next.

- Click on the **Intersection** tool from the **Curves** panel in the **Surfacing** tab of **Ribbon**. The **Intersection** command box will be displayed; refer to Figure-37.

Figure-37. Intersection command box

- Select the first set of surfaces or faces to be used for intersection curve while holding the **CTRL** key.
- Select the second set of surfaces or faces which are intersecting with first set. Preview of intersection curve will be displayed.
- Click on the **Finish** button from command box. You will notice that a curve is created at intersection of two bodies; refer to Figure-38. Press **ESC** to exit the tool.

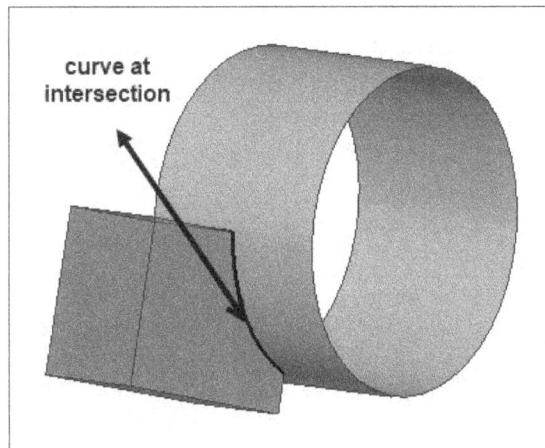

Figure-38. Intersection curve

PROJECT

The **Project** tool is used to project one or more curve on to a surface or set of surfaces. You can also project the curve along a vector or along surface normals and a point onto a surface using this tool. The procedure to use this tool is discussed next.

- Click on the **Project** tool from the **Curves** group in the **Surfacing** tab of **Ribbon**. The **Project** command box will be displayed; refer to Figure-39.

Figure-39. Project command box

- The **Options** button is used to set the projection to be either along a vector or normal to the selected surface. On clicking this button, the **Project Curve Options** dialog box will be displayed; refer to Figure-40. Select desired radio button from the dialog box and click on the **OK** button.

Figure-40. Project Curve Options dialog box

- Select desired option from the **Select** drop-down to specify the curve or point that you want to projected.
- Set desired option in the **Select Surface Step** section to define the plane or face onto the curve will be projected and right-click to apply selection.
- Set desired option in the **Projection Plane Step** section to define the plane or face whose normal will determine the direction of the projection.
- Define the direction from the **Direction Step** rollout after defining projection plane and click on the **Finish** button from command box; refer to Figure-41.

Figure-41. Curve projected

WRAPPING SKETCH CURVES ON SURFACE/FACE

The **Wrap Sketch** tool in **Project** drop-down is used to wrap selected sketch curve on a surface/face. The procedure to use this tool is given next.

- Click on the **Wrap Sketch** tool from the **Project** drop-down in the **Curves** group of **Surfacing** tab in the **Ribbon**. The **Wrap Sketch** command box will be displayed; refer to Figure-42 and you will be asked to select a face on which curve will be wrapped.

Figure-42. Wrap Sketch command box

- Select desired surfaces/faces on which curves will be wrapped and right-click in the drawing area. You will be asked to select the curves to be projected.
- Select desired sketch then the Preview of the wrap curve will be displayed; refer to Figure-43. Note that the sketch curve should be on a plane tangent to selected face/surface.

Figure-43. Preview of wrap curve

- Click on the **Finish** button from the command box to create wrap curve feature.

ETCH

The **Etch** tool is used to create etch feature using selected sketch. The procedure to use this tool has been discussed in Chapter 12.

HELICAL CURVE

The **Helical Curve** is used to create helical or spiral curves. The procedure to use this tool is discussed next.

- Click on the **Helical Curve** tool from the **Curves** group in the **Surfacing** tab of the **Ribbon**. The **Helical Curve** command box will be displayed; refer to Figure-44.

Figure-44. Helical Curve command box

- Select desired edge, circular face, cone face, or line to define axis of helical curve. The outer edge of helical curve will get attached to cursor and you will be asked to specify diameter of helical curve.
- Click at desired location to specify diameter of curve. The **Helical Curve Parameters** dialog box will be displayed; refer to Figure-45.

Figure-45. Helical Curve Parameters dialog box

- Select desired option from the **Type** drop-down of the dialog box to define type of helical curve. Select the **Constant Pitch** option from the drop-down if gap between two consecutive coils of helical curve are same; refer to Figure-46. Select the **Variable Pitch** option from the drop-down if you want to specify different pitch at start and end of the helical curve; refer to Figure-47. Select the **Compound** option from the drop-down if you want to specify varying pitch, number of turns, and diameter at different length points; refer to Figure-48. Select the **Spiral** option from the drop-down if you want to create spiral curve; refer to Figure-49.

Figure–46. Constant pitch helical curve

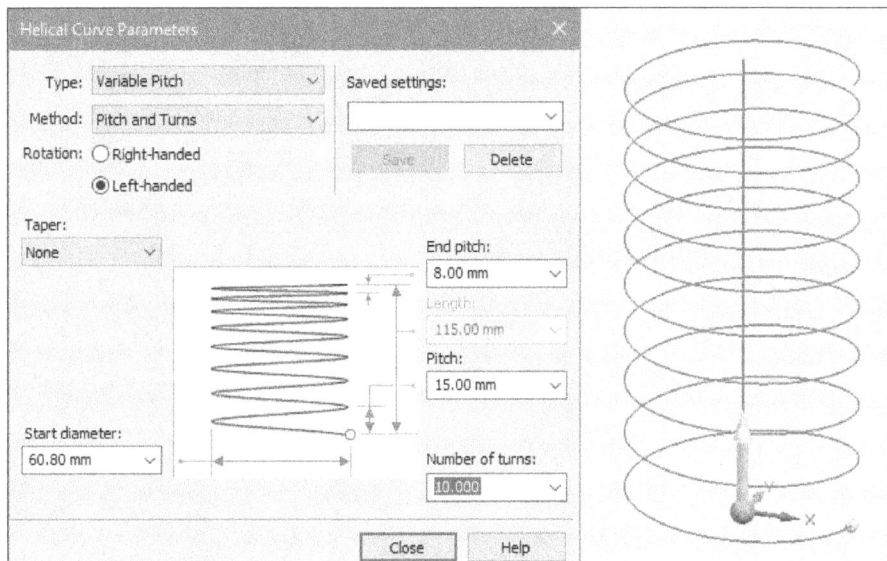

Figure–47. Variable pitch helical curve

Figure-48. Compound helical curve

Figure-49. Spiral curve

- Select desired option from the **Method** drop-down to define method for specifying parameters of helical/spiral curve. Select the **Length and Turns** option from the drop-down to specify total height of curve and number of turns in the helical curve. Select the **Length and Pitch** option from the drop-down to specify the total height of curve and gap between two consecutive coils of the curve. Select the **Pitch and Turns** option from the drop-down if you want to specify gap between two consecutive coils and number of turns in the curve.

- If you are creating spiral curve then in place of length parameter, you will be asked to specify End diameter and in place of pitch, you will be asked to specify radial pitch.

- Select desired radio button from the **Rotation** section to define left handed or right-handed rotation in curve.
- Specify the parameters of helical/spiral curve in edit boxes and click on **Close** button. The preview of helical/spiral curve will be displayed; refer to Figure-50.

Figure-50. Preview of helical curve

- Click on the **Finish** button from the command box to create the feature.

CROSS

The **Cross** tool is used to create a curve at intersection of virtual extrusion of two curves. The procedure to use this tool is given next.

- Click on the **Cross** tool from the **Curves** group in the **Surfacing** tab of the **Ribbon**. The **Cross** command box will be displayed; refer to Figure-51.

Figure-51. Cross command box

- Select desired sketch curve to preview the cross step will be displayed; refer to Figure-52.

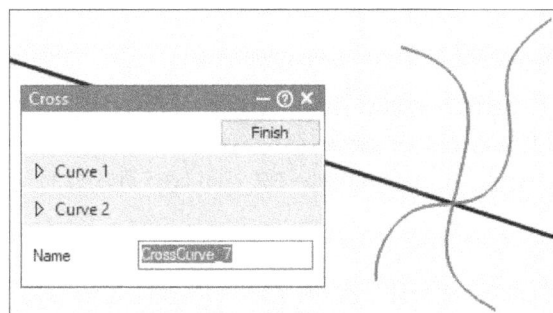

Figure-52. Selecting the curve by cross

- Click on the **Finish** button from the command box to create the feature. Press **ESC** to exit the tool.

CONTOUR

The **Contour** tool is used to draw a curve directly on the selected surface. The procedure to use this tool is given next.

- Click on **Contour** tool from the **Curves** group in the **Surfacing** tab of the **Ribbon**. The **Contour** command box will be displayed; refer to Figure-53.

Figure-53. Contour command box

- Select the surface on which you want to create curve and press **ENTER**. You will be asked to specify points of the curve.
- There are two type of curves that can be created on the surface, open curve and close curve. Select the **Open** button from the command box to create open curves and select the **Close** button from the command box to create closed curves.
- Click to specify the points on the surface. Preview of the curve will be displayed as you specify the points; refer to Figure-54.

Figure-54. Contour curve created

- If you want to insert a point anywhere after creating the curve then click on the **Insert Point** button from the command box.
- After specifying the points, click on the **OK** button from the command box to create the curve.
- Specify desired name of curve in the **Name** edit box of the command box displayed and click on the **Finish** button. Press **ESC** to exit the tool.

ISOCLINE

The **Isocline** tool is used to create a curve on selected surface at specified angle from selected plane. The procedure to use this tool is discussed next.

- Click on the **Isocline** tool from the **Curves** group in the **Surfacing** tab of the **Ribbon**. The **Isocline** command box will be displayed; refer to Figure-55.

Figure-55. Isocline command box

- Selects a reference plane or a planar face to be used for specifying vector angle. The plane should be perpendicular and intersecting with the surface.
- Select face or body on which isocline curve is to be created. The direction should be towards the face or bodies. The preview of isocline curves will be displayed; refer to Figure-56.

Figure-56. Preview of Isocline

- Right-click to accept the selection and click on the **Finish** button to create curve.
- Press **ESC** to exit the tool.

DERIVED

The **Derived** tool is used to create a curve using one or more selected curves/edges. The procedure to use this tool is given next.

- Click on the **Derived** tool from the **Curves** group in the **Surfacing** tab of the **Ribbon**. The **Derived** command box will be displayed; refer to Figure-57.

Figure-57. Derived command box

- Selects one or more curves or edges as the base for creating derived curves. Click **RMB** button to accept. Refer to Figure-58.

Figure-58. Derived curve created

- **Single Curve :** If this option is turn **"ON"** from command box then it creates a single curve from end-connected, tangent input elements.
- Click on the **Finish** button from the command box to create the curves. Press **ESC** to exit the tool.

SPLIT

The **Split** tool is used to split a curve using intersecting geometry. The procedure is given next.

- Click on the **Split** tool from the **Curves** group in the **Surfacing** tab of the **Ribbon**. The **Split** command box will be displayed; refer to Figure-59.

Figure-59. Split command box

- Select the curves to be split and right-click in the drawing area. You will be asked to select the intersecting geometry.
- Select desired option from the **Select** drop-down and then select respective intersecting geometry.
- Click on the **OK** button from the command box. The curve(s) will split at intersection points; refer to Figure-60.

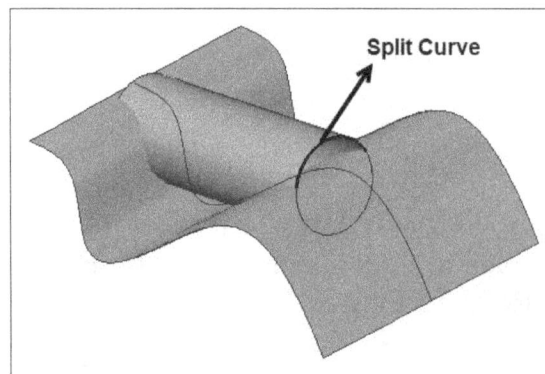

Figure-60. Split curve created

• Click on the **Finish** button to create feature and press **ESC** to exit the tool.

INTERSECTION POINT

The **Intersection Point** tool is used create points at the intersection of selected curves and edges. The procedure to use this tool is given next.

• Click on the **Intersection Point** tool from the **Curves** group in the **Surfacing** tab of the **Ribbon**. The **Intersection Point** command box will be displayed; refer to Figure-61.

Figure-61. Intersection Point command box

• Select a continuous primary curve to be used for generating intersection points and right-click in the drawing area. You will be asked to select an intersecting plane/axis/body.
• Select desired intersecting object and right-click in the drawing area. The intersection points will be generated; refer to Figure-62.

Figure-62. Intersection point created

• Click on the **Finish** button from the command box to create the feature and press **ESC** to exit the tool.

SECTION SKETCHES

The **Section Sketches** tool is used to create multiple 2D cross-section sketches. The procedure to use this tool is discussed next.

• Click on the **Section Sketches** tool from **Curves** group in the **Surfacing** tab of the **Ribbon**. The **Section Sketches** command box will be displayed; refer to Figure-63. You will be asked to select the body to create the section sketches.

Figure-63. Section Sketches command box

- Select desired body from which you want to create the section sketches; refer to Figure-64. You will be asked to select plane step.

Figure-64. Selecting the body to create section sketches

- Select desired plane or face to create the section sketches; refer to Figure-65. You will be asked to specify number of planes and offset distance.

Figure-65. Selecting the plane

- Specify desired number of section planes in the **Number of offset Planes** edit box of **Pattern of Planes** rollout.
- Specify desired distance between the planes in the **Offset** edit box; refer to Figure-66.

Figure-66. Specifying number of planes and distance

- After specifying desired parameters in the command box, click on the **OK** button and then **Close** button to create the section sketches; refer to Figure-67.

Figure-67. Section sketches created

CREATING SURFACES

The tools to create surfaces are available in the **Surfaces** group of the **Surfacing** tab in the **Ribbon**; refer to Figure-68. These tools are discussed next.

Figure-68. Surfaces panel

Creating Surface using BlueSurf tool

The **BlueSurf** tool is used to create surface by using selected edges or curves. You can also use a guide curve to control the shape of surface. Note that the guide curve should intersect with the cross-section curves and should not have sharp turns. The procedure to use this tool for creating surface is discuss next.

- Click on the **BlueSurf** tool from the **Surfaces** group in the **Surfacing** tab of the **Ribbon**. The **BlueSurf** command box will be displayed; refer to Figure-69.

Figure-69. BlueSurf command box

- Select the first curve and then right-click in drawing area. You will be asked to select the second curve.
- Select the second curve and right-click in drawing area. Preview of surface will be displayed; refer to Figure-70.

Figure-70. BlueSurf Preview

- **Tangency Control**   : Click on the down arrow of the **Tangency Control** box. The options to control the shape of surface at the ends of the feature will be displayed; refer to Figure-71. Select the **Natural** option if you want to continue natural flow in surface based on selected curves. Select the **Normal to Section** option if you want the surface to be perpendicular to edge at start/end point. Select the **Parallel to Section** option to make the surface edge parallel to selected cross-section.

Figure-71. Tangency Control drop down

- Click on the **Guide Curve** button from the command box and select the curve that you want to use as guide curve. Preview of the surface will be displayed.
- Right-click in the drawing area to accept the selection.

- Click on the **BlueSurf Options** button to modify standard and advanced parameters of the surface. The **BlueSurf Options** dialog box will be displayed; refer to Figure-72.

Figure-72. BlueSurf Options dialog box

- Set the tangency options as discussed earlier in the dialog box.
- Select desired option from the **Curve Connectivity** area to define how curves of surface will be connected. Select the **Use Pierce Points** radio button to use connect relationship at intersections of cross-section and guide curves. When you move cross section curves or guide curves then connections will be automatically updated for new intersection points. Select the **Use BlueDots** radio button to modify the shape of surface by using the blue dots displayed on curves.
- Select the **Open ends** radio button from the **End Capping** area of the dialog box if you want to keep the end faces of surface open. Select the **Closed ends** radio button if you want to cap the end faces of the surface; refer to Figure-73.

Figure-73. Closed end capping

- Select desired option from the **Extent Type** area to define whether you want to create closed surface or open surface. If selected curvatures can form surface then select the **Closed** radio button from this area and click on the **Apply** button; refer to Figure-74.

Figure-74. Closed surface

- Click on the **Advanced** tab in the dialog box if you want to reorder sections in the surface; refer to Figure-75.

Figure-75. Advanced tab

- To reorder a section, select it from the list and click on the **Up** or **Down** button from the dialog box.
- After setting desired parameters in the dialog box, click on the **OK** button.
- Click on the **Insert Sketch** button from the command box if you want to add another sketch in the surface at intersection of a plane and curve.
- After setting desired parameters for surface, click on the **Next** button from the command box and press **ENTER**. The surface will be created. Press **ESC** to exit the tool.

Creating Surface using Bounded tool

The **Bounded** tool is used to create a surface using selected closed boundaries. The procedure to use this tool is discuss next.

- Click on the **Bounded** tool from the **Surfaces** group in the **Surfacing** tab of the **Ribbon**. The **Bounded** command box will be displayed; refer to Figure-76.

Figure-76. Bounded command box

- Select the closed boundary to be used for creating surface. The preview of surface will be displayed; refer to Figure-77.

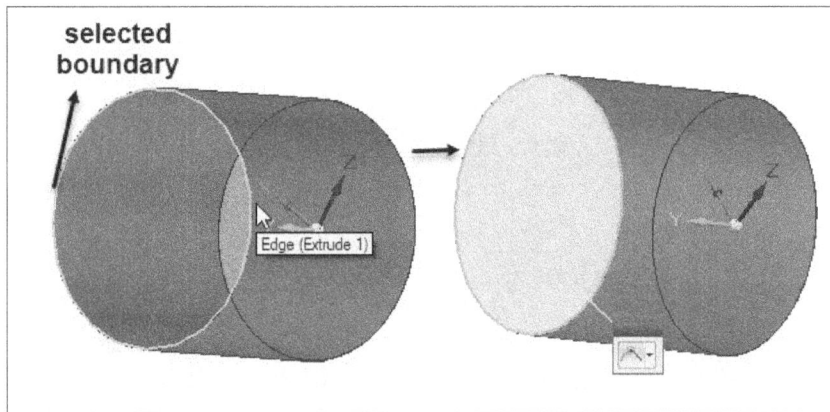

Figure-77. Bounded surface created

- Click on the **Bounded Surface Options** button from the command box to define how created surface will be connected to curves and adjoining faces. The **Bounded Surface Options** dialog box will be displayed; refer to Figure-78.

Figure-78. Bounded Surface Options dialog box

- Select the **No Merge** radio button from the **Face Merging** area of the dialog box to create new surface as new body. Select the **Merge (as many as possible)** radio button to merge new surface with connected surfaces. Select the **Full Merge** radio button to merge new surface with connected surfaces and form one single surface body.

- Select desired radio button from the **Curve Connectivity** area of the dialog box to define which points will be used to control shape of surface. The function of these radio buttons has been discussed earlier. After setting desired parameters in the dialog box, click on the **OK** button.
- Click on the **Guide Curve** button from the command box if you want to use a guide curve to define shape of bounded surface. The method to use a guide curve is same as discussed for **BlueSurf** tool.
- After setting desired parameters, click on the **Finish** button from the command box to create surface. Press **ESC** to exit the tool.

Creating Surface using Redefine tool

The **Redefine** tool is used to create a surface by using existing model faces and surfaces. The procedure is given next.

- Click on the **Redefine** tool from the **Surfaces** group in the **Surfacing** tab of the **Ribbon**. The **Redefine** command box will be displayed; refer to Figure-79.

Figure-79. Redefine command box

- Click on the **Options** button from the command box to display the redefine surface options. The **Redefine Surface** dialog box will be displayed; refer to Figure-80. Specify the parameters as discussed earlier. Select the **Replace faces on solid body** check box from the **Replace** area to replace the selected faces with a new surface. After setting desired parameters, click on the **OK** button from the dialog box.

Figure-80. Redefine Surface dialog box

- Select the faces of the model which you want to use for creating surfaces and right-click in the drawing area. The surfaces will be created; refer to Figure-81.

Figure-81. Redefine surface created

- Click on the **Finish** button from the command box to create the surface and press **ESC** to exit the tool.

Creating Surface using Swept tool

The **Swept** tool in surfacing is similar to the **Sweep** tool used in solid creation. The **Swept** tool in surfacing is used to sweep a section along the selected path to form a surface. The steps to use this tool are given next.

- Click on the **Swept** tool from **Surfaces** group in the **Surfacing** tab of **Ribbon**. The **Sweep Options** dialog box will be displayed; refer to Figure-82. Set desired parameters as discussed in chapter related to 3D Modeling and click on the **OK** button. The **Swept** command box will be displayed; refer to Figure-83.

Figure-82. Sweep Options dialog box

Figure-83. Swept command box

- Click on the **Surface Visualization Options** button, the **Swept-Surface Visualization options** will be displayed; refer to Figure-84.

Figure-84. Surface visualization options dialog box

- Select the **Show surface visualization** check box and modify default colors for various visualizations of the surfaces, if needed. Click on the **Close** button to exit.
- Select desired curve to be used as path for sweep. You will be asked to select a sketch/curve for section of sweep.
- Click on the **Swept- Select Sketch Step** button from the **Path** drop-down to select the sketch for the feature. Select desired option from the Select drop-down. Select **Swept-Start Design Point** from the drop-down to specify a point to the cross section will correspond with the start point on adjacent cross section. Select **Edit** button to select the profile within the feature of editing.
- Select the open or close sketch for the section. The procedure is similar to **Sweep** tool for solid protrusion. Preview of the swept surface will be displayed; refer to Figure-85.
- Click on the **Finish** button to create the feature.

Figure-85. Swept Surface

Creating Surface using Extruded tool

The **Extruded** tool is used to extrude a closed or open sketch to the specified height to from a surface. The **Extruded** tool of surfacing is similar to **Extrude** tool in solid modeling. The steps to use this tool are given next.

- Click on the **Extruded** tool from the **Surfaces** group in the **Surfacing** tab of the **Ribbon**. The **Extruded** command box will be displayed; refer to Figure-86. You are asked to select the existing sketch or a plane to draw sketch.

Figure-86. Extruded command box

- Select **Single** option from the **Selection Type** drop-down of the **Sketch** rollout if you want to select single or multiple curves one by one for extrude feature. Select **Chain** option from the drop-down to select complete chain of curves connected to selected curve.
- Select desired closed or open sketch and right-click in the drawing area. The preview of the surface will be displayed with extent of extrude feature attached to the cursor; refer to Figure-87.

Figure-87. Creating extruded surface

- Specify desired value of extent in the **Distance** edit box of the command box to define size of feature.
- Select desired value of **Step** edit box to set the increment by which the value will change.
- You will be asked to specify the side on which you want to create feature with respect to selected section. Click on desired side. The preview of feature will be displayed; refer to Figure-88.

Figure-88. Extruded Surface

- Click on the **Finish** button from the command box to create the feature. Press **ESC** to exit the tool.

Revolved

The **Revolved** tool is used to revolve a close or open sketch up to specified angle with respect to selected reference to create a surface. The **Revolved** tool in surfacing is similar to **Revolve** tool in solid creation. The steps to use this tool are given next.

- Click on the **Revolved** tool from the **Surfaces** group in the **Surfacing** tab of the **Ribbon**. The **Revolved** command box will be displayed; refer to Figure-89 and you will be asked to select an existing sketch or a plane to draw sketch.

Figure-89. Revolved command box

- Select **Single** option from the **Selection Type** drop-down of the **Sketch** rollout and select single or multiple lines or edges to be revolved. Select **Chain** option from the drop-down to select single line or edge to be revolved. Click on the Tick mark displayed in the Selected area. Select desired line or edge to be used as revolution axis. You will be asked to specify extent of feature.
- Specify the angle of revolution in the **Angle** edit box from the **Extend** drop-down of the command box and click in the drawing area. Preview of the surface will be displayed; refer to Figure-90.
- Click on the **Finish** button from the command box to create the feature and press **ESC** to exit the tool.

Figure-90. Revolved Surface

Offset

The **Offset** tool is used to create a surface at an offset distance from selected face/ surface. The steps to create offset surface are given next.

- Click on the **Offset** tool from the **Surfaces** group in the **Surfacing** tab of the **Ribbon**. The **Offset** command box will be displayed; refer to Figure-91.

Figure-91. Offset command box

- Select the face/surface which you want to use for creating the offset surface and right-click in the drawing area. You will be asked to specify direction of offset and distance value.
- Specify desired value of distance in the **Distance** edit box of command box.
- Select the **Show Boundaries** button from the command box if you want to include internal cuts of surface while offsetting. By default, the **Remove Boundaries** button is selected in command box and so, the cuts on surface are not created in the offset surface.
- After specifying the parameters, click on desired side of base surface to create offset surface. Preview of the surface will be displayed; refer to Figure-92.

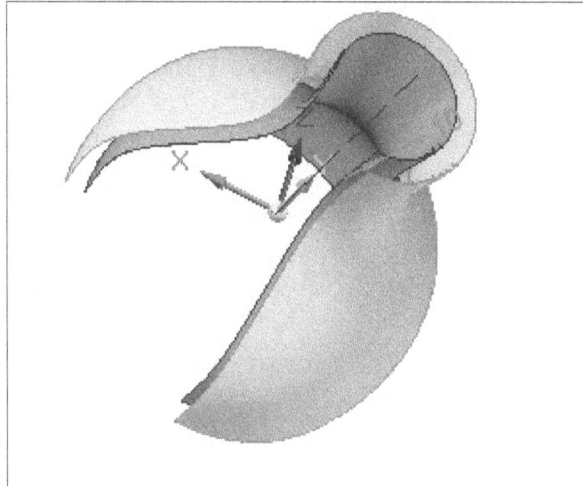

Figure-92. Offset Surface

- Click on the **Finish** button to create the offset surface. Press **ESC** to exit the tool.

Copy

The **Copy** tool is used to create surface derived from one or more input faces. The select faces may or may not be adjacent to each other. You can specify whether any internal or external boundaries are removed on the new copy of the surface. The procedure to use this tool is given next.

- Click on **Copy** tool from the **Surfaces** group in the **Surfacing** tab of the **Ribbon**. The **Copy** command box will be displayed; refer to Figure-93.

Figure-93. Copy command box

- Select the faces/surfaces to be copied and right-click in the drawing area to accept selection. Preview of faces will be displayed; refer to Figure-94.

Figure-94. Faces copied as surfaces

Ruled

The **Ruled** tool is used to create a surface that extends out in a specified direction from the selected edges. This type of surface becomes very important when creating parting surface for molding/casting. The steps to use this tool are given next.

- Click on the **Ruled** tool from the **Surfaces** group in the **Surfacing** tab of the **Ribbon**. The **Ruled** command box will be displayed; refer to Figure-95.

Figure-95. Ruled command box

- Select desired edge of surface to be used for creating ruled surface. The preview of surface will be displayed; refer to Figure-96.

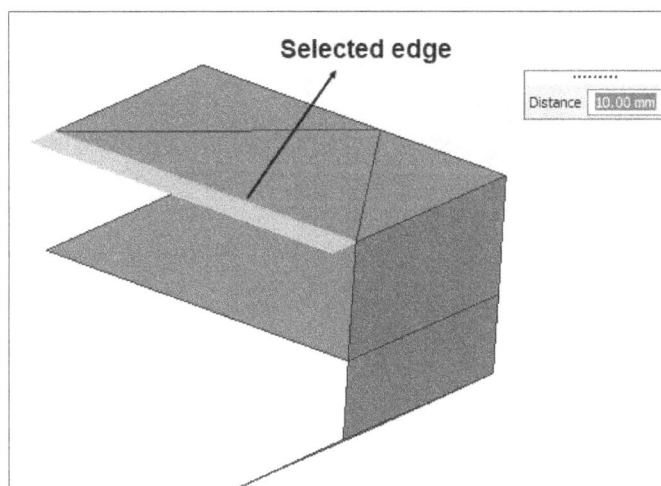

Figure-96. Ruled

- **Ruled Options** drop-down : Click on the down arrow of the **Ruled Options** button in the command box. The **Ruled Options** drop-down will be displayed; refer to Figure-97.

Figure-97. Ruled Options drop-down

- Select the **Tangent Continuous** option from the drop-down if you want to create an extension of surface tangent to base surface at selected edge.
- Select the **Normal to face** option from the drop-down if you want to create new surface perpendicular to base face at selected edge.
- Select the **Tapered to plane** option from the drop-down if you want to create surface at an angle to reference plane at selected edge. After selecting this option, select desired reference plane for specifying angle. Preview of surface will be displayed; refer to Figure-98.

Figure-98. Preview of surface at angle

- Select the **Natural** option to create extension surface in natural flow of base surface.
- Select the **Along an Axis** option to create extension surface along selected axis. You will be asked to select an axis. Select desired axis or edge to define direction of surface extension. Preview of the surface will be displayed; refer to Figure-99.

Figure-99. Preview of surface at angle using axis

- After setting desired parameters, click on the **OK** button from the command box to create the surface.
- Press **ESC** to exit the tool.

Blank Surface

The **Blank Surface** is used to creates a flattened surface from the selected model faces. The steps to use this tool are given next.

- Click on the **Blank Surface** tool from **Surfaces** group in the **Surfacing** tab of the **Ribbon**. The **Material Table** dialog box will be displayed if material has not been defined earlier.
- Select desired material from the table and click on the **Apply to Model** button from the dialog box. The **Blank Surface** command box will be displayed; refer to Figure-100.

Figure-100. Blank Surface command box

- Select desired faces/surfaces from the model and specify the direction to the model.
- Click on the **Preview** button from the command box. The preview of surface will be displayed; refer to Figure-101.

Figure-101. Blank Surface

MODIFYING SURFACES

The tools in the **Modify Surfaces** group of **Surfacing** tab in the **Ribbon** are used to modify surfaces; refer to Figure-102. These tools are discussed next.

Figure-102. Modify Surfaces group ordered

Intersect

The **Intersect** tool is used to extend or trim selected surfaces to a common intersection. Using this tool, you can also create bodies using the areas fully enclosed by surfaces. The procedure to use this tool is discuss next.

- Click on the **Intersect** tool from the **Modify Surfaces** group in the **Surfacing** tab of **Ribbon**. The **Intersect** command box will be displayed; refer to Figure-103.

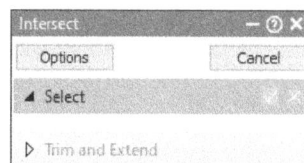

Figure-103. Intersect command box

- Select the two or more intersecting surfaces and then right-click in the drawing area. You will be asked to select the sections to be removed.
- Click on desired sections of surfaces to be removed; refer to Figure-104. You can also extend the surfaces to intersect by using this tool; refer to Figure-105.

Figure-104. Selecting region to trim

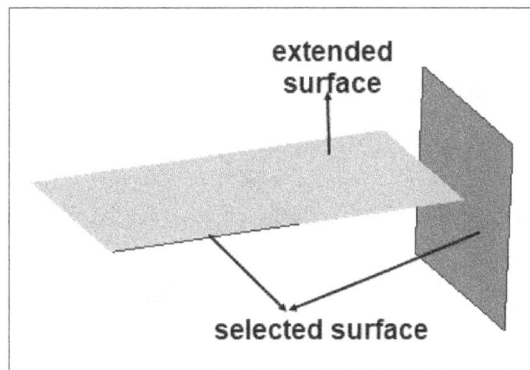

Figure-105. Intersect Extend surface

- If selected surfaces form a closed region then you can use this tool to convert selected region into solid. After selecting all the surfaces, select the **Create design bodies** option from the **Intersect Options** drop-down in the command box; refer to Figure-106 and right-click in the drawing area. The **Volume Regions** dialog box will be displayed with preview of solid region; refer to Figure-107.

Figure-106. Create design bodies option

Figure-107. Volume Regions dialog box

- Select the **Unite Regions** check box to unite all the selected surfaces as single body.
- Select the **Consume Surfaces** check box if you want to remove selected surfaces after creating design body.
- Click on the **Close** button from the **Volume Regions** dialog box to exit and click on the **OK** button from the command box to accept selection. The design body will be created.
- Click on the **Finish** button from the command box and press **ESC** to exit the tool.
- Similarly, you can use the **Auto-trim** option from the **Intersect Options** drop-down of the command box to automatically trim surfaces to form closed region; refer to Figure-108. Note that using this option will not form solid.

Figure-108. Auto trim option

Replace Face

The **Replace Face** tool is used to replace selected faces of a solid with a surface. When replacing more than one face, the faces being replaced cannot touch each other. The procedure is discuss next.

- Click on the **Replace Face** tool from the **Modify Surfaces** group in the **Surfacing** tab of the **Ribbon**. The **Replace Face** command box will be displayed; refer to Figure-109.

Figure-109. Replace Face command box

- Select the part face to be replaced and right-click in the drawing area.
- Select the surface which should replace the part face and click on **Finish** button. The replacement of face will be applied; refer to Figure-110. Press **ESC** to exit the tool.

Figure-110. Replace Face

Trim

The **Trim** tool is used to trim a surface with selected curve, surface, or plane. The procedure is discuss next.

- Click on the **Trim** tool from the **Modify Surfaces** group in the **Surfacing** tab of Ribbon. The **Trim** command box will be displayed; refer to Figure-111.

Figure-111. Trim command box

- Select the surfaces to be trimmed and right-click in the drawing area.
- Select curves, planes, surfaces, and sketch elements to be used as trim tool, and right-click in the drawing area. You will be asked to select the regions to be removed.
- Select the regions formed by trim tool that will be trimmed and right-click in the drawing area. The preview of trimming will be displayed; refer to Figure-112.
- Click on the **Finish** button to trim surfaces and press **ESC** to exit the tool.

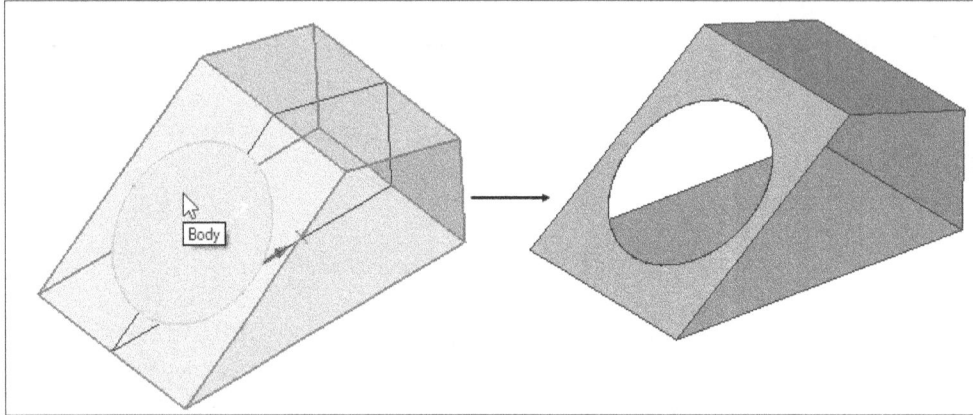

Figure-112. Preview of Trim

Extend

The **Extend** tool is used to extend a surface along selected edges. The procedure is given next.

- Click on the **Extend** tool from the **Modify Surfaces** group in the **Surfacing** tab of the **Ribbon**. The **Extend** command box will be displayed; refer to Figure-113.

Figure-113. Extend command box

- Select edges of surface to be extend and right-click in the drawing area. Selected end of surface will get attached to cursor and you will be asked to specify the extent of surface.
- Click at desired location or enter the value in the **Distance** edit box of the command box. After specifying the value in edit box, click in the drawing area. The preview of extend surface will be displayed; refer to Figure-114.

Figure-114. Extend preview

- Click on the **Finish** button from the command box to create feature and press **ESC** to exit the tool.

Split

The **Split** tool is used to split faces/surfaces using sketch, curve, surface, or reference planes. The procedure to use this tool is discuss next.

- Click on the **Split** tool from **Modify Surface** group in the **Surfacing** tab of the **Ribbon**. The **Split** command box will be displayed; refer to Figure-115.

Figure-115. Split command box surfacing

- Select desired option from the **Select** drop-down. You will be asked to select geometry being used as splitting tool.
- Select the geometry to be used as splitting tool for the surface/face and right-click in the drawing area. Preview of split faces/surfaces will be displayed; refer to Figure-116.

Figure-116. Preview of split surface

- Click on the **Finish** button from the command box to perform splitting and press **ESC** to exit the tool.

Stitched

The **Stitched** tool is used to creates a construction surface by stitching multiple sheet surfaces. The procedure to use this tool is discuss next.

- Click on the **Stitched** tool from the **Modify Surfaces** group in the **Surfacing** tab of the **Ribbon**. The **Stitched Surface Options** dialog box will be displayed; refer to Figure-117. Select the **Heal stitched surfaces** check box to stitch and heal the surfaces which have gap within specified tolerance range. Select the **Show this dialog box when the command begins** check box to display **Stitched Surface Options** dialog box every time you invoke this tool.
- After setting desired parameters in the dialog box, click on the **OK** button. The **Stitched** command box will be displayed; refer to Figure-118.

Figure-117. Stitched Surface Options

Figure-118. Stitched command box

- Select the surfaces to be stitched together and click on the **Preview** button from the command box. The preview of stitch operation will be displayed. Note that if selected surfaces form a closed region then a message box will be displayed asking you whether you want to convert stitched surfaces into solid; refer to Figure-119. Click on the **Yes** button if you want to create a solid body and click on the **No** button if you want to create a surface body after stitching.
- Click on the **Finish** button from the command box to create the feature and press **ESC** to exit the tool.

Figure-119. Message box for stitched surfaces

Show Non-Stitched Edges

The **Show Non-Stitched Edges** tool is used to display all non-stitched edges of selected surfaces. The procedure to use this tool is discuss next.

- Click on the **Show Non-Stitched Edges** from **Stitched** drop-down. The **Show Non-Stitched Edges** command box will be displayed; refer to Figure-120.

Figure-120. Show Non-Stitched Edges command box

- Select the model to observe the non-stitched edges; refer to Figure-121. Click on the **Reset** button from command box to start fresh selection and click on the **Close** button from the command box to exit the tool.

Figure-121. Show non stitched edges preview

Parting Split

The **Parting Split** tool is used to create parting lines along the silhouette edges of the part. This tool is useful in creating mold die design for the part. The procedure to use this tool is discuss next.

- Click on the **Parting Split** tool from the **Modify Surfaces** group in the **Surfacing** tab of the **Ribbon**. The **Parting Split** command box will be displayed; refer to Figure-122 and you will be asked to select a plane for creating split lines on part.

Figure-122. Parting Split command box

- Select desired option from the **Create From** drop-down in the command box and select/define the plane to be used for creating split line. You will be asked to select the faces/surfaces to be split.
- Selects the face/surface on which split lines will be created and click on the **Preview** button. Preview of the lines will be displayed; refer to Figure-123.

Figure-123. Preview of Parting split

- Click on the **Finish** button from the command box to create the splitting lines and press **ESC** to exit the tool.

Parting Surface

The **Parting Surface** is used to create parting surface for mold tools. The procedure to use this tool is discuss next.

- Click on the **Parting Surface** tool from the **Parting Split** drop-down in the **Modify Surfaces** panel of the **Surfacing** tab in the **Ribbon**. The **Parting Surface** command box will be displayed refer to Figure-124 and you will be asked to select a reference plane.

Figure-124. Parting Surface command box

- Select the plane to be used as direction reference for split surface. Note that surface will be created along the selected plane/face. You will be asked to sketch, edge, or curve to be used for creating split surface.
- Select desired curve from the drawing area. Generally, we use parting split curve for creating split surface. After selecting curve, right-click in the drawing area. You will be asked to specify the extent of surface.
- Specify desired direction value in the **Distance** edit box of the command box and click on the direction curve to generate face. The preview of parting surface will be displayed; refer to Figure-125.

Figure-125. Parting Surface preview

• Click on the **Finish** button from the command box to create surface and press **ESC** to exit the tool.

Offset Edge

The **Offset Edge** tool is used to split a face by offsetting an existing edge. The procedure to use this tool is discuss next.

• Click on the **Offset Edge** tool from the **Parting Split** drop-down in the **Modify Surfaces** group of **Surfacing** tab in the **Ribbon**. The **Offset Edge** command box will be displayed; refer to Figure-126.

Figure-126. Offset Edge command box

• Select desired option from the **Selection Type** drop-down to define what type of entity you want to select for offsetting.
• Select the face/edge/loop to offset edges. You will be asked to specify the distance up to which the edges will be offset in dynamic input box. The preview of offset will be displayed; refer to Figure-127.

Figure-127. Preview of Offset Edge

• Right-click in the drawing area to create the edge offset and press **ESC** to exit the tool.

BlueDot (Ordered Environment)

⌾ The **BlueDot** tool is used to create bi-directional connect point. You can create a BlueDot by identifying the two curves to connect. No more than two curves can be BlueDot connected. The first curve you select will move to intersect the second curve. The second curve is unchanged (it maintains its location and shape). There are several solutions possible when creating a BlueDot. The solution is dependent on how you select the curves. Each curve has four select zones (two endpoints, a midpoint, curve). The endpoints and midpoint display a red dot for selection. If you select a midpoint or endpoint, that point on the curve will move to the zone you select on the second curve. You should experiment with the BlueDot command to better understand the several solutions. The procedure is given next.

• Click on the **BlueDot** tool from the **Modify Surfaces** group in the **Surfacing** tab of the **Ribbon**. You will be asked to select the curves.
• One by one select the two curves to be joined by BlueDot. The curves will be joined by BlueDot; refer to Figure-128.

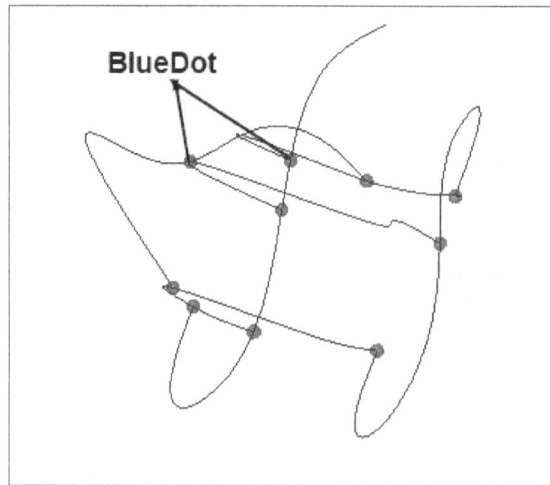

Figure-128. Creating BlueDot by Using curve

Press **ESC** to exit the tool. If you want to modify the position of joint then click on the blue dot and select the **Dynamic Edit** button from the mini toolbar displayed; refer to Figure-129. The **Dynamic Edit** command box will be displayed where you can specify the respective new location of joint or you can dynamically drag the joint. After changing position, press **ESC** to exit the tool.

Figure-129. Dynamic Edit button

SURFACE VISUALIZATION SETTINGS

The **Surface Visualization** tool is used to specify settings related to how the surface will be displayed in the drawing area. Using this tool, you can define parameters related to mesh density and curvature comb of surface. The procedure to use this tool is given next.

- Click on the **Surface Visualization** tool from the **Inspect** group in the **Surfacing** tab of the **Ribbon**. The **Surface Visualization Settings** dialog box will be displayed; refer to Figure-130.

Figure-130. Surface Visualization Settings dialog box

- Select the **Show surface visualization** check box to display specified surface visualization on selected surfaces.
- Select the **Direction 1** and **Direction 2** check boxes from the **Mesh density** area of the dialog box to display mesh density in X and Y directions. After selecting a check, you can define color of mesh lines in the drop-down below it.
- Using the sliders, you can increase or decrease the density of mesh lines.
- Similarly, you can set parameters for curvature comb in the **Curvature comb** area of the dialog box.
- After setting desired parameters in the dialog box, select the surfaces to display surface visualization; refer to Figure-131.

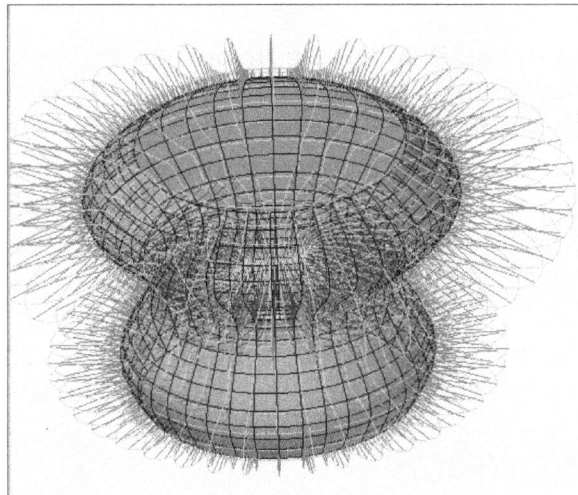

Figure-131. Displaying surface visualization

- Click on the **Clear All Surfaces** button to clear all selected surfaces and start a new selection set.
- Click on the **Close** button from the dialog box to exit tool.

DISPLAYING SECTION CURVATURE

The Section curvature is used to display curvature of surface at intersection with selected reference plane. The procedure to display section curvature is discussed next.

- Click on the **Section Curvature Settings** tool from the **Inspect** group of **Surfacing** tab in the **Ribbon**. The **Section Curvature Settings** dialog box will be displayed with **Section Curvature Settings** command box; refer to Figure-132.

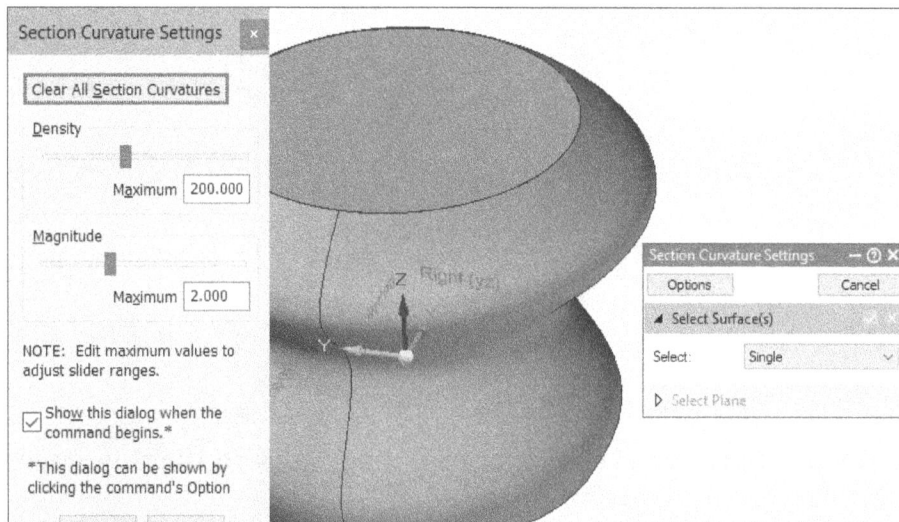

Figure-132. Section Curvature Settings dialog box and command box

- Set desired value of density and magnitude of curvature using the sliders in the dialog box. After setting parameters, click on the **Close** button from the dialog box.
- Select desired surfaces/faces for which you want to display curvature and right-click in the drawing area. You will be asked to select a section plane that intersects with the surfaces/faces.
- Select desired planar face or plane. Preview of curvature will be displayed and you will be asked to specify distance of curvature plane with respect to selected reference plane/face.
- Specify desired value in the **Distance** edit box of the command box and click at desired location. If you have specified value in the edit box then you need to click on desired side of reference plane to place the curvature comb.
- Click on the **Finish** button from the command box to create plot.
- Select or clear the **Show** check box from the **Section Curvature** area of the **Inspect** group in the **Surfacing** tab of the **Ribbon** to display or hide the section curvature respectively.

Similarly, you can use the **Reflective Plane** options in the **Inspect** group of **Surfacing** tab in **Ribbon** to display reflection of surfaces. The other tools of **Surfacing** tab have been discussed earlier.

PRACTICAL 1

Create the model of helmet glass as shown in Figure-133. The dimensions of the model are given in Figure-134.

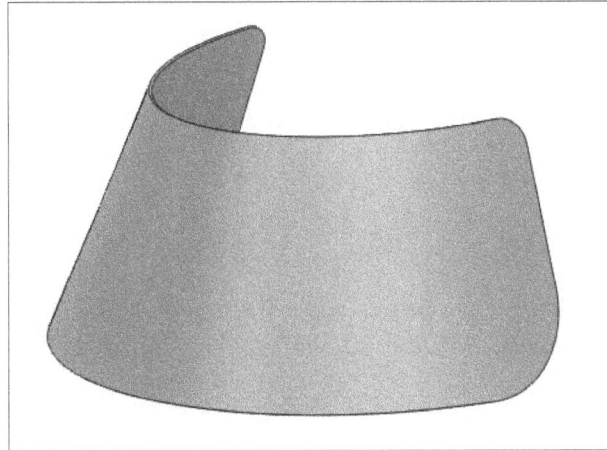

Figure-133. Practical 1 model

Figure-134. Practical 1 drawing

Starting New Document

- Start Solid Edge if not started yet.
- Click on the **ISO Metric Part** tool from the **New** cascading menu of the **Application** menu. The Part modeling environment will be displayed in **Synchronous** mode.
- Select the **Ordered** radio button from the **Model** group in the **Tools** tab of the **Ribbon**. The ordered modeling mode will be activated.

Creating Sketches

- Click on the **Sketch** tool from the **Sketch** group in the **Home** tab of **Ribbon**. You will be asked to select a sketching plane.
- Select the XY plane from the drawing area and create the sketch as shown in Figure-135.

Figure-135. Sketch 1 for practical

- Similarly, create the sketch at plane parallel to XY plane by **40** distance as shown in Figure-136 and Figure-137.

Figure-136. Sketch 2 for practical

Figure-137. Sketch for loft surface

Creating Surface using sketches

- Click on the **BlueSurf** tool from the **Surfaces** group in the **Surfacing** tab of the **Ribbon**. The **BlueSurf** command box will be displayed.
- Select the bottom curve near the end point as shown in Figure-138 and right-click in the drawing area. You will be asked to select the next curve.

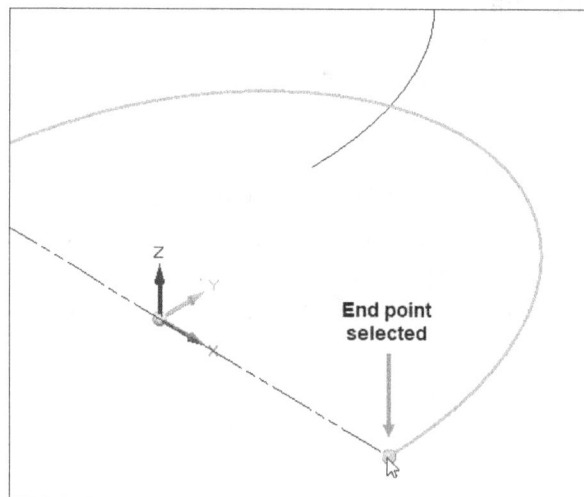

Figure-138. End point of curve selected

- Select the corresponding point on next curve; refer to Figure-139 and right-click in the drawing area. Preview of the surface will be displayed; refer to Figure-140.

Figure-140. Preview of surface

Figure-139. Selecting next curve point

- Click on the **Next** button and then **Finish** button from the command box. The surface will be created. Press **ESC** to exit the tool.

Applying Thickness to Surface

- Click on the **Thicken** tool from the **Add** drop-down in the **Solids** group of **Home** tab in the **Ribbon**. The **Thicken** command box will be displayed and you will be asked to select the surface body to be thickened.
- Select the surface body recently created and specify thickness value as **0.50** in the edit box of command box; refer to Figure-141.

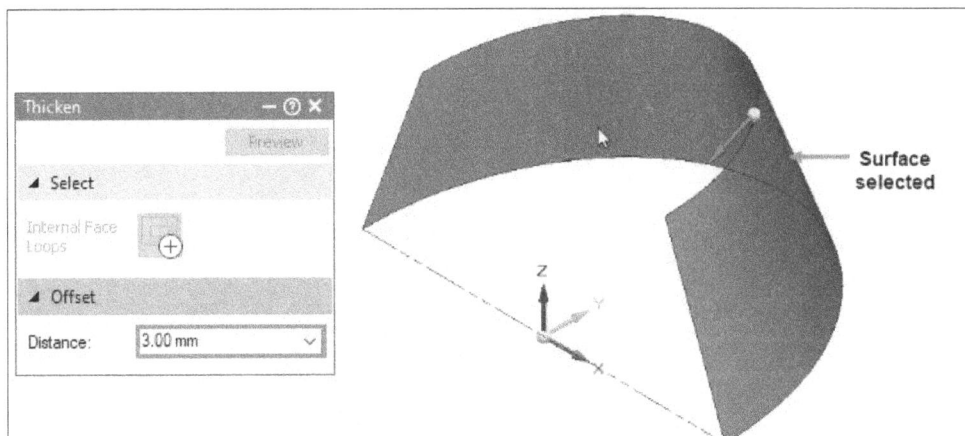

Figure-141. Specifying parameters for thicken

- Click towards the inside of surface to define direction of thickening. Preview of thickened feature will be displayed.
- Click on the **Finish** button from the command box to create the feature and press **ESC** to exit the tool.

Applying Round

- Click on the **Round** tool from the **Round** drop-down in the **Solids** group of **Home** tab in the **Ribbon**. The **Round** command box will be displayed.

- Specify the radius value as **10** in the **Radius** edit box of command box and select the corner edges of model. Preview of round will be displayed; refer to Figure-142.

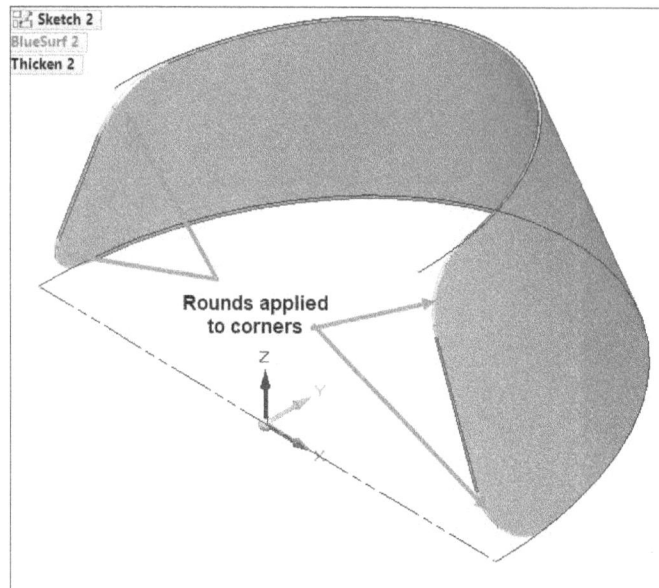

Figure-142. Preview of rounds

- Right-click in the drawing area to accept selection.
- Click on the **Preview** button to check preview of round and then click on the **Finish** button to create round feature. Press **ESC** to exit the tool.

PRACTICE 1

Create the surface model of tank as shown in Figure-143. The dimensions of the model are given in Figure-144.

Figure-143. Practice 1 model

Figure-144. Practice 1 drawing

PRACTICE 2

Create the surface model of car bumper as shown in Figure-145. The dimensions of the model are given in Figure-146. **Assume the missing dimensions**. **(Hint:** You will need vertex pulling in freestyle after making the base component.**)**

Figure-145. Practice 2 model

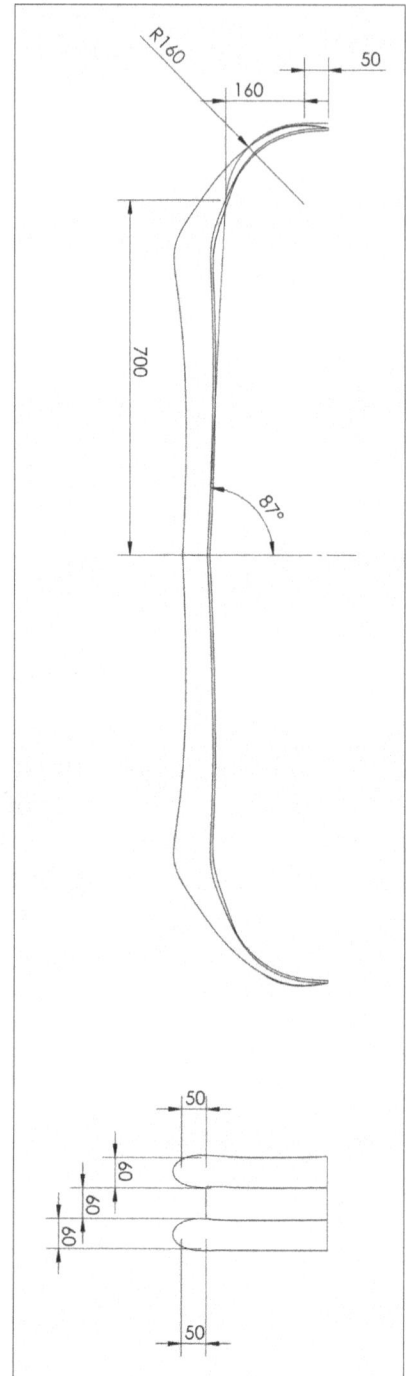

Figure-146. Car Bumper

SELF ASSESSMENT

Q1. What is the primary function of the Box tool?

A) To create a cylindrical-shaped feature
B) To create a box-shaped primitive feature
C) To smooth out edges of a model
D) To connect edges of a model

Q2. What does the Start Symmetry tool allow you to do?

A) Create a symmetrical feature on both sides of the model
B) Stop symmetrical editing
C) Scale the model uniformly
D) Blend edges of the model

Q3. What is the purpose of the Blend tool?

A) To fill missing faces in a model
B) To connect edges with a lofted structure
C) To sharpen or smooth the hard edges of a subdivision model
D) To project curves onto surfaces

Q4. What is the function of the Fill tool?

A) To connect two edges with a lofted structure
B) To fill missing faces in the subdivision model
C) To align a shape to a curve
D) To project curves onto a surface

Q5. Which of the following tools is used to connect the edges of selected cage elements?

A) Bridge
B) Split
C) Scale
D) Fill

Q6. How do you specify the segments used for the major diameter of a torus cage?

A) Major Diameter Segments edit box
B) Base Segments edit box
C) Side Segments edit box
D) Z-segments edit box

Q7. What is the primary purpose of the Isocline tool?

A) To create a helical curve
B) To create a curve at a specified angle from a selected plane
C) To split curves
D) To generate offset surfaces

Q8. What does the Intersection Point tool do?

A) It creates curves on surfaces
B) It generates points at the intersection of selected curves and edges
C) It creates a helical curve
D) It defines a plane for sketches

Q9. What is the purpose of the Section Sketches tool?

A) To create multiple 2D cross-section sketches
B) To create helical curves
C) To generate isocline curves
D) To split curves

Q10. What does the Swept tool do in surfacing?

A) It sweeps a section along a selected path to form a surface
B) It creates a closed surface from open curves
C) It extrudes a sketch to a specified height
D) It redefines existing model faces

Q11. What is the function of the Revolved tool in surfacing?

A) To create a helical curve
B) To revolve a sketch up to a specified angle
C) To create section sketches
D) To sweep a section along a path

Q12. What is the purpose of the Offset tool in surfacing?

A) To create a curve at a specified angle
B) To create a surface at an offset distance from a selected face/surface
C) To create multiple section sketches
D) To split existing curves

Q13. What is the primary function of the Copy tool in the Surfacing tab?

A) To create a surface from existing model features.
B) To derive a surface from one or more input faces.
C) To replace selected surfaces with new materials.
D) To extend surfaces along specified edges.

Q14. What does the Blank Surface tool accomplish?

A) Creates a surface that extends out from edges.
B) Flattens selected model faces into a surface.
C) Modifies existing surfaces to fit new parameters.
D) Joins multiple surfaces into a single entity.

Q15. What is the purpose of the Intersect tool?

A) To create surfaces from selected edges.
B) To split faces using curves or planes.
C) To extend or trim surfaces to a common intersection.
D) To replace one surface with another.

Q16. What does the Stitched tool do?

A) Creates a new design body from intersecting surfaces.
B) Splits surfaces using sketch curves.
C) Creates a construction surface by stitching multiple sheet surfaces.
D) Extends surfaces to specified distances.

Q17. Which tool is specifically used to create a surface that extends from selected edges?

A) Blank Surface
B) Extend
C) Ruled
D) Trim

Q18. Which option in the Replace Face tool allows you to substitute a surface for a part face?

A) Select face to be replaced
B) Choose material
C) Modify surface properties
D) Select curve for trimming

Q19. In the Split tool, what is required to perform the split operation?

A) Select geometry for trimming only
B) Use curves or reference planes as splitting tools
C) Choose a material to apply
D) Define the angle of split

Q20. What does the Stitched tool do?

A) Creates a new design body from intersecting surfaces.
B) Splits surfaces using sketch curves.
C) Creates a construction surface by stitching multiple sheet surfaces.
D) Extends surfaces to specified distances.

Q21. The Show Non-Stitched Edges tool is used to:

A) Create parting lines along silhouette edges.
B) Display all non-stitched edges of selected surfaces.
C) Replace surfaces with new material.
D) Visualize surface curvature.

Q22. Which command allows you to visualize the curvature of surfaces?

A) Surface Visualization
B) Section Curvature
C) Show Non-Stitched Edges
D) BlueDot

Q23. Which option allows you to specify the angle of a surface created using the Ruled tool?

A) Tangent Continuous
B) Normal to face
C) Tapered to plane
D) Natural

Q24. What feature does the Trim tool provide?

A) It merges multiple surfaces into one.
B) It removes sections of a surface based on a selected curve or plane.
C) It extends a surface beyond its existing boundaries.
D) It creates a new surface from selected edges.

Q25. When using the BlueDot tool, how many curves can be connected at once?

A) One
B) Two
C) Three
D) Four

Q26. What must you select to visualize the surface properties using the Surface Visualization tool?

A) The edges of the surface
B) The surfaces you want to apply visualization settings
C) The type of material used
D) The density of mesh lines only

Q27. In the Stitched tool, what happens if selected surfaces form a closed region?

A) A solid body is automatically created.
B) The operation cannot be performed.
C) The surfaces will be stitched but not converted.
D) You will need to select a new surface.

Q28. Which of the following tools can be used to display non-stitched edges?

A) Surface Visualization
B) Show Non-Stitched Edges
C) Parting Split
D) Intersect

Q29. What is required when specifying a plane for the Parting Split tool?

A) The plane must be curved.
B) The plane must be selected for creating split lines.
C) The plane must be a solid face.
D) The plane must be offset from the model.

Q30. To modify the joint position of a BlueDot connection, which button should you select?

A) Finish
B) Reset
C) Dynamic Edit
D) Close

Q31. In the BlueSurf tool, what does the Natural option in the Tangency Control box ensure?

A) The surface will not change shape.
B) The surface flows naturally based on selected curves.
C) The surface ends will be capped.
D) The surface will be drawn in a straight line.

Q32. Which parameter defines how curves of the surface will be connected in the BlueSurf Options dialog?

A) Open ends
B) Curve Connectivity
C) Tangency Control
D) Extent Type

Q33. What is the function of the Insert Sketch Step button in the BlueSurf command box?

A) To start a new sketch
B) To add another sketch at the intersection of a plane and curve
C) To remove a sketch
D) To preview the surface

Q34. What is the main purpose of the Bounded tool?

A) To create open sketches
B) To generate surfaces from closed boundaries
C) To extend existing surfaces
D) To modify surface textures

Q35. What does the Finish button do when using the Extruded tool?

A) Deletes the surface created
B) Saves the current sketch
C) Creates the extruded surface feature
D) Opens the Extrude Options dialog

Q36. In the context of the Offset tool, what does the Distance edit box specify?

A) The length of the sketch
B) The height of the surface
C) The offset distance from the selected face or surface
D) The width of the feature

Q37. What do you need to do to create a closed surface using the Bounded tool?

A) Select open boundaries only
B) Ensure that selected boundaries are closed
C) Click the Merge option
D) Specify open ends

Q38. What does the Ruled Options drop-down allow you to do when creating a ruled surface?

A) Set the color of the surface
B) Choose the type of extension for the surface
C) Delete existing edges
D) Merge surfaces

Q39. Which tool would you use to create a surface that extends in a specified direction from selected edges?

A) Redefine tool
B) Ruled tool
C) Swept tool
D) Bounded tool

Q40. What parameter do you need to specify when using the Revolved tool?

A) The distance of the offset
B) The angle of revolution
C) The color of the surface
D) The material properties

Chapter 7

Product Manufacturing Information

Topics Covered

The major topics covered in this chapter are:

- *Introduction to Product Manufacturing Information*
- *Setting Planes for PMI*
- *Applying PMI Dimensions*
- *Dimension Styles*
- *Applying Annotations*
- *Setting Annotation Font and Size*
- *Generating Model and Section Views*

INTRODUCTION

PMI (Product Manufacturing Information) is non geometric data included with 3D model for critical digital inspection reporting. Rather than relying on 2D drawings or digital documents to convey engineering or manufacturing data, it can all be embedded within the original CAD file. In some software, it is also called Model Based Definition. PMI may include the following information:

*   GD&T (Geometric dimensions & tolerances).
*   Bill of materials (BOM).
*   Surface finish.
*   Weld symbols.
*   Material specifications.
*   Metadata & notes.
*   History of engineering change orders.
*   Legal/proprietary/export control notices.
*   Other definitive digital data.

The tools to apply PMI are available in the **PMI** tab of the **Ribbon**; refer to Figure-1. Various tools in this tab are discussed next.

Figure-1. PMI tab of Ribbon

LOCKING DIMENSION PLANE (ORDERED ENVIRONMENT)

When we annotate a 3D model using dimensions then there is a need of plane to be used for placing the annotations. We select the planes which are parallel to feature for annotating the model. The procedure to lock a plane for annotation is given next.

*   Click on the **Lock Plane** tool from the **Tools** group in the **PMI** tab of the **Ribbon**. The **Lock Dimension Plane** command box will be displayed; refer to Figure-2.

Figure-2. Lock Dimension Plane command box

*   Select desired option from the **Create From** drop-down in the command box to select or create lock plane. Create a plane as discussed earlier or select a planar face to be used as annotation plane. The selected plane will be assigned as dimension plane.

SETTING DIMENSION AXIS (ORDERED ENVIRONMENT)

The **Set Axis** tool is used to define axis for PMI dimensions at specified angle. This tool is useful when create inclined dimensions.

* Click on the **Set Axis** tool from the **Tools** group in the **PMI** tab of the **Ribbon**. You will be asked to select a line or key points to define the dimension axis.
* Select two key points on the model or select desired line to define dimension axis. The axis will be created. You can check the axis when you next time click on the **Set Axis** tool; refer to Figure-3.

Figure-3. Dimension axis created

Other tools of **Tools** group will be discussed later in this chapter.

CREATING DIMENSION

The tools to create dimensions for PMI are similar to the tools discussed earlier for dimensioning in sketches. Here, we will discuss the use of **Smart Dimension** tool. You can apply the other tools in the same way.

* Click on the **Smart Dimension** tool from the **Dimension** group in the **PMI** tab of the **Ribbon**. You will be asked to select the entities to be dimensioned and the **Smart Dimension** command box will be displayed; refer to Figure-4.

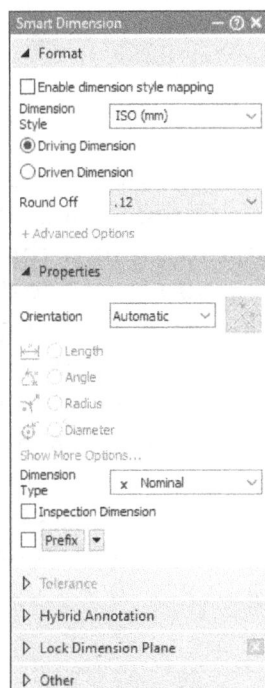

Figure-4. Smart Dimension command box

- Select **Enable dimension style mapping** check box from **Format** rollout to specify the **Dimension style** as specified on the Dimension Style page of **Solid Edge Options** dialog box.

- Select desired option from the **Dimension Style** drop-down to define unit and style of dimensions. By default, **ANSI(mm)** option is selected in our case.

- On selecting the **Driving Dimension** radio button, its value controls the size, orientation, or location of connected sketch elements.

- On selecting the **Driven Dimension** radio button, its value is controlled by the element it refers to, or by a formula or variable you define.

- Select desired option from the **Round-Off** drop-down to define accuracy level of dimension.

- If you want to define accuracy for primary and secondary dimensions separately then click on the **Advanced options** button from the command box. The advance option will be displayed; refer to Figure-5. Set desired precision levels in drop-downs of this dialog box and click on the **OK** button.

Figure-5. Advanced Option displayed

- Select desired option from the **Orientation** drop-down from **Properties** rollout to define how dimension will be oriented. Select the **Horizontal/Vertical** option from the drop-down if you want to create a dimension aligned horizontally or vertically. Select the **By 2 Points** option from the drop-down to create a dimension directly measuring selected two points; refer to Figure-6. Select the **Use Dimension Axis** option from the drop-down if you want to create dimension along dimension axis created earlier. You can create a dimension axis while the **Smart Dimension** tool is active by clicking on the **Dimension Axis** button in the **Smart Dimension** command box. The procedure to create dimension axis has been discussed earlier. When you have selected the **Use Dimension Axis** option from the drop-down then any dimension you create will be aligned to the axis; refer to Figure-7. Select the **Automatic** option from the drop-down to automatically orient the dimensions based on selection.

Figure-6. Dimension by 2 points

Figure-7. Creating dimension using dimension axis

- Select the **Tangent** toggle button to dimension objects using their tangent points. On activating this button, when you select a circle or arc then its tangent point will be used to dimensioning in place of default center point.
- If you are creating a radial dimension then **Radius**, **Diameter**, and **Concentric Diameter** buttons will be active in the command box; refer to Figure-8. Select the **Radius** button if you want to apply radius dimension to selected entity. Select the **Diameter** button to create diameter dimension.

Figure-8. Buttons for radial dimensioning

- Select the **Diameter Normal** button from the **Smart Dimension** command box. The diameter dimension will be created perpendicular to face of model; refer to Figure-9.
- Select the **Jog** button from the command box to convert radial dimension into jogged line dimension; refer to Figure-10.

Figure-9. Normal diameter dimension

Figure-10. Jogged line dimension

- Select the **Diameter - Half/Full** button from the command box to display diameter dimension in half or full length.
- If you are dimensioning an arc in the model then you can create the arc length or arc angle dimension by using the **Length** or **Angle** buttons from the command box; refer to Figure-11.

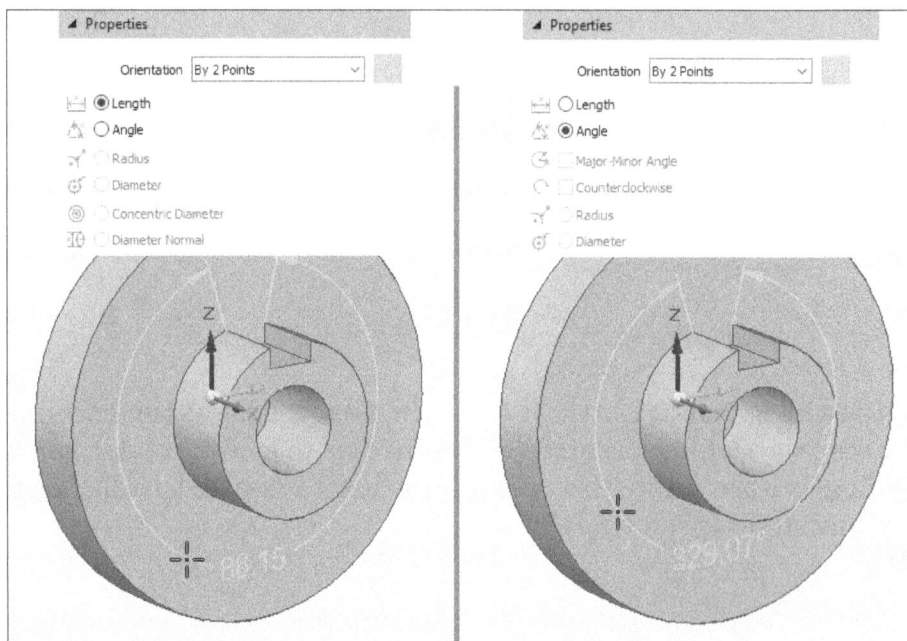

Figure-11. Length and angle dimension of arc

- If you are creating an angular coordinate dimension then you can use **Counterclockwise** button to measure dimensions counterclockwise. Similarly, you can use the **Major-Minor** button from command box to switch between major and minor angle dimension.
- If you want to add symbols and prefixes to the dimension then click on the **Prefix** button from the command box. The **Dimension Prefix** dialog box will be displayed; refer to Figure-12. Click in the **Prefix**, **Suffix**, **Superfix**, and **Subfix** edit boxes of the dialog box and specify desired values. You can also add symbols in these edit boxes by selecting the respective buttons from the **Symbol** area of the dialog box. After specifying desired parameters, click on the **OK** button from the dialog box.

Figure-12. Dimension Prefix dialog box

- Click on the **Dimension Type** button from the command box to define type of dimension. A list of options will be displayed as shown in Figure-13. These options have been discussed in next sections.

Figure-13. Dimension Type options

- Select **Inspection Dimension** check box to add an inspection bubble around the dimension text.
- Select **Feature Control Frame** check box from the **Hybrid Annotation** section to add the feature control frame to the dimension. Click on the **Smart Dimension-Hybrid Feature Control Frame** button, the dialog box will be displayed; refer to Figure-14.
- Specify the Geometric Dimensioning and Tolerance (GD&T) data in the dialog box as needed and click on the **OK** button.

- Select the **Dimension** or the **Feature Control Frame** radio button from **Attach Dimension Line** section to attach dimension line to dimension or feature control frame, respectively.
- After setting desired parameters, click at desired location in graphics area to place the dimension.

Figure-14. Feature control frame properties dialog box

Nominal Dimension Type

Select the **Nominal** option from the list if you want to specify only nominal value of dimension. This option is selected by default.

Unit Tolerance Dimension Type

- Select the **Unit Tolerance** option from the list if you want to specify upper and lower tolerances of dimension. The options to specify upper and lower limit of dimension will be displayed in the command box; refer to Figure-15.

Figure-15. Options to specify upper and lower limits of dimension

- If you want to change the sign of tolerance value then click on the button for sign.
- Specify desired values in tolerance edit boxes of the command box.

Alpha Tolerance Dimension Type

- Select the **Alpha Tolerance** option from the **Dimension Type** list. The options to specify alphanumeric tolerance will be displayed in the command box.
- Specify the parameters as discussed for **Unit Tolerance** dimension type. Note that you can also specify alphabets along with numerical values in the tolerance edit boxes.

Class Dimension Type

- Select the **Class** option from the list of dimension type if you are applying dimensions to holes and shafts.
- Select desired option from the **Type** drop-down to define how tolerance value will be displayed with dimension.
- Select desired tolerance class for shaft/hole from respective drop-down in the command box; refer to Figure-16.

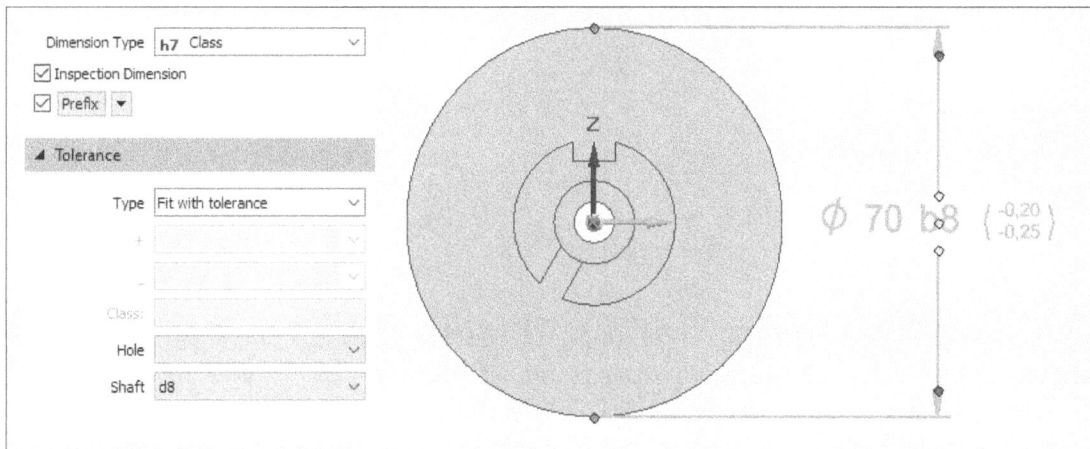

Figure-16. Applying class tolerance

Limit Dimension Type

- Select the **Limit** option from the **Dimension Type** list if you want to display dimensions by their maximum and minimum values after adding/subtracting the tolerance values from nominal value.
- Specify desired tolerance values in the **Upper Tolerance** and **Lower Tolerance** edit boxes of the command box; refer to Figure-17.

Figure-17. Limit tolerance type dimensions

Basic Dimension Type

- Select the **Basic** option from the **Dimension Type** list if you want to specify theoretical value of dimension. Note that theoretical value is impossible to achieve repeatedly in manufacturing so by specifying basic value, you want to say that all the other dimensions should be measured in reference to specified basic value dimension. The title block tolerance does not apply to basic dimensions. The tolerance associated with a basic dimension usually appears in a feature control frame or a note. Basic dimensions have a box around their values.

Reference Dimension Type

- Select the **Reference** option from the **Dimension Type** list if you want to add a dimension which can be accessed already by other dimensions in the drawing; refer to Figure-18.

Figure-18. Reference dimension

Feature Callout Dimension Type

- Select the **Feature Callout** option from the **Dimension Type** list if you want to dimension features like holes which have callouts associated with them.
- Select the round edge of hole or other feature which has callout associated with it. The callout dimension will be attached to cursor; refer to Figure-19.

Figure-19. Callout dimension attached to cursor

Blank Dimension Type

- Select the **Blank** option from the **Dimension Type** list if you want to create a dimension without dimension value. Note that although the dimension value will not be displayed in blank dimension type but prefix and other parameters specified in the **Dimension Prefix** dialog box will be displayed.
- Select the entities to be dimensioned as discussed earlier.

- Select the **Inspection** button from the command box if you want the selected dimension inspected after manufacturing.
- Click on the **Lock Dimension Plane** rollout from the command box and select a plane/face used as base reference for placing dimensions. Select desired plane/face for locking dimensions.
- Click on the **Keypoint for Dimension** button from the command box and select desired points while creating the dimension. The list of keypoints will be displayed; refer to Figure-20.

Figure-20. Keypoints options

DIMENSION STYLES

The **Styles** tool in **Dimension** group of the **PMI** tab in **Ribbon** is used to create and manage dimension styles and parameters for various annotation features. The procedure to use this tool is given next.

- Click on the **Styles** tool from the **Dimension** group in the **PMI** tab of the **Ribbon**. The **Styles** dialog box will be displayed. The options of this dialog box have been discussed earlier in Chapter 2.

MODEL SIZE AND PIXEL SIZE PMI

There are two methods to define size of annotations in PMI, using model text size or using Pixel size. Select the **Model Size PMI** radio button from the **Dimension** group in the **Ribbon** if you want to use size parameters specified in the **Styles** dialog box. Select the **Pixel Size PMI** radio button from the **Dimension** group of **PMI** tab in **Ribbon** if you want to increase or decrease the size of dimension dynamically. On selecting the **Pixel Size PMI** radio button, the **Increase PMI Font** and **Decrease PMI Font** buttons will become active in the **Dimension** group of **Ribbon**. Using these buttons, you can increase or decrease the size of annotations and dimensions. Note that size of all the dimensions will increase or decrease in same proportion.

CREATING CALLOUT ANNOTATION

The **Callout** tool in **Annotation** group is used to create user defined text annotation with leader. The procedure to use this tool is given next.

- Click on the **Callout** tool from the **Annotation** group in the **PMI** tab of the **Ribbon**. The **Callout Properties** dialog box will be displayed; refer to Figure-21.

Figure-21. Callout Properties dialog box

General Tab

- Specify desired text in the **Callout text** and **Callout text 2** edit boxes. You can also add symbols in the edit boxes by using the buttons in **Special characters** area of the dialog box.

- Click on the **Property Text** button from the **Reference** area to add properties used in current document in the edit boxes. On clicking this button, the **Select Property Text** dialog box will be displayed; refer to Figure-22.

Figure-22. Select Property Text dialog box

- Double-click on desired property from the **Properties** list of the dialog box. The property parameter will be added in the edit box. Click on the **OK** button from the dialog box.
- Click on the **Reference Text** button from the **Reference** area of the dialog box. The **Select Reference Text** dialog box will be displayed; refer to Figure-23.

Figure-23. Select Reference Text dialog box

- Select desired option from the **Source type** drop-down to select source of reference text.
- Double-click on the reference text you want to add in the callout and click on the **OK** button.
- Click on the **Select Symbols and Values** button from the **Reference** area to add GD&T symbols and other drawing parameters in the callout. The **Select Symbols and Values** dialog box will be displayed; refer to Figure-24.

Figure-24. Select Symbols and Values dialog box

- Expand desired category of symbol or parameter in the dialog box and double-click on the symbol/parameter to be added.
- You can select the standard feature references, smart depths of features, and bend parameters by selecting respective buttons from the **General** tab of **Callout Properties** dialog box.
- Select desired option from the **Taper symbol** drop-down to add a taper symbol in the callout if selected feature is taper. Apart from the **Off** option which hides taper symbol, you can select the **Left** or **Right** option to place symbol at the left or right in the callout respectively.
- Select the **All around symbol with leader** check box to apply the callout annotation all around selected entity.
- Select the **All over symbol with leader** check box to add all over symbol with the leader.

Text and Leader Tab

The options in the **Text and Leader** tab of the dialog box are used to specify parameters related to text size, leader lines, projection lines, color, terminator, and so on; refer to Figure-25. Set the parameters as desired in this tab. The options in this tab have been discussed earlier.

Figure-25. Text and Leader tab of Callout Properties dialog box

Smart Depth Tab

The options in the **Smart Depth** tab are used to add smart depth parameters in the callout; refer to Figure-26. Click in desired edit box and specify desired parameters. You can use the buttons to add parameters.

Figure-26. Smart Depth tab

Feature Callout Tab

The options in the **Feature Callout** tab are used to define the parameters to be added in the standard hole and slot callouts; refer to Figure-27.

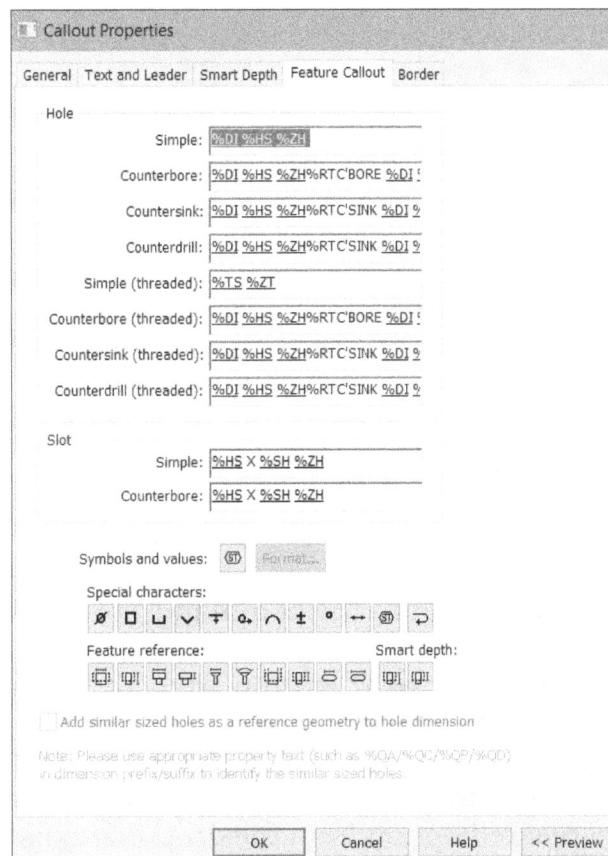

Figure-27. Feature Callout tab

Border Tab

The options in the **Border** tab are used to define thickness of callout outlines, gap between border and text, and text position in the callout; refer to Figure-28.

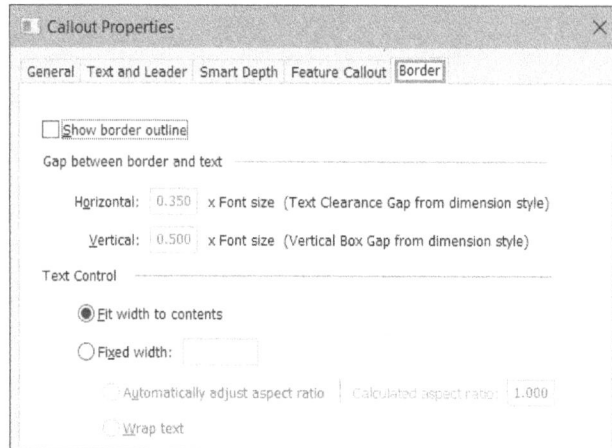

Figure-28. Border tab

- Select the **Show border outline** check box to display border line in callout; refer to Figure-29.

Figure-29. Callout border line preview

- Specify desired parameters in the edit boxes of the **Gap between border and text** area to define gap between border of callout and text inside it.
- Select desired radio button from the **Text Control** area of the dialog box to define the width of frame.
- After specifying the parameters, click on the **OK** button from the dialog box. The **Callout** command box will be displayed; refer to Figure-30.

- Specify desired angle value in the **Leader** edit box to define orientation of callout with respect to leader landing.
- Select the **Break Line** button from the command box to add a landing line with leader in annotation.
- Click on the **Horizontal Alignment** button from the command box and select desired option from the drop-down list displayed to define where text will be placed in callout box.
- Select the **Position** button to define position of text with respect to landing line in annotation. Select desired option from the drop-down list displayed on clicking this button. Note that the **Position** button is active only when **Break Line** button is selected in the command box.

Figure-30. Callout command box

- Select the **Underline Text** button to underline the text in callout.
- Select the **Parallel Text** button from the command box to make callout box parallel to selected face and edge.
- Click on the **Invert Text** button from the command box to invert the text of callout.
- Toggle on/off the **Show Border** button to display or hide the borders of callout.
- Click on the **Text Control** button from the command box and select desired option from the drop-down list to define how text will be arranged in the callout box.
- Click on the **Lock Dimension Plane** button from the command box to lock the active dimension plane for PMI dimension and annotation creation.
- Click on the **Referenced Geometry** button from the command box to select reference geometry callout.
- After setting desired parameters, click on the edge or face to which you want to add callout. The callout will get attached to the cursor; refer to Figure-31.

Figure-31. Callout attached to cursor

- Click at desired location to place the callout. Press **ESC** to exit the tool.

CREATING BALLOON ANNOTATION

The **Balloon** tool in the **Annotation** group is used to create balloon annotation on the model. The procedure to use this tool is given next.

- Click on the **Balloon** tool from the **Annotation** group in the **PMI** tab of the **Ribbon**. The **Balloon** command box will be displayed; refer to Figure-32.

Figure-32. Balloon command box

- Click on the **Shape** button from the command box and select desired option from the list to define shape of balloon annotation.
- Specify desired alphanumeric values in the **Upper**, **Lower**, **Prefix**, and **Suffix** edit boxes to define balloon parameters.
- Set the other parameters as discussed earlier and click on the face to which you want to apply annotation. The balloon annotation will get attached to the cursor; refer to Figure-33.

Figure-33. Balloon annotation attached to cursor

- Click at desired location to place the balloon. Press **ESC** to exit the tool.

CREATING SURFACE TEXTURE SYMBOL

The **Surface Texture Symbol** tool is used to apply surface texture symbol on selected face. The procedure to use this tool is given next.

- Click on the **Surface Texture Symbol** tool from the **Annotation** group of the **PMI** tab in the **Ribbon**. The **Surface Texture Symbol Properties** dialog box will be displayed; refer to Figure-34.

Figure-34. Surface Texture Symbol Properties dialog box

- Click on the button from the **Symbol type** area to select symbol for surface finish. The list of symbols will be displayed as shown in Figure-35.

Figure-35. Surface finish symbols

- Select desired symbol from the list and specify the parameters in edit boxes next to the button.
- Click on the **Surface lay** button from the dialog box and select desired option to define direction of surface finish.
- Set desired parameters of surface finish in the edit boxes, and click on the **OK** button from the dialog box. You will be asked to specify the location for symbol.
- Click at desired face, surface, edge, or other entity. The surface symbol will get attached to cursor.
- Click at desired location to place the symbol. Press **ESC** to exit the tool.

CREATING WELD SYMBOL

The **Weld Symbol** tool is used to create weld symbol in PMI and drawing. The procedure to use this tool is given next.

*   Click on the **Weld Symbol** tool from the **Annotation** group in the **PMI** tab of the **Ribbon**. The **Weld Symbol Properties** dialog box will be displayed; refer to Figure-36.

Figure-36. Weld Symbol Properties dialog box

*   Specify desired parameters in the edit boxes of dialog box for weld symbol and click on the **OK** button.
*   The **Weld Symbol** command box will be displayed; refer to Figure-37 and you will be asked to specify the location of weld symbol.

Figure-37. Weld Symbol command box

*   Click at desired location to specify weld bead location. The symbol will get attached to cursor.
*   Click at desired location to place the symbol.

CREATING EDGE CONDITION

The **Edge Condition** tool is used to annotation tolerance conditions for edges. The procedure to use this tool is given next.

- Click on the **Edge Condition** tool from the **Annotation** group in the **PMI** tab of **Ribbon**. The **Edge Condition Properties** dialog box will be displayed; refer to Figure-38.

Figure-38. Edge Condition Properties dialog box

- Specify desired tolerances values in the edit boxes and set the other parameters as discussed earlier.
- Click on the **OK** button from the dialog box. The **Edge Condition** command box will be displayed; refer to Figure-39.

Figure-39. Edge Condition command box

- You can set desired parameters in the command box as discussed earlier and click at desired edge to which you want to apply annotation. The symbol will get attached to cursor; refer to Figure-40.

Figure-40. Symbol attached to cursor

• Click at desired location to place the symbol and press **ESC** to exit the tool.

CREATING DATUM FRAME AND FEATURE CONTROL FRAME ANNOTATION

The **Feature Control Frame** tool is used to apply GD&T annotations to selected features. Geometrical tolerance is defined as the maximum permissible overall variation of form or position of a feature.

Geometrical tolerances are used,
(i) to specify the required accuracy in controlling the form of a feature.
(ii) to ensure correct functional positioning of the feature.

Figure-41 shows a geometric symbol with meaning of each box.

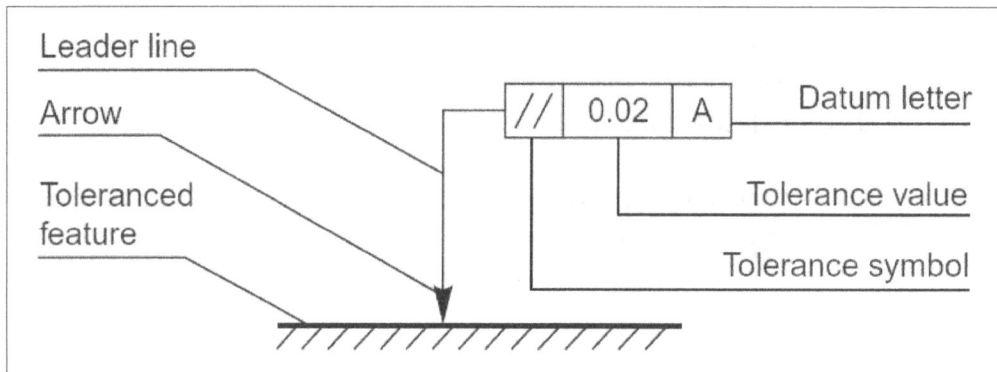

Figure-41. Geometric Tolerance symbol

Before we start using the tools in Solid Edge, it is important to understand the meaning of each Tolerance Symbol.

Meaning of various geometric tolerance symbols are given in Figure-42.

Characteristics to be toleranced		Symbols
Form of single features	Straightness	—
	Flatness	▱
	Circularity (roundness)	○
	Cylindricity	⌀
	Profile of any line	⌒
	Profile of any surface	⌓
Orientation of related features	Parallelism	//
	Perpendicularity (squareness)	⊥
	Angularity	∠
Position of related features	Position	⊕
	Concentricity and coaxiality	◎
	Symmetry	≡
	Run-out	↗

Figure-42. Meaning of geometric Tolerance symbol

Figure-43 and Figure-44 shows the use of geometric tolerances in real-world.

Figure-43. Use of geometric tolerance 1

4. Cylindricity tolerance	8. Concentricity and coaxiality tolerance
⌀ 0.1	Ⓞ φ 0.01 A A
5. Parallelism tolerance	
∥ 0.01 D D	9. Symmetry tolerance
	A = 0.08 A
6. Perpendicularity tolerance	10. Radial run-out
⊥ 0.08 A A	⟋ 0.1 D D
7. Angularity tolerance	11. Axial run-out
∠ 0.08 A 40° A	⟋ 0.1 D D

Figure-44. Use of geometric tolerance 2

Note that in applying most of the Geometrical tolerances, you need to define a datum plane. So, first we will create a datum Frame and then we will apply Feature Control Frame.

Creating Feature Control Frame

The **Feature Control Frame** tool is used to create GD&T frame. The procedure to use this tool is given next.

- Click on the **Feature Control Frame** tool from the **Annotation** group in the **PMI** tab of the **Ribbon**. The **Feature Control Frame Properties** dialog box will be displayed; refer to Figure-45.

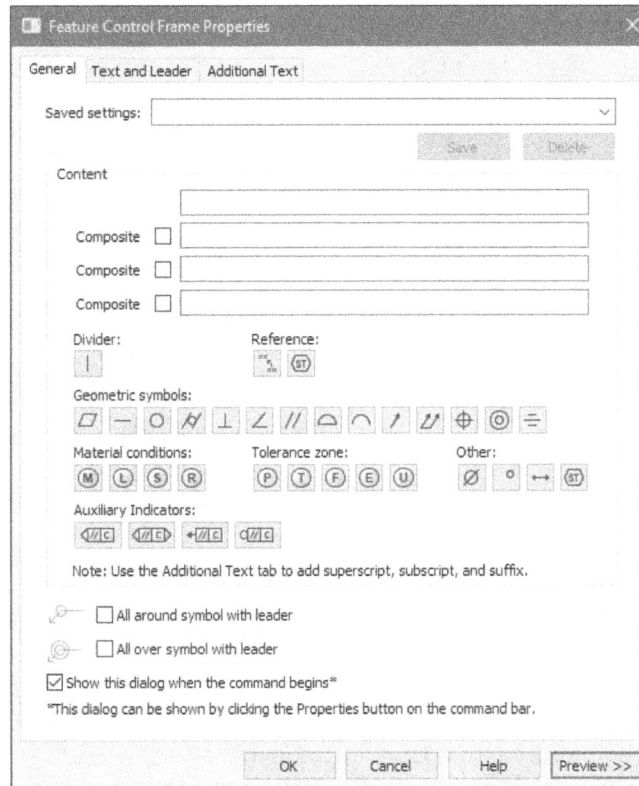

Figure–45. Feature Control Frame Properties dialog box

- Click in the first edit box of the **Content** area and specify GD&T symbol.
- Click on the **Divider** button from the dialog box to create division in frame box.
- Specify the value of tolerance in the edit box and create the divider.
- Click on the **Reference Text** button from the dialog box. The **Select Reference Text** dialog box will be displayed.
- Double-click on desired datum reference value from the list and click on the **OK** button from the dialog box. The preview of symbol will be displayed; refer to Figure-46.

Figure–46. Preview of Feature Control Frame

- You can add material conditions, tolerance zones, and other symbols in the frame, in the same way as discussed earlier.

- Click on the **OK** button from the dialog box. The **Feature Control Frame** command box will be displayed and you will be asked to select the entity to be annotated.
- Click at desired face/surface/entity. The control frame will be attached to cursor; refer to Figure-47.

Figure-47. Feature control frame attached to cursor

- Click at desired location to place the symbol and press **ESC** to exit the tool.

Creating Datum Frame

The **Datum Frame** tool is used to define reference face/plane for applying geometric dimensioning and tolerance. The procedure to use this tool is given next.

- Click on the **Datum Frame** tool from the **Annotation** group in the **PMI** tab of the **Ribbon**. The **Datum Frame** command box will be displayed; refer to Figure-48.

Figure-48. Datum Frame command box

- Specify desired text for datum frame in the **Text** edit box of command box.
- Click on the **Datum Frame Shape** button from the command box. A list frame shapes will be displayed. Select desired shape for boundary of datum frame.
- Select desired face/surface/element to be marked as datum frame. The frame will get attached to cursor; refer to Figure-49.

Figure-49. Datum frame attached to cursor

- Click at desired location to place the symbol.
- Press **ESC** to exit the tool.

CREATING DATUM TARGET

The **Datum Target** tool is used to create datum target annotation for a datum point. The procedure to use this tool is given next.

- Click on the **Datum Target** tool from the **Annotation** group in the **PMI** tab of the **Ribbon**. The **Datum Target** command box will be displayed; refer to Figure-50.

Figure-50. Datum Target command box

- Specify desired parameters in the edit boxes of the command box.
- Click on the **Show Datum Area** button to show dimension with datum target symbol.
- Click at desired entity to apply the annotation. The datum target symbol will get attached to cursor.
- Click at desired location to place the annotation. Press **ESC** to exit the tool.

CREATING LEADER

The **Leader** tool is used to create leader without text. The leader is used to mark target for a text or annotation. The procedure to use this tool is given next.

- Click on the **Leader** tool from the **Annotation** group in the **PMI** tab of the **Ribbon**. The **Leader** command box will be displayed; refer to Figure-51.

Figure-51. Leader command box

- Click at desired location to specify start point of leader. The end point will get attached to cursor.
- Click at desired location to specify end point of leader. Press **ESC** to exit the tool.

CREATING SECTION BY PLANE

The **Section by Plane** tool is used to create 3D section of model using section plane. The procedure to use this tool is given next.

- Click on the **Section by Plane** tool from the **Model Views** group of the **PMI** tab in the **Ribbon**. The **Section by Plane** command box will be displayed with preview of section; refer to Figure-52 and Figure-53.

Figure-52. Section by Plane command box

Figure-53. Preview of section

- Click on the **Options** button from the command box. The **Section by Plane Options** dialog box will be displayed; refer to Figure-54.

Figure-54. Section by Plane Options dialog box

- Specify desired parameters in the dialog box to define name, plane color, edge color, and so on. Click on the **OK** button from the dialog box.
- Select or deselect the **Bounded** button from the command box to define scope of section.
- Click on the **Add Cutting Plane** button from the command box to add more standard cutting plane. If you want to delete a cutting plane then select the plane from the **Cutting Plane List** drop-down and click on the **Delete Cutting Plane** button from the command box.
- You can flip the cut direction by double-clicking on the arrow shown in preview of section in graphics area.
- After setting desired parameters, click on the **OK** button from the command box and then click on the **Finish** button. The 3D section will be created; refer to Figure-55.

Figure-55. 3D section created

CREATING USER DEFINED SECTION

The **Section** tool is used to create section using selected/created planes. The procedure to use this tool is given next.

- Click on the **Section** tool from the **Model Views** group in the **PMI** tab of the **Ribbon**. The **Section** command box will be displayed; refer to Figure-56.

Figure-56. Section command box

- Select desired plane that you want to use for creating section. You will be asked to create sketching boundary and the sketching environment will be activated.
- Click on the **Draw Profile Step** rollout to create a closed sketch section; refer to Figure-57 and click on the **Close Sketch** button from the **Ribbon**. You will be asked to specify whether you want to section inside or outside of the sketch.

Figure-57. Sketch section created

- Click inside the sketch to create section. You will be asked to specify extent of section region.
- Enter desired distance value in the **Distance** edit box or click at desired location; refer to Figure-58.

Figure-58. Section extent

- Click on the **OK** button from the command box. The preview of section will be displayed; refer to Figure-59.

Figure-59. Preview of section by plane

- Click on the **Finish** button from the command box to create the section.

CREATING MODEL VIEWS

The **Model View** tool is used to create model views using current model display. The procedure to use this tool is given next.

- Click on the **Model View** tool from the **Model Views** group in the **PMI** tab of the **Ribbon**. The **Model View** command box will be displayed; refer to Figure-60.

Figure-60. Model View command box

- Select desired option from the **Render Mode** drop-down.
- Select **Include in Publish 3D PDF** check box to publish the model view while creating a 3D PDF.
- Select **Plane** option from the **Selection Mode** drop-down to select the plane or planar face for PMIs to be displayed. Only PMIs created parallel to selected plane will be displayed. Select **PMI** option from the drop-down to display and select the PMIs that are to be displayed based on options selected PMI section of dialog box.

- Click on the **Model View- Select All Displayed** button from the command box to add all display PMIs to the model view. Select **Model View- Deselect All** button from the command box to remove all PMIs from the model view.
- Click on the **Model View- Show All** button from the command box to display all PMI in active document. Click on the **Model View- Hide All** button to hide all PMI in active document. Click on the **Model View- Show Parallel to Plane** button to display all PMI parallel to the view orientation or selected PMI plane. Click on the **Model View- Show Coplanar to Plane** button to display all PMI coplanar to the view orientation or selected PMI plane.
- Click on the **Sections** rollout, select desired option from **Cut** display drop-down to display the cut section.
- Select desired option from the **Cut Display** drop-down in the **Sections** rollout how cut faces will be displayed in the model view.
- After specifying desired parameters in the command box, click on the **Finish** button from the command box. The model view will be created and added in the **Model Views** node of **Feature Tree**; refer to Figure-61.

Figure-61. Model view created

PRACTICAL 1

Create the model and apply the model based annotations as shown in Figure-62.

Figure-62. Practical

Starting a Document

- Start Solid Edge by using Start menu or desktop icon, if not started yet.
- Click on the **ISO Metric Part** option from the **New** cascading menu of the **Application** menu. A new part file document will open.
- Select the **Ordered** radio button from the **Model** group in the **Tools** tab of the **Ribbon** to activate ordered mode.

Creating Extrude Feature

- Click on the **Extrude** tool from the **Solids** group in the **Home** tab of **Ribbon**. The **Extrude** command box will be displayed and you will be asked to select a sketching plane.
- Select the Front (XZ) plane and create the sketch as shown in Figure-63.
- Apply the dimension and then modify them as per the drawing to apply Basic dimension box; refer to Figure-64.

Figure-63. Base sketch for Extrude

Figure-64. Sketch after applying basic dimension symbols

- After creating sketch, click on the **Close Sketch** tool from the **Close** group in the **Home** tab of **Ribbon**. You will be asked to specify extent of extrude feature.
- Specify the value **38** in the **Distance** edit box of command box and press **ENTER**. You will be asked to specify the side on extrude feature will be created with respect to sketching plane.

- Click behind the sketching plane to specify side of feature and click on the **Finish** button. The feature will be created.

Creating Extrude Cut Feature

- Click on the **Extrude - Add/Cut** button from **Extrude** command box and select the **Cut** option from the list displayed. You will be asked to select a sketching plane.
- Select the back face of model as shown in Figure-65 and orient the sketching plane parallel to screen.

Figure-65. Face selected for cut feature

- Create the sketch for cut as shown in Figure-66 and apply the dimensions.

Figure-66. Sketch for cut feature

- Click on the **Close Sketch** button from the **Ribbon** to exit sketching environment. You will be asked to specify extent of cut feature.
- Change the orientation of model for clear view, select the **Through All** button from the command box, and click on desired side to create the feature. Preview of cut feature will be displayed.
- Click on the **Finish** button from the command box to create the feature and press **ESC** to exit the tool.

Creating Holes

- Click on the **Hole** tool from the **Hole** drop-down in the **Home** tab of **Ribbon**. The **Hole** command box will be displayed and you will be asked to select a plane for creating hole.

- Select the back face of model as shown in Figure-67 and place the sketch circle of hole as shown in Figure-68.

Figure-67. Face selected

Figure-68. Sketch for hole

- Click on the **Close Sketch** tool from the **Close** group in the **Home** tab of **Ribbon**. Preview of the hole will be displayed.
- By default, the holes will be created through. Click on the **Finish** button from the command box to create the hole and press **ESC** to exit the tool.
- Similarly, create the hole of diameter **4** as per the drawing.

Creating PMI Annotations

- Click on the **Copy to PMI** tool from the **Tools** group in the **PMI** tab of the **Ribbon**. The **Copy to PMI** command box will be displayed.
- Select the model whose 2D dimensions are to be copied. Preview of dimensions will be displayed; refer to Figure-69.

Figure-69. Dimensions displayed on selecting model

- Select all the dimensions to be copied and right-click in the drawing area.
- Repeat the steps to copy all the dimensions and then press **ESC** to exit the tool.
- Click on the **Datum Frame** tool from the **Annotation** group in the **PMI** tab of **Ribbon**. The **Datum Frame** command box will be displayed.
- Click on the **Feature Control Frame** tool from the **Annotation** group in the **PMI** tab of **Ribbon** and place the control frames as shown in Figure-70 using left orientation of model.

Figure-70. Placing annotations in left orientation

- Similarly, apply feature control frames and datum frames in Front orientation to the model; refer to Figure-71.

Figure-71. Placing annotations in front orientation

PRACTICE 1

Create the model and apply annotations as per the drawing shown in Figure-72.

Figure-72. Drawing for practice

SELF ASSESSMENT

Q1. What is the purpose of the Prefix button in the Smart Dimension command box?

A) To create a diameter dimension
B) To add symbols and prefixes to the dimension
C) To set the accuracy level of the dimension
D) To select inspection dimensions

Q2. In the Smart Dimension command box, what option allows you to specify the style of dimensions?

A) Enable dimension style mapping
B) Select Dimension Axis
C) Automatic Option
D) Tangent Toggle Button

Q3. Which dimension type allows you to specify upper and lower tolerances of a dimension?

A) Nominal Dimension Type
B) Unit Tolerance Dimension Type
C) Reference Dimension Type
D) Feature Callout Dimension Type

Q4. Which tool is used to create and manage dimension styles in the PMI tab?

A) Callout Tool
B) Styles Tool
C) Lock Dimension Plane Tool
D) Surface Texture Symbol Tool

Q5. What is the function of the Callout tool in the PMI tab?

A) To apply surface texture symbols on selected faces.
B) To create user-defined text annotations with leaders.
C) To create balloon annotations on models.
D) To define reference faces for geometric dimensioning.

Q6. What does the Edge Condition tool annotate?

A) Surface texture conditions.
B) Tolerance conditions for edges.
C) Geometric tolerances for features.
D) Callouts for specific dimensions.

Q7. What must you specify when creating a Feature Control Frame in the PMI tab?

A) The color of the annotation.
B) The GD&T symbol and tolerance value.
C) The position of the text relative to the leader.
D) The height ratio of the balloon.

Q8. What is the primary purpose of the Leader tool in the Annotation group?

A) To create a 3D model.
B) To create a leader without text for marking a target for annotations.
C) To create sections using selected planes.
D) To define dimensions for a model.

Q9. What does the Section by Plane tool allow you to do?

A) Create leaders with text.
B) Generate 3D sections of the model using a section plane.
C) Create annotations with symbols.
D) Manage model view styles.

Q10. What does the Model View tool allow you to create?

A) 3D sections of a model.
B) Various model views using the current model display.
C) Annotations and leaders for models.
D) Cutting planes for sections.

Q11. Which command would you use to show all PMIs in the active document?

A) Model View - Hide All
B) Model View - Show Parallel to Plane
C) Model View - Show All
D) Model View - Deselect All
Answer: C) Model View - Show All

Q12. What is the first step to create a Feature Control Frame?

A) Select the desired entity to annotate
B) Click on the Feature Control Frame tool from the Annotation group
C) Specify the tolerance value in the edit box
D) Create a datum reference

Q13. What is the purpose of clicking on the Divider button in the Feature Control Frame Properties dialog box?

A) To set the material conditions
B) To create divisions in the frame box
C) To specify the text for the datum frame
D) To add multiple GD&T symbols

Q14. Which of the following symbols can you place using the Surface Texture Symbol tool?

A) Datum frame symbols
B) Geometric tolerance symbols
C) Surface finish symbols
D) Edge condition symbols

Q15. What is the function of the Automatic option in the Orientation drop-down?

A) To manually adjust the dimension size
B) To automatically orient dimensions based on the selection
C) To lock the dimension plane
D) To hide dimension values

Q16. In the context of dimensioning, what does the Limit Dimension Type specify?

A) The theoretical value of a dimension
B) Dimensions by their maximum and minimum values
C) Upper and lower tolerances only
D) Nominal values without tolerances

Q17. Which of the following is true about Reference Dimension Type?

A) It displays dimensions with no value
B) It represents theoretical values
C) It adds dimensions accessible by other dimensions
D) It specifies tolerances for dimensions

Q18. What does the Blank Dimension Type allow you to create?

A) Dimensions with no associated values
B) Dimensions with tolerances
C) Dimensions only for holes and shafts
D) Dimensions with theoretical values

Q19. What function does the Leader button serve in the Callout command box?

A) To specify the text style in the callout
B) To define the orientation of the callout text
C) To add a landing line with leader in annotation
D) To select the reference geometry for callouts

Q20. In the Callout Properties dialog box, what do the Special Characters buttons allow you to do?

A) Change the font size of the text
B) Add GD&T symbols and other drawing parameters
C) Create new geometry
D) Define the background color of the callout

Q21. When using the Balloon tool, what does the Height edit box specify?

A) The width of the balloon annotation
B) The angle of the balloon orientation
C) The balloon height as a ratio of text height
D) The size of the text in the balloon

Q22. What is the first step to lock a dimension plane for annotation?

A) Select a planar face
B) Click on the Lock Plane tool in the PMI tab
C) Create a new model
D) Open the Dimension Styles dialog

Q23. In the context of dimension styles, what does the Styles tool in the PMI tab allow you to do?

A) Create and manage dimension styles and parameters
B) Change the model size of annotations
C) Add symbols and prefixes to dimensions
D) Define the default color of dimensions

Q24. What does the Callout tool in the Annotation group allow you to create?

A) Radial dimensions for circles
B) User-defined text annotations with leaders
C) Lock dimension planes
D) Automatic dimensions

Q25. What is the purpose of the 'Jog' button in the Smart Dimension command box?

A) To create a curved dimension
B) To convert radial dimensions into jogged line dimensions
C) To delete existing dimensions
D) To change dimension units

Q26. What is the purpose of selecting the 'Add a leader line' check box when creating balloon annotations?

A) To define the angle of the balloon
B) To display a leader line connecting the balloon to the geometry
C) To change the shape of the balloon
D) To enable dimension style mapping

Q27. What is the purpose of the Text and Leader tab in the dialog box?

A) To define surface texture parameters
B) To specify parameters related to text size, leader lines, and color
C) To create geometric dimensioning and tolerancing (GD&T) annotations
D) To generate balloon annotations

Q28. Which tab allows you to add smart depth parameters to a callout?

A) Text and Leader
B) Border
C) Smart Depth
D) Feature Callout

Q29. What should you do to display a border line in a callout?

A) Adjust the gap parameters
B) Select the Show border outline check box
C) Specify text parameters
D) Click on the Lock Dimension Plane button

Q30. What action is taken after setting parameters in the Feature Control Frame tool?

A) The callout is automatically placed
B) The user is asked to specify the location for the weld symbol
C) The user selects the entity to be annotated
D) The command box disappears

Q31. Which tool is used to create balloon annotations on the model?

A) Feature Control Frame
B) Surface Texture Symbol
C) Weld Symbol
D) Balloon tool

Q32. What is the primary purpose of the Weld Symbol tool?

A) To create geometric dimensioning annotations
B) To apply surface texture symbols
C) To create a weld symbol in PMI and drawing
D) To define reference planes for GD&T

Q33. What does the Datum Target tool allow you to create?

A) A leader without text
B) A datum target annotation for a datum point
C) A weld symbol
D) A feature control frame

Q34. Which command is used to lock the active dimension plane for PMI dimension and annotation creation?

A) Show Border
B) Lock Dimension Plane
C) Position Button
D) Invert Text

Q35. What does the Break Line button do in the callout command box?

A) Adjusts the text size
B) Adds a landing line with a leader in annotation
C) Defines the balloon height
D) Creates a surface texture symbol

Q36. What should you do after setting parameters in the Weld Symbol tool?

A) Click on the surface texture symbol
B) Specify the location for the weld symbol
C) Select the datum frame
D) Choose the border outline option

FOR STUDENTS NOTES

FOR STUDENTS NOTES

Chapter 8

Assembly-I

Topics Covered

The major topics covered in this chapter are:

- *Starting Assembly Document*
- *Inserting Component*
- *Assembly Constraint*
- *Modification Tools*
- *Applying Motors*
- *Simulation of Motors*
- *Pattern in Assembly*
- *Inspection Tools*
- *Practical*
- *Practice*

ASSEMBLY

In Engineer's language, assembly is the combination of two or more components and these components are constrained to each other in a specified manner called assembly constraints. In Solid Edge, assembly design is a separate environment. To start the Assembly design, click on **New** button from the menu bar. The **New** dialog box will display; refer to Figure-1. Double-click on the **iso metric assembly.asm** template from the dialog box. The assembly design environment will be displayed as shown in Figure-2. The tools related to assembly are available in the **Ribbon**.

Figure-1. New Solid Edge document dialog box

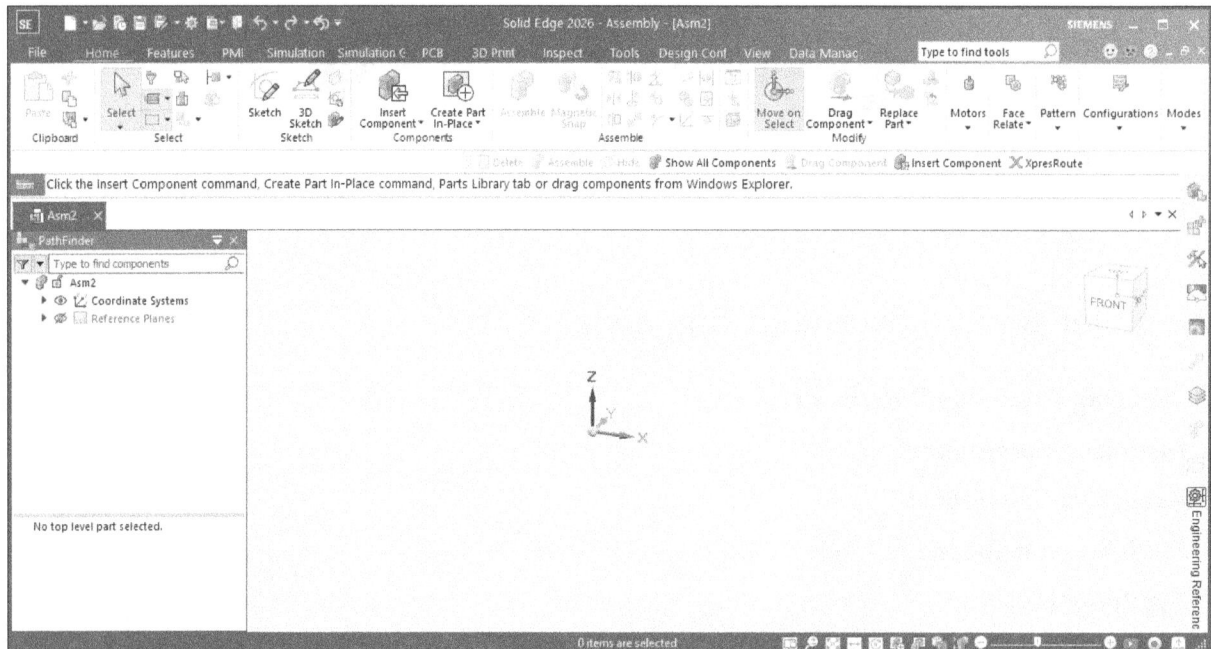

Figure-2. Assembly Design Environment

INSERTING GROUND COMPONENT

After starting Assembly environment, select desired part file from the **Parts Library** box which is located on the left side in application window; refer to Figure-3. If you want to fix the **Parts Library** at the left in the application window then click on the **Auto Hide** (Pin) button at the top in the box; refer to Figure-3. Drag desired part to drawing area, by default the first inserted part is grounded automatically. You can locate the newly inserted component in the **Path Finder**. When you hover the cursor on part name in the **Path Finder**, the part is highlighted in the drawing area; refer to Figure-4.

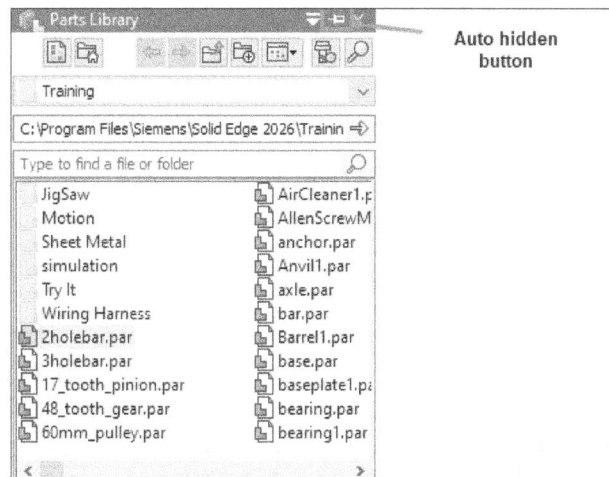

Figure-3. Parts library

Figure-4. Part ground

INSERTING COMPONENTS IN ASSEMBLY

Once, you have inserted the base component in the assembly, follow the steps given next to insert more components.

- Click on the **Insert Component** tool from the **Components** group in the **Home** tab of the **Ribbon**. The **Parts Library** box will be displayed.
- Browse to the location of desired part file from **Part Library** box, click on it, and drag to the drawing area. The **FlashFit** command box will be displayed; refer to Figure-5.

Figure-5. FlashFit command box

- The **List** drop-down in the **Relationships** rollout display the existing relationships for the selected occurrence. Select desired relationship type to be applied from **Types** drop-down to define constraint applied between base component and inserted new component. Various constraints will be discussed later.
- Select desired offset type from **Type** area of the **Offset** rollout and specify respective parameters.
- Click on the **Direction** rollout and select **Lock axial rotation** toggle button to lock the relative rotation of the components. Click on the **Flip** button to flip the direction of the relationships.
- Click on the **Show** rollout and turn on desired entity toggle button that you want to display in assembly. Select **Designed** or **Simplified** radio button to show design or simplified models, respectively.
- Click on the **Occurrence Properties** rollout and click on the **Open** button. The **Occurrence Properties** dialog box will be displayed; refer to Figure-6. Set desired parameters in the dialog box like quantity of component and other properties to be added in the assembly. After setting desired parameters, click on the **OK** button from the dialog box.

Figure-6. Occurrence Properties dialog box

- Click on the **Settings** rollout. The options will be displayed; refer to Figure-7.

Figure-7. Options displayed

- Turn on **Use FlashFit as the default placement method** toggle button to automatically apply FlashFit constraint when a new component is inserted in the assembly.
- Turn on toggle buttons from **Locate** area to define which elements will be used for placing component when FlashFit constraint is used.
- Turn on **Use Reduced Steps when placing parts** toggle button to place parts with minimum number of steps needed like selecting the part and then applying placement constraints.
- Turn on **Automatically Capture Fit when placing parts** toggle button to automatically fit holes, shafts, fasteners, and so on which need to be repeatedly inserted in the assembly.
- Turn on **Use distance between faces as default offset** toggle button to automatically apply offset distance constraint when two faces are selected from assembly.
- Turn on **Place as Adjustable** toggle button to make the inserted sub-assembly adjustable.
- Turn on **Disperse after placement** toggle button to add inserted sub-assemblies as individual parts.
- Turn on **Match click points on the parts when creating first assembly relationship (pre-ST8 behavior)** toggle button to use match behavior used in Solid Edge ST8 and previous versions.
- Turn on **Show all dimensions** toggle button to display dimensions when placing components.
- After setting desired parameters in dialog box, click on the **OK** button. Various constraints used in assembly are discussed next.

ASSEMBLY CONSTRAINT

The assembly constraints are used to combine various parts in a single assembly. Click on the **Assemble** button from the **Assemble** group or insert the component, the **Assemble** command box will be displayed. Or, you can use the respective tool from the **Assemble** group in the **Ribbon**. These constraints are discussed next.

Magnetic Snap

The **Magnetic Snap** tool is used to automatically detect and apply geometric relationships between parts such as faces, edges, axes, and holes. The procedure to use this tool is discussed next.

- Click on the **Magnetic Snap** tool from the **Assemble** group in the **Ribbon**. The **Magnetic Snap** dialog box will be displayed; refer to Figure-8.

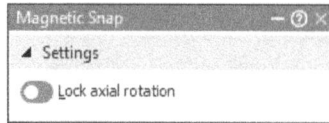

Figure-8. Magnetic Snap command box

- Select **Lock axial rotation** toggle button to stop the rotation of part being dragged for snapping. When this toggle button is OFF then depending on nearby geometries, the part being dragged can change orientation.
- Select the axis of child part and drag it to the desired hole of parent part. The child part will be snapped to the hole of parent part due to magnetic snapping; refer to Figure-9. Similarly, you can drag part by face or edge to apply related constraints via magnetic snapping.

Figure-9. Magnetic Snap created

FlashFit

The **FlashFit** tool is active by default when you insert a new component. Using this tool, you can automatically apply mate, planar align, axial align, and connect constraints based on selected entities.

Mate

The **Mate** tool is used to mate two parts by making their faces/planes coincident. The procedure to use this mate is given next.

- After inserting desired components, click on the **Mate** tool from the **Assemble** group in the **Ribbon**. The **Mate** command box will be displayed as shown in Figure-10.

Figure-10. Mate command box

- Select the face of inserted part. You will be asked to select face of target part.
- Select desired face of target part. The constraint will be applied; refer to Figure-11.

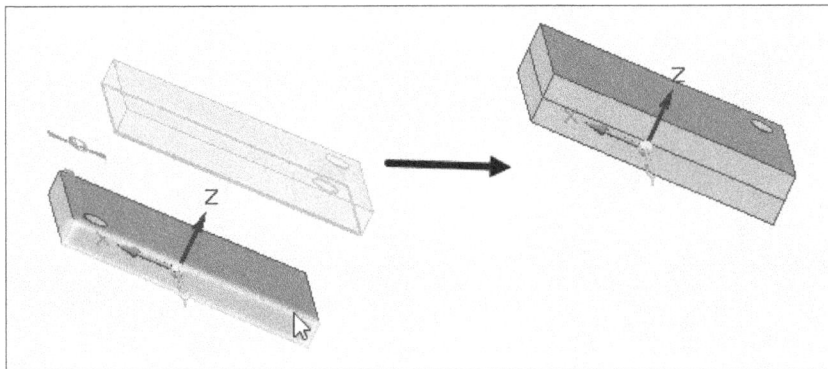

Figure-11. Mate constraint applied

- By default, the **Fixed Offset** option is selected in the **Type** area of **Offset** rollout and specify offset distance in the **Value** edit box to define gap between selected faces.
- Select the **Floating Offset** option from the **Type** area of **Offset** rollout to allow free lateral movement of inserted part.
- Select the **Range** option from the **Type** area of **Offset** rollout to define minimum and maximum gap allowed between selected faces.
- Select **Lock rotation** check box from **Direction** rollout if you want to stop rotation of inserted part in the assembly.
- After setting desired parameters, right-click in the drawing area. The constraint will be applied.

Planar Align

The **Planar Align** tool is used to make faces of inserted part and base part aligned to each other. The procedure to use this tool is discussed next.

- Click on the **Planar Align** tool from the **Assemble** group in the **Home** tab of the **Ribbon**. The **Planar Align** command box will be displayed; refer to Figure-12.

Figure-12. Planar Align command box

- One by one select the faces of base part and inserted part. The selected faces will be aligned; refer to Figure-13. Right-click in the drawing area to create the constraint.

Faces selected for aligning

After aligning faces

Figure-13. Planar align face

Axial Align

The **Axial Align** tool is used to align two cylindrical axes, a cylindrical axes and a linear element, or two linear elements by using their axes. The procedure to use this tool is given next.

- Click on the **Axial Align** tool from the **Assemble** group of **Home** tab in the **Ribbon**. The **Axial Align** command box will be displayed; refer to Figure-14.

Figure-14. Axial Align command box

- Select the axes of base component and inserted component one by one. The constraint will be applied; refer to Figure-15.

Figure-15. Axial Align

- You can lock and unlock rotation of components from the command box as discussed earlier.

Insert

The **Insert** tool is used to apply insert relationship to place axial-symmetric parts such as nuts and bolts into holes or onto cylindrical protrusions. The procedure to apply insert constraint is given next.

- Click on the **Insert** tool from the **Assemble** group in the **Home** tab of the **Ribbon**. The **Insert** command box will be displayed; refer to Figure-16.

Figure-16. Insert command box

- Select cylindrical surface of first entity and circular axis of second entity to align the axes.
- Select the face of first entity and face of second entity to make faces coincident. You can also specify offset value to put gap between selected faces; refer to Figure-17.

Figure-17. Insert

Parallel

The **Parallel** tool is used to make two components parallel with respect to each other by selected cylindrical faces, edges, or axes. The procedure to use this tool is given next.

- Click on **Parallel** tool from the **Assemble** group in the **Home** tab of the **Ribbon**. The **Parallel** command box will be displayed; refer to Figure-18.

Figure-18. Parallel command box

- Select the two faces/axes/planes of two components. The components will be parallel with respect to these references; refer to Figure-19.

Figure-19. Parallel Constraint

Connect

The **Connect** tool is used to connect a keypoint of one part to another keypoint, line or plane of another part. The procedure to apply the constraint is discuss next.

- Click on the **Connect** tool from the **Assemble** group in the **Home** tab of the **Ribbon**. The **Connect** command box will be displayed; refer to Figure-20.

Figure-20. Connect command box

- Select the key points of the two components. The selected points will become coincident; refer to Figure-21.

Figure-21. Connect constraint

- You can also specify offset value from the command box.

Angle

The **Angle** tool is used to constrain angle between faces of two components. The procedure to apply this constraint is given next.

- Click on the **Angle** tool from the **Assemble** group in the **Home** tab of the **Ribbon**. The **Angle** command box will be displayed as shown in Figure-22.

Figure-22. Angle command box

- Select desired angle format from **Format** drop-down and specify the angle value in **Value** edit box of **Properties** rollout.
- Select the faces of first and second entity. You will be asked to select the plane along which angle will be measured.
- Select desired plane along which angle measurement will be performed; refer to Figure-23. Right-click in the drawing area to apply constraint.

Figure-23. Angle constraint applied

Tangent

The **Tangent** tool is used to make two components tangent to each other. The procedure to use this tool is given next.

- Click on the **Tangent** tool from the **Assemble** group in the **Home** tab of the **Ribbon**. The **Tangent** command box will de displayed; refer to Figure-24.

Figure-24. Tangent command box

- Select two faces that you want to make tangent. The preview of tangent constraint will be displayed; refer to Figure-25.

Figure-25. Tangent constraint

- Right-click in the drawing area. The constraint will be applied.

Cam

The **Cam** tool is used to create cam-follower mate between two entities. The procedure to apply cam constraint is given next.

- Click on the **Cam** tool from the **Path** drop-down in the **Assemble** group of **Home** tab in the **Ribbon**. The **Cam** command box will be displayed; refer to Figure-26.

Figure-26. Cam command box

- Select a face or point of follower which will run over cam. You will be asked to select the face chain of cam.
- Select the face chain of cam and click on the **OK** button to accept the selection.
- Right-click in the drawing area to create the constraint. Figure-27 shows an example of cam mate.

Figure-27. Cam

Note that you need to limit the motion of follower so that it moves only in vertical direction and do not deviate from cam path.

Gear

The **Gear** tool is used to create a joint between two gears with specified gear ratio. The procedure to use this tool is given next.

- Click on the **Gear** tool from the **Assemble** group in the **Home** tab of the **Ribbon**. The **Gear** command box will be displayed; refer to Figure-28.

Figure-28. Gear command box

- Select the cylindrical faces of two mating gears. Note that by default, the **Rotation-Rotation** option is selected in the **Gear type** area of command box and hence two cylindrical gears are assembled. Select the **Rotation-Linear** option from the **Gear type** area of command box if you are assembling a cylindrical gear and rack then select the cylindrical face of gear and linear edge of rack; refer to Figure-29. Select the **Linear-Linear** option if you want to assemble two linear gears and select the linear edges of two gears.

Figure-29. Gear

- Specify desired values of gear ratio and click on the **OK** button from the command box to accept the selection.
- Right-click in the drawing area to create the constraint.

Match Coordinate Systems

The **Match Coordinate Systems** tool is used to mate two component with their coordinate system. You can use the default coordinate systems of parts as well as user defined coordinates for applying this mate. If coordinate system of active component is not displayed by default then select the part from **Path Finder** and right-click on it. A shortcut menu will be displayed; refer to Figure-30. Click on the **Show/Hide Component** option from the shortcut menu. The **Show/Hide Component** dialog box will be displayed as shown in Figure-31. Select the **On** check box for **Coordinate Systems** from the dialog box and click on the **OK** button. The coordinate system will be displayed. The procedure to apply Match Coordinate Systems mate is discussed next.

Figure-30. Cascade menu

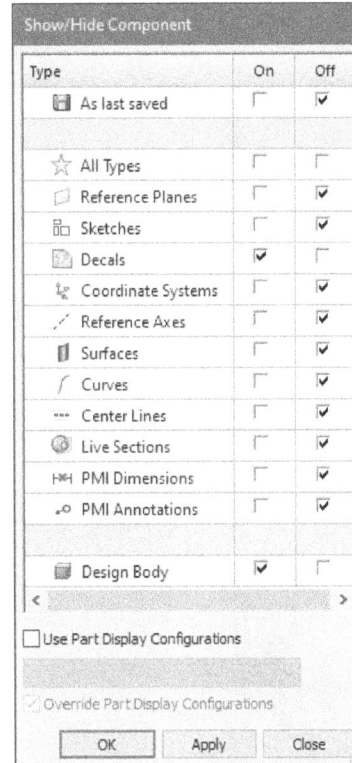

Figure-31. Show hide component

- Click on the **Match Coordinate Systems** tool from the **Assemble** group in the **Home** tab of the **Ribbon**. The **Match Coordinate Systems** command box will be displayed; refer to Figure-32.

Figure-32. Match Coordinate Systems command box

- Select the coordinate system of first and second component. The selected coordinate systems will become coincident; refer to Figure-33.

Figure-33. Match coordinate system

Center-Plane

The **Center-Plane** tool is used to place a component at the center of another component. The procedure to use this tool is given next.

- Click on the **Center-Plane** tool from the **Assemble** group in the **Home** tab of the **Ribbon**. The **Center-Plane** command box will be displayed; refer to Figure-34 and you will be asked to select the faces of components.

Figure-34. Center Plane command box

- Select the start and end faces of the component which is to be centered.
- Select the faces of another component to be used as boundary for component; refer to Figure-35. The component will be placed.
- Right-click in the drawing area to apply mate.

Figure-35. Applying Center-Plane mate

Path

The **Path** tool is used to move part along the selected path. The procedure to use this tool is given next.

- Click on the **Path** tool from the **Path** drop-down in the **Assemble** group of the **Home** tab in the **Ribbon**. The **Path** command box will be displayed; refer to Figure-36.

Figure-36. Path command box

- Note that to use this constraint, there must be a path on the target plane so that another part can follow that path. For that you need to draw path by using sketching tools in Assembly environment. You can also use an edge of model as path for creating the constraint.
- Select desired keypoint of inserted part and an edge or sketch path; refer to Figure-37.

Figure-37. Creating path constraint

Rigid Set

The **Rigid Set** tool is used to make selected assembly components rigid or fix with respect to each other, so if we move one component then other will also move. The procedure to use this tool is given next.

- Click on the **Rigid Set** tool from the **Assemble** group in the **Home** tab of the **Ribbon**. The **Rigid Set** command box will be displayed; refer to Figure-38.

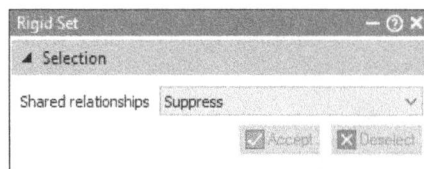

Figure-38. Rigid Set command box

- Select desired option from the **Shared relationships** drop-down to specify whether to delete, suppress, or ignore the applied constraints when applying rigid set.

- Select the faces of components that you want to make rigid set and right-click in the drawing area. Press **ESC** to exit the tool.

Ground

The **Ground** tool is used to apply a ground relationship to the selected part in an assembly. A ground relationship is automatically applied to the first part placed in an assembly.

Assembly Relationships Manager

The **Assembly Relationships Manager** tool provides the list of assembly relationships applied between various components of assembly and allow you to modify/repair the relationships. The procedure to use this tool is given next.

- Click on the **Assembly Relationships Manager** tool from the **Assemble** group in the **Home** tab of the **Ribbon**. The **Assembly Relationships** dialog box will be displayed; refer to Figure-39. In this dialog box, you can locate all the relationships applied in the assembly.

Figure-39. Assembly Relationships Manager

- Right-click on the relationship you want to modify and select the **Edit Definition** option from the shortcut menu. The command box will be displayed with options related to modifying the constraint.
- You can also use the **Suppress** and **Unsuppress** options in the shortcut menu to suppress or activate the selected constraint.
- After setting desired parameters, click on the **Close** button to close the dialog box.

Assembly Relationships Assistant

The **Assembly Relationship Assistant** is used to apply assembly relationships between selected parts and subassemblies based on their current geometric orientation. This command is useful when working with assemblies whose parts are oriented correctly, but do not have assembly relationships, such as assemblies that were imported into Solid Edge from another CAD system. The procedure to use this tool is given next.

- Click on the **Assembly Relationship Assistant** tool from the **Assemble** group in the **Home** tab of the **Ribbon**. The **Relationship Assistant Options** dialog box will be displayed; refer to Figure-40.

Figure-40. Relationship Assistant Options dialog box

- Select desired radio button from the **Check Select Set1 Against** area to define scope of components to be checked for applying constraints automatically.
- Select the **Remove relationships on parts in Select Set1** check box to remove existing relationships when applying new one. By default, the **Remove grounded part relationships** radio button is selected which deletes the ground relationship automatically. Select the **Remove all relationships** radio button to remove all existing relationships of selected components when applying new one.
- Similarly, set the other parameters and click on the **OK** button. The **Assembly Relationship Assistant** command box will be displayed; refer to Figure-41.

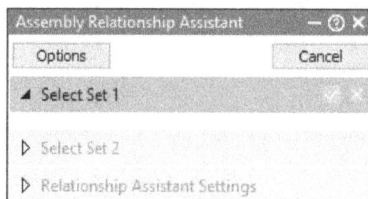

Figure-41. Assembly Relationship Assistant command box

- Select the components of first set and right-click in the drawing area. You will be asked to select components of second set.
- Select the components of second set and right-click in the drawing area. The **Relationship Assistant Settings** dialog box will be displayed; refer to Figure-42.

Figure-42. Relationship Assistant Settings dialog box

- Select the check boxes for relationships you want to apply and click on the **Process** button. The constraints that can be applied will be displayed in the list box; refer to Figure-43.

Figure-43. List of available relationships

- Click on the **Accept** button from the dialog box to apply new relationships.
- Click on the **Close** button from the dialog box to exit the tool.

CAPTURE FIT

The **Capture Fit** tool is used to copy already existing relationships of a component so that when next time you insert the same component, the relationships will be applied automatically. The procedure to use this tool is given next.

- Select the assembly component whose relationships are to be copied and click on the **Capture Fit** tool from the **Assemble** group in the **Home** tab of the **Ribbon**. The **Capture Fit** dialog box will be displayed; refer to Figure-44.

Figure-44. Capture Fit dialog box

- Select the relationships that you do not want to copy from the **Learn these relationships** area of the dialog box and click on the **<< Remove** button.

- Similarly, select the relationships that you want to copy from the **Do not learn these relationships** area of the dialog box and click on the **Add >>** button.
- After setting desired parameters, click on the **OK** button from the dialog box to make the software learn which relationships will be applied automatically when next time this component is inserted. Note that when next time you will insert the same component, you will be asked to select the similar reference entities on other base component as you specified while inserting the original component.

MOVE ON SELECT

The **Move on Select** tool is used to allow moving the component in the specific direction after selecting it. When a component is selected a steering wheel appears on it; refer to Figure-45. Select the arrow of direction from steering wheel in which you want to move or rotate the component.

Figure-45. Steering Wheel

DRAG COMPONENT

The **Drag Component** tool is used to drag component which is under constraint or grounded. The step of using this tool is discussed next.

- Click on the **Drag Component** tool from the **Drag Component** drop-down in the **Modify** group of **Home** tab in the **Ribbon**. The **Analysis Options** dialog box will be displayed; refer to Figure-46.

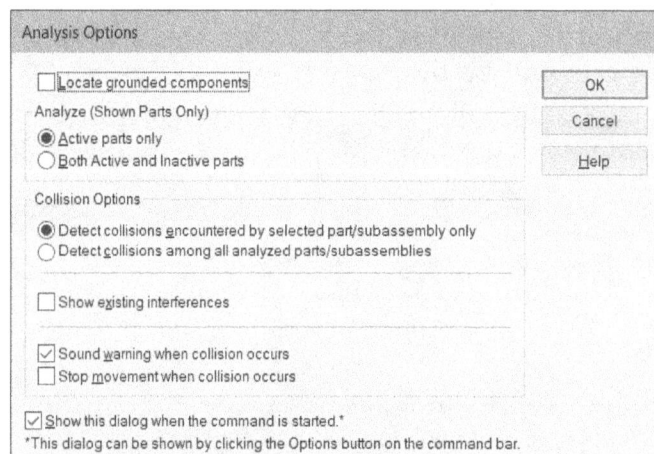

Figure-46. Analysis Options dialog box

- Select the **Locate grounded components** check box to select only the components and subassemblies which are grounded.
- Select the **Active parts only** radio button to drag only displayed parts. Select the **Both Active and Inactive parts** radio button to allow dragging both active and inactive parts of assembly.

- Select desired radio button from the **Collision Options** area to define the scope for checking collision.
- Similarly, set the other parameters and click on the **OK** button. The **Drag Component** command box will be displayed as shown in Figure-47.

Figure-47. Drag Component command box

- Select desired button from the command box to define motion of components. Select the **Move** button from **Drag Type** area of the command box to move components in linear directions. Select the **Rotate** button from the command box to rotate the components about selected edge/axis. Select the **Freeform Move** button from the command box to freely move the selected components in any direction by dragging.
- Select desired option from the **Motion Analysis** drop-down to define how motion of components will be analyzed while dragging.
- Select object and drag component by holding Left Mouse Button; refer to Figure-48. Press **ESC** to exit the tool.

Figure-48. Drag component under constraint

MOVE COMPONENT

The **Move Components** tool is used to move or copy selected components. The procedure to use this tool is given next.

- Click on the **Move Components** tool from **Drag Component** drop-down in the **Modify** group of **Home** tab in the **Ribbon**. The **Move Options** dialog box will be displayed; refer to Figure-49.

Figure-49. Move Options dialog box

- Select desired radio button from the dialog box to define whether relationships will be maintained or deleted while moving. Click on the **OK** button from the dialog box. The **Move Components** command box will be displayed; refer to Figure-50.

Figure-50. Move Components command box

- Set desired options in the command box and select the components you want to move, rotate, or copy.
- Click on the **OK** button from the command box. Depending on button selected in the command box earlier, the options to specify linear or rotational movement will be displayed. (These buttons are discussed in next topic).
- Click on desired point to specify starting reference point for the model. You will be asked to point where you want to move the selected component.
- Specify the coordinate values in the edit boxes of command box or click at desired location.
- Press **ESC** to exit the tool.

Linear Move

The **Move Components Option** button is used to move the component in specified direction or path. By default, this button is active in the **Move Components** command box on activating the tool. The step to use this option are given next.

- Select the component(s) and click right mouse button in the drawing area to accept selection. The options to specify start point will be displayed; refer to Figure-51.

Figure-51. Options for reference point of movement

- Select desired keypoint or specify the coordinates in command box and press **ENTER**. You will be asked to specify the location where you want to move the components.
- Click at desired keypoint or enter the coordinates in the command box; refer to Figure-52. Press **ESC** to exit the tool.

Figure-52. Move component by path

Rotational Move

The **Rotate Components Option** button is used to move component(s) along selected axis of rotation. Click on the **Rotate Components Option** button in the **Move Components** command box to activate rotation mode. The steps to use this option are given next.

- Select the components that you want to rotate and right-click in the drawing area to accept selection. You will be asked to specify axis of rotation.
- Select desired edge or axis from the model about which you want to rotate the selected components. You will be asked to specify the rotation angle.
- Specify desired rotation angle value in the edit box of command box and press **ENTER**. The components will rotate accordingly; refer to Figure-53. Press **ESC** to exit the tool.

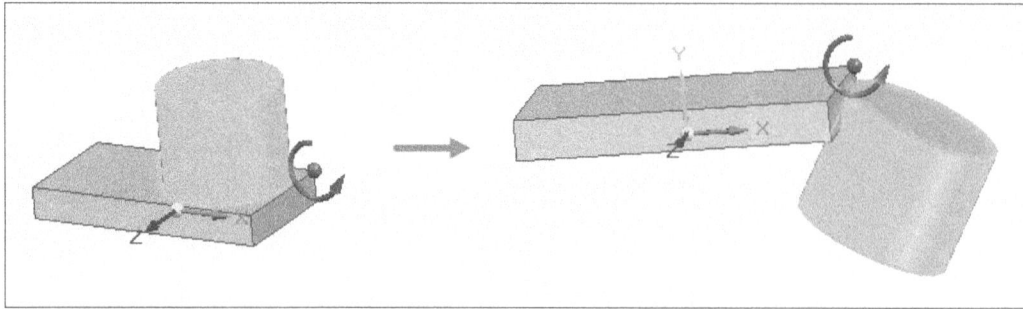

Figure-53. Rotating objects using Move tool

REPLACE PART

The tools in the **Replace Part** drop-down are used to replace existing part with another one; refer to Figure-54. Various tools in this drop-down are discussed next.

Figure-54. Replace part drop-down

Replace Part

The **Replace Part** tool is used to replace selected part using an existing part. The procedure to use this tool is given next.

- Click on the **Replace Part** tool from the **Replace Part** drop-down in the **Modify** group of **Home** tab in the **Ribbon**. The **Replace Part** command box will be displayed; refer to Figure-55.

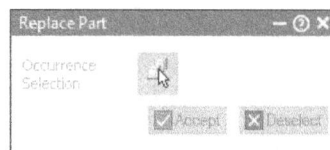

Figure-55. Replace Part command box

- Select part of assembly which you want to replace and right-click in the drawing area to accept the selection. The **Replacement Part** dialog box will be displayed; refer to Figure-56.

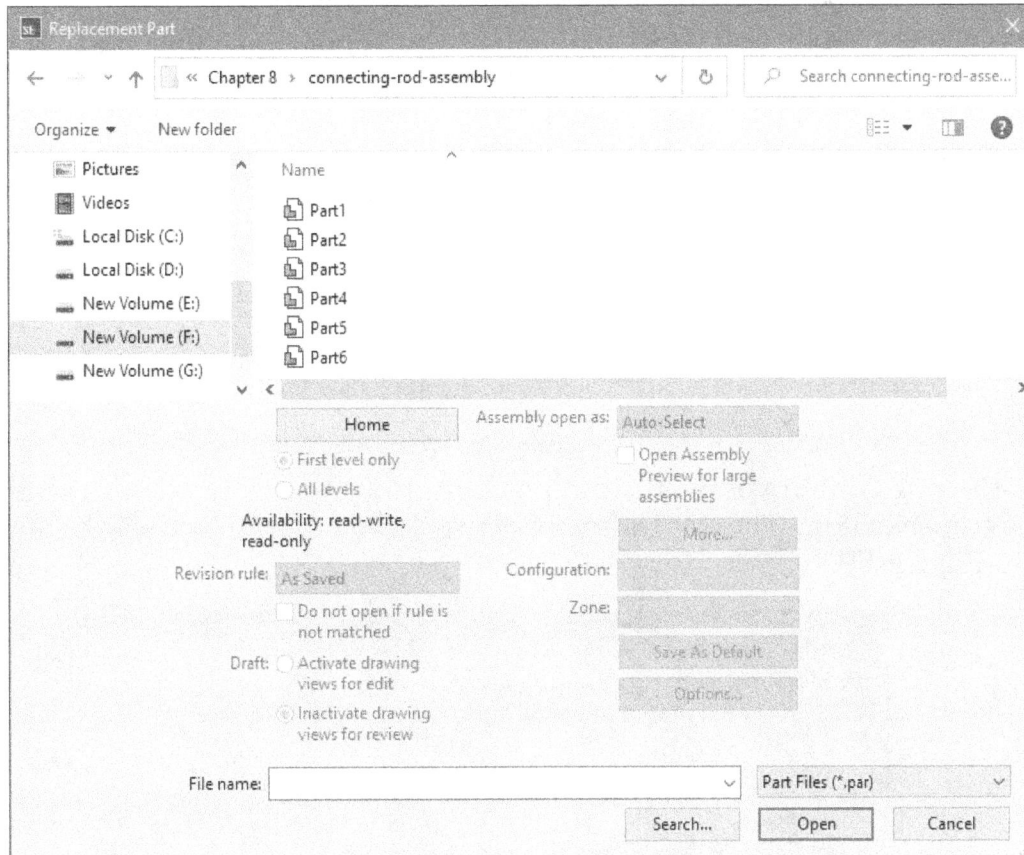

Figure-56. Replacement Part dialog box

- Select desired part from the dialog box and click on the **Open** button. The part will be replaced by new one; refer to Figure-57.

Figure-57. Replacing part

Replace Part With Standard Part

The **Replace Part with Standard Part** tool is used to replace an existing part with a standard part from library. This tool is also available in the **Replace Part** drop-down. The procedure to use this tool is same as discussed for previous tool.

Replace Part with New Part

The **Replace Part with New Part** tool is used to replace existing part by creating new part in the existing assembly. The procedure is discussed next.

- Click on the **Replace Part with New Part** tool from the **Replace Part** drop-down. The **Replace Part with New Part** command box will be displayed; refer to Figure-58.

Figure-58. Replace Part with New Part command box

- Select the part which you want to replace and right-click in the drawing area to accept the selection. The **Replacement Part - New** dialog box will be displayed; refer to Figure-59.

Figure-59. Replacement Part New dialog box

- Specify the name of part, template, and location of the new component. After setting desired parameters, click on the **Create Part** button. The new part will be added in the list.
- Now, select the newly created part from the **Path Finder** in the left of application window and right-click on it. A shortcut menu will be displayed; refer to Figure-60.

Figure-60. Shortcut menu displayed

- Select the **Open** option from the shortcut menu to create part in separate window or select the **Edit** option to create part in assembly environment. The tools to create part model have been discussed earlier.

Replace Part with Copy

The **Replace Part with Copy** tool is used to replace a part with another copy of part. The procedure to use this tool is given next.

- Click on the **Replace Part with Copy** tool from the **Replace Part** drop-down in the **Ribbon**. The **Replace Part with Copy** command box will be displayed.
- Select desired part you want to replace from the assembly and right-click in the drawing area. The **Replacement Part - Copy** dialog box will be displayed.
- Specify desired name of new copy of component and click on the **Save** button from the dialog box. Now, you can edit the new copy of component without modifying the original component.

Replace Part with 3DfindIT.com Part

The **Replace Part with 3DfindIT.com Part** tool is used to replace selected part with similar parts available on 3Dfind.it.com portal. The procedure to use this tool is given next.

- Click on the **Replace Part with 3DfindIT.com Part** tool from the **Replace Part** drop-down in the **Modify** group of **Ribbon**. The **Replace Part with 3DfindIT Part. com** command box will be displayed.
- Select the part to be replaced and right-click in the drawing area. The parts similar to selected part will be displayed in the browser window; refer to Figure-61.

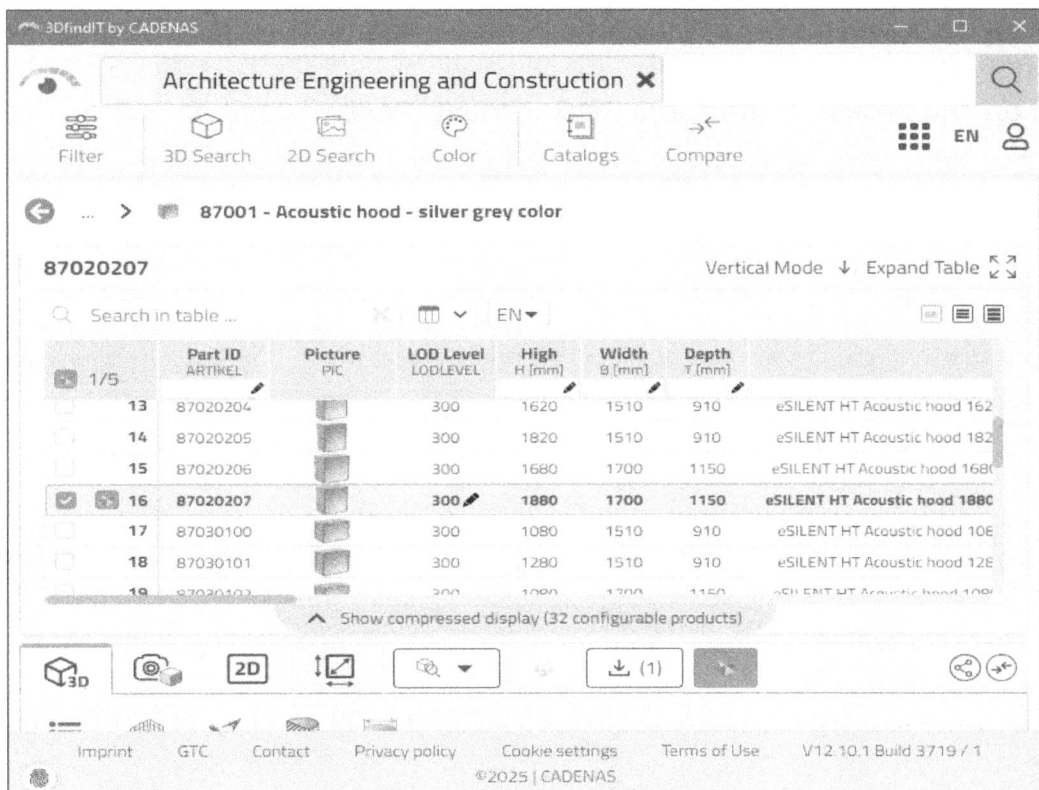

Figure-61. 3Dfind it parts

• Select the part from browser and download it to replace the part.

Define Alternate Components

The **Define Alternate Components** tool is used to specify and manage alternate parts or subassemblies that can replace existing components in an assembly. The procedure to use this tool is discussed next.

• Select desired alternate component and click on the **Define Alternate Components** tool from the **Replace Part** drop-down in the **Ribbon**. The **Define Alternate Components** dialog box will be displayed; refer to Figure-62.

Figure-62. Define alternate component dialog box

• Click on the **Search** button to search for documents in a specific location to find potential alternate components.
• Click on the **Browse** button, the **Replacement Part** dialog box will be displayed; refer to Figure-63.

Figure-63. Replacement Part dialog box

- Select desired file then click on the **Open** button from the dialog box. The file will be displayed in the **Define Alternate Components** edit box.
- Select desired file and click on the **Add** button to move the file from the alternate candidate list.
- After specifying desired file then click on the **OK** button from the dialog box. The alternate component will be displayed.
- Press **ESC** to exit the tool.

TRANSFERRING PART TO ANOTHER SUBASSEMBLY

The **Transfer** tool in the **Modify** group of **Ribbon** is used to transfer selected part into another sub-assembly. The procedure to use this tool is given next.

- Click on the **Transfer** tool from the **Modify** group in the **Home** tab of the **Ribbon** after selecting the component. The **Transfer to Assembly Level** dialog box will be displayed; refer to Figure-64.

Figure-64. Transfer to Assembly Level dialog box

- Select desired sub-assembly from the dialog box and click on the **OK** button. The selected part will be transferred to new sub-assembly.

DISPERSING SUB-ASSEMBLY

The **Disperse** tool is used to transfer all the components of selected sub-assembly into one higher level assembly. The procedure to use this tool is given next.

- Select the sub-assembly you want to disperse and click on the **Disperse** tool from the **Modify** group in the **Home** tab of **Ribbon**. The **Disperse Assembly** message box will be displayed.
- Click on the **Yes** button from the dialog box. The selected sub-assembly will be deleted and all its components will be added in the main assembly.

ROTATIONAL MOTORS

The **Rotational Motor** tool is used to create a motor that rotates an under constrained component. Motors help you to observe the motion of various interrelated parts in assembly. The procedure to use this tool is given next.

- Click on the **Rotational Motor** tool from the **Motors** group in the **Home** tab of **Ribbon**. The **Rotational Motor** command box will be displayed as shown in Figure-65.

Figure-65. Rotational Motor command box

- Select the component which you want to rotate using the motor. You will be asked to select the axis about which you want to rotate the part.
- Select the axis of component. The options to specify speed and limit will be displayed in the command box; refer to Figure-66.

Figure-66. Component selected for rotational motor

- Click on the **Flip Direction** button from the command box to reverse the rotation direction.
- Specify desired values in **Speed** and **Limit** edit boxes of command box to specify the rotational speed of component and total angular span of rotation, respectively.
- After setting desired parameters, click on the **Finish** button from the command box. Press **ESC** to exit the tool.

LINEAR MOTOR

The **Linear Motor** tool is used to move under constrained component in linear direction. The procedure to use this tool is given next.

- Click on the **Linear Motor** tool from **Motors** group in the **Home** tab of **Ribbon**. The **Linear Motor** command box will be displayed; refer to Figure-67.

Figure-67. Linear Motor command box

- Select the component which you want to move linearly using the motor. You will be asked to specify the direction of movement.
- Select desired edge or axis to define direction of movement.
- Specify the speed and translation limit for component in respective edit boxes.
- After setting desired parameters, click on the **Finish** button to apply motor. Press **ESC** to exit the tool.

VARIABLE TABLE MOTOR

The **Variable Table Motor** tool is used to create a motor that drives the component through the table which is created by the user. The procedure is given next.
- Click on the **Variable Table Motor** tool from **Motor** group in the **Ribbon**. The **Variable Table Motor** dialog box will be displayed; refer to Figure-68.

Figure-68. Variable Table Motor dialog box

- Click on the **Variable Table Motor** [icon] button from the dialog box. The **Variable Table** dialog box will be displayed; refer to Figure-69.

Figure-69. Variable Table Spreadsheet

- Select desired unit from the **Unit type** drop-down and set the parameters in table like name of variable, its value, and so on to create a new variable. After selecting/creating the variable, click on the **Close** button from the **Variable Table** dialog box to exit. The **Variable Table Motor** dialog box will be active again.
- Specify desired value in **Speed** and **Limit** edit boxes of the dialog box to specify the speed and limits of movement.
- Click on the **OK** button from the dialog box. The motor will be applied.

SIMULATE MOTOR

The **Simulate Motor** tool is used to observe simulation of motor through animation. It is active when any kind of motor is applied to the Assembly. The procedure to use this tool is given next.

- Click on the **Simulate Motor** tool from the **Motor** group in the **Home** tab of the **Ribbon**. The **Motor Group Properties** dialog box will be displayed; refer to Figure-70.

Figure-70. Motor Group Properties dialog box

- Select desired radio button from the dialog box to define whether you want to perform a collision detection, physical motion, or no analysis.
- Select the **Use motor limits as duration if defined** check box to convert limits to duration during analysis. Specify desired value of motor duration in the **Default motor duration** edit box.
- Select all the motors you want to include for checking motion from the **Available Motors** list box of the dialog box and click on the **Add>>** button.

- After setting desired parameters, click on the **OK** button from the dialog box.
- The animation box view in the bottom of screen will be displayed as shown in Figure-71. Click on the **Play** button to start the animation.

Figure-71. Animation

- Press **ESC** to exit the tool. The **Animation Editor** message box will be displayed.
- Click on the **Yes** button to save changes in animation or click **No** button to exit without saving changes.

PATTERN

We have already discussed about **Pattern** tool for part modeling but in assembly, we use this tool to create copies of the components. The procedure to create assembly pattern is discussed next.

- Click on the **Pattern** tool from the **Pattern** group in the **Home** tab of the **Ribbon**. The **Pattern** command box will be displayed as shown in Figure-72.

Figure-72. Pattern command box

- Select the component which you want to pattern.
- Select the base part on which have pattern created for assembling the component.
- Select the reference feature on base part to be used as starting reference point for creating the pattern. Preview of pattern will be displayed; refer to Figure-73.
- Click on the **Finish** button from the command box to create the pattern.

Figure-73. Preview of pattern

MIRROR COMPONENTS

We have already discussed about the **Mirror** tool in Part modeling, here we are going to discuss about mirroring components in assembly. The procedure to use this tool is given next.

• Click on the **Mirror** tool from the **Pattern** group in the **Home** tab of **Ribbon**. The **Mirror Components** command box will be displayed and you will be asked to select the components to be mirrored; refer to Figure-74.

Figure-74. Mirror Components command box

• Select the components to be mirrored and right-click in the drawing area. You will be asked to select the mirror plane.
• Select desired mirror plane. Preview of the mirrored components will be displayed; refer to Figure-75 and the **Mirror Components** dialog box will be displayed; refer to Figure-76 .

Figure-75. Mirror Assembly

Figure-76. Mirror Components dialog box

- Set desired action and adjustment parameters for mirror copies in the dialog box and click on the **OK** button.
- Click on the **Finish** button from the command box to create the mirror copies.

DUPLICATE COMPONENT

The **Duplicate Component** tool in assembly is same as discussed earlier in Part modeling.

PATTERN ALONG CURVE

The **Along Curve** tool is used to create pattern of existing components along guided curve. To use this tool, you must have a curve on the body which will be followed while creating pattern. The procedure to use this tool is given next.

- Click on the **Along Curve** tool from the **Pattern** group in the **Home** tab of the **Ribbon**. The **Along Curve** command box will be displayed; refer to Figure-77 and you will be asked to select the components, you want to pattern.

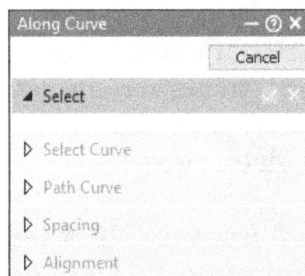

Figure-77. Along Curve command box

- Select the components that you want to pattern and right-click in the drawing area. You will be asked to select curve for 1st direction.
- Select desired curve or edge and right-click in drawing area. You will be asked to specify start point for pattern.
- Click at desired end point of curve to specify start point of pattern.

- Click in the **Offset** edit box and enter the distance between start point and first instance of component in pattern. The options to specify number of instances and distance between two consecutive instances of pattern.
- Specify desired parameters in the command box and click on the **Next** button. The **Along Curve** command box will be displayed with options to define orientation and rotation of part in pattern.
- Specify desired parameters in drop-downs of command box and click on the **Preview** button. Preview of pattern will be displayed; refer to Figure-78.
- Click on the **Finish** button from the command box to create pattern.

Figure-78. Along Curve

CLONE COMPONENT

The **Clone Component** tool is used to place multiple occurrences of one or more components at various locations in an assembly based on geometry recognition. The procedure to use this tool is same as discussed for **Duplicate Component** tool.

INSPECTION TOOLS

The tools in the **Inspect** tab are used to inspect the assembly properties like material, dimension, and so on. Various tools in this tab are discussed next.

Smart Measure

The **Smart Measure** tool is used to measure dimensions of various elements. It is useful when you are assembling components whose dimensions are not annotated. The procedure to use this tool is given next.

- Click on the **Smart Measure** tool from the **2D Measure** group in the **Inspect** tab of **Ribbon**. The **Smart Measure** command box will be displayed; refer to Figure-79.

Figure-79. Smart Measure command box

- You can measure the entities in the same way as discussed earlier for **Smart Dimension** tool; refer to Figure-80.

Figure-80. Smart Measure

Measure

The **Measure** tool is used to measure 3D surfaces, faces, angles, lines, and others entities. The procedure to use this tool is given next.

- Click on the **Measure** tool from the **3D Measure** group in the **Inspect** tab of the **Ribbon**. The **Measure** command box will be displayed; refer to Figure-81.

Figure-81. Measure command box

- Select the entity from the part or body like face, edge, vertex, and so on; refer to Figure-82.

Figure-82. Measuring entities

Measure Distance

The **Measure Distance** tool is used to measure the distance between two or more keypoints. The procedure to use this tool is given next.

- Click on the **Measure Distance** from the **3D Measure** group in the **Inspect** tab of **Ribbon**. The **Measure Distance** command box will be displayed; refer to Figure-83.

Figure-83. Measure Distance command box

- Select the keypoints of the part or body in the order by which you want to measure them. A spreadsheet will be displayed with measurement in the **Measure Distance** dialog box; refer to Figure-84.

Figure-84. Measure Distance

Measure Minimum Distance

The **Measure Minimum Distance** tool is used to measure minimum distance between two keypoints. The procedure to use this tool is given next.

- Click on the **Measure Minimum Distance** tool from the **3D Measure** group. The **Measure Minimum Distance** command box will be displayed; refer to Figure-85.

Figure-85. Measure Minimum Distance command box

- Select the keypoints on part or body; refer to Figure-86.

Figure-86. Measure Minimum Distance

Measure Normal Distance

The **Measure Normal Distance** tool is used to measure a normal distance from a plane or line to a keypoints. The procedure is given next.

• Click on the **Measure Normal Distance** tool from the **3D Measure** group in the **Inspect** tab of **Ribbon**. The **Measure Normal Distance** command box will be displayed; refer to Figure-87.

Figure-87. Measure Normal Distance command box

• Select desired face, line, and then key point to measure distance of keypoint perpendicular to selected face/line; refer to Figure-88.

Figure-88. Measure Normal Distance

• Press **ESC** to exit the tool.

Measure Angle

The **Measure Angle** tool is used to measure the angle between selected faces/planes/lines/keypoints. The procedure to use this tool is given next.

- Click on the **Measure Angle** tool from the **3D Measure** group in the **Inspect** tab of **Ribbon**. The **Measure Angle** command box will be displayed; refer to Figure-89.

Figure-89. Measure Angle command box

- Select the faces/edges/keypoints. A spread sheet will be displayed with measurement; refer to Figure-90.

Figure-90. Measuring angle

- Press **ESC** to exit the tool.

Inquire Element

The **Inquire Element** tool is used to get information of selected element. The procedure to use this tool is given next.

- Click on the **Inquire Element** tool from the **3D Measure** group in the **Inspect** tab of **Ribbon**. The **Inquire Element** command box will be displayed; refer to Figure-91.

Figure-91. Inquire Element command box

- Select the element about which you want to get the information. The **Inquire Element** dialog box will be displayed; refer to Figure-92.

Figure-92. Inquire Element

PROPERTIES

The **Properties** tool is used to display all the physical properties of selected parts or assembly. The procedure to use this tool is given next.

- Click on the **Properties** tool from the **Physical Properties** group of the **Inspect** tab in the **Ribbon**. The **Physical Properties** dialog box will be displayed; refer to Figure-93.

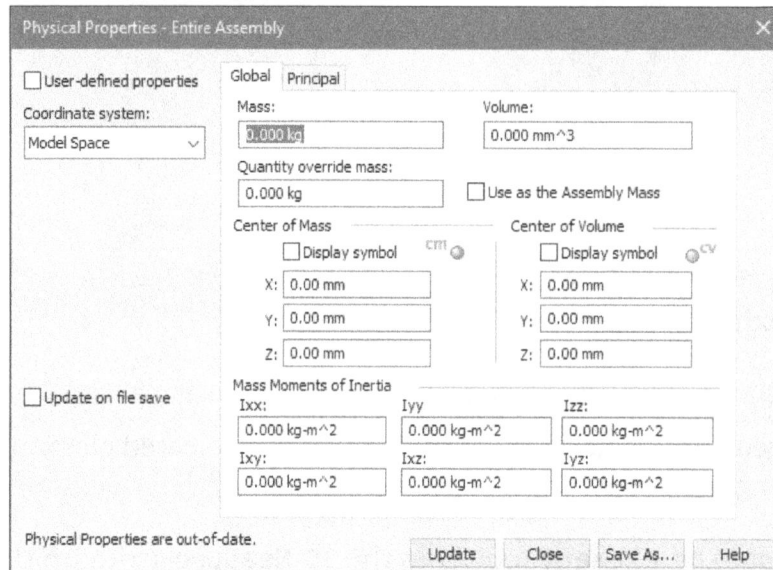

Figure-93. Physical Properties dialog box

- Select desired option from the **Coordinate system** drop-down and click on the **Update** button. The mass, volume, and other properties of assembly/selected elements will be displayed in the dialog box.
- After checking the parameters, click on the **Close** button from the dialog box to exit.

PROPERTIES MANAGER

The **Properties Manager** tool is used to modify physical properties for the top level assembly and associated component. The procedure to use this tool is given next.

- Click on the **Properties Manager** tool from the **Physical Properties** group of the **Inspect** tab in the **Ribbon**. The **Physical Properties Manager** dialog box will be displayed; refer to Figure-94.

- Select desired material from the **Material** column of component which parameters are to be checked. After selecting the material, click on the **Update All** button from the dialog box.
- Click on the **OK** button from the dialog box to exit the dialog box.

Figure-94. Spread Sheet of Properties Manager

CHECK INTERFERENCE

The **Check Interference** tool is used to find parts that physically interfere with one another in assembly. The procedure to use this tool is given next.

- Click on the **Check Interference** tool from the **Evaluate** group in the **Inspect** tab of the **Ribbon**. The **Check Interference** command box will be displayed; refer to Figure-95.

Figure-95. Check Interference command box

- Select the components of first set of parts and right-click in the drawing area. You will be asked to select parts of second set.
- Select the second set of parts used for interference checking and click on the **Process** button from the command box. If there is an interference then it will be displayed in graphics; refer to Figure-96. If there is no interference then a message box will be displayed.
- Press **ESC** to exit the tool.

Figure-96. Check Interference

ASSEMBLY STATISTICS

The **Assembly Statistics** tool is used to display the assembly statistics like;

- Assembly file name
- Total number of parts
- Number of unique parts
- Total number of subassemblies
- Number of unique subassemblies

On clicking the **Assembly Statistics** tool from the **Evaluate** group in the **Inspect** tab of the **Ribbon**. The **Assembly Statistics** dialog box will be displayed with all the parameters; refer to Figure-97.

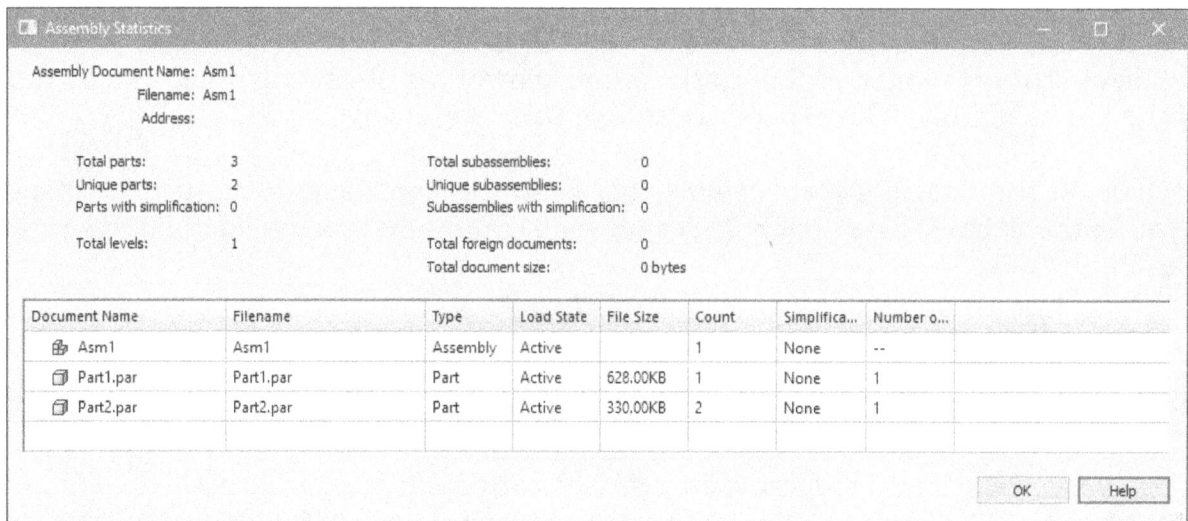

Figure-97. Assembly Statistics

GEOMETRY INSPECTOR

The **Geometry Inspector** tool is used to check errors and small entities in the model. The procedure to use this tool is given next.

- Click on the **Geometry Inspector** tool from the **Evaluate** group in the **Inspect** tab in the **Ribbon**. The **Geometry Inspector** dialog box will be displayed; refer to Figure-98.

Figure-98. Geometry Inspector dialog box

- Select desired check boxes from the dialog box and parameters.
- After setting parameters, click on the **Go** button from the dialog box. The results will be displayed in the **Results** list box.
- Click on the **Close** button from the dialog box to exit.

PRACTICAL - 1

Assemble the parts of Knuckle Joint as shown in Figure-99. The exploded view of assembly is shown in Figure-100.

Figure-99. Knuckle joint

Figure-100. Exploded View Practical 1

Part files for this assembly are available in the resource kit. The steps to assemble components are given next.

Inserting Ground Component

- Start Solid Edge if not started yet and click on the **New** button from **Quick Access Toolbar**. The **New** dialog box will be displayed.
- Start the assembly environment by selecting **iso metric assembly.asm** template from the **New** dialog box and click on the **OK** button. The assembly environment will be displayed.
- Click on the **Insert Component** tool from the **Components** group in the **Home** tab of the **Ribbon**. The **Parts Library** box will be displayed.
- Select the Fork part from the box and drag the component in drawing area; refer to Figure-101.

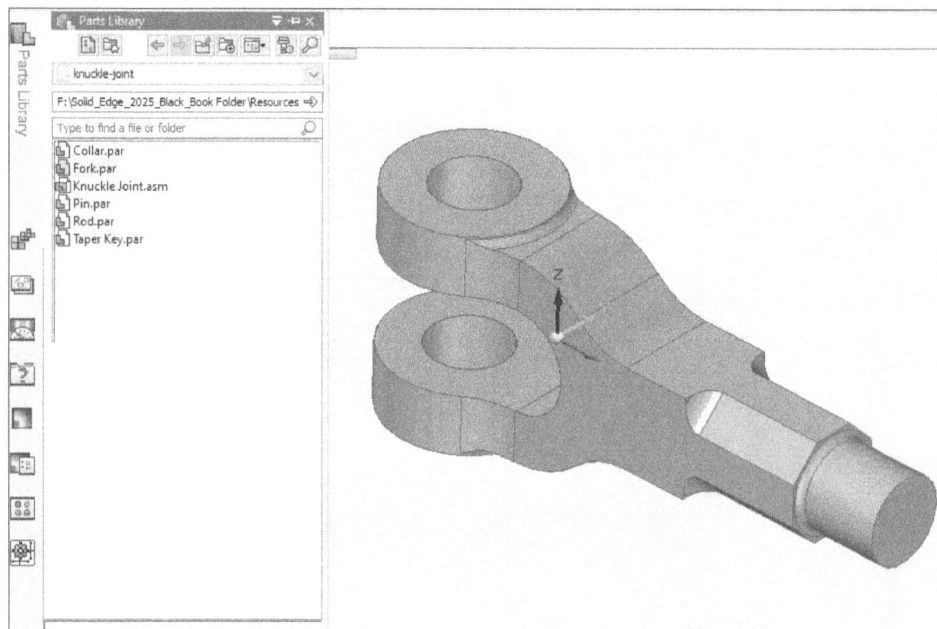

Figure-101. Inserting Componenet

Inserting and Constraining Component

- Now, select next component from the part library that is Rod and drag it to assembly environment.
- Select the **Planar Align** option from the **Relationship Types** drop-down in the command box to make the face of Fork and Rod aligned. Select the inner face of fork end and outer face eye end. Refer to Figure-102. Note that you may need to flip the part.

Figure-102. Applying planar align constraint

- Select the **Axial Align** option from the **Relationship Types** drop-down to make components aligned axially. Select the circular face of fork end and eye end; refer to Figure-103.

Figure-103. Components aligning axially

- Press **ESC** to exit the command box.
- Now, insert Pin from the **Parts Library** box. Select the radial axis of pin and the radial axis of Fork end; refer to Figure-104. The axes of components will be aligned.

Figure-104. Axial align of pin

- Select the flat inner face of pin and top flat face of fork end; refer to Figure-105.

Figure-105. Selecting faces of pin and fork

- Drag the Collar part from the **Parts Library** box into the drawing area. Select the round faces of collar and pin to align their axes.
- Select the collar axial align with the hole created on the pin; refer to Figure-106.

Figure-106. Collar axial align

- Insert Tapper pin and make the hole of collar aligned with pin. Move the taper pin by dragging inside the hole. This completes the assembly; refer to Figure-107.

Figure-107. Assembly completed

PRACTICE - 1

Assemble the Connecting Rod parts as shown in Figure-108. The parts of assembly are available in the resource kit.

Figure-108. Connecting Rod Assembly

PRACTICE - 2

Assemble the Pipe Vice as shown in Figure-109. The components of assembly are available in the resource kit.

Figure-109. Pipe vice assembly

SELF ASSESSMENT

Q1. What does "Assembly" mean in engineer's language?

a) Combining unrelated parts without constraints
b) Combination of two or more components with specific constraints
c) Joining components without any constraints
d) Assembling parts randomly

Q2. What tool in Solid Edge is primarily used to start an assembly design?

a) Open File
b) New button
c) Save As
d) Template Manager

Q3. Which tool is used to drag a desired part into the drawing area?

a) FlashFit
b) Drag Component
c) Insert Component
d) Move Components

Q4. Which tool automatically applies constraints like mate, align, or connect?

a) Mate
b) FlashFit
c) Insert
d) Planar Align

Q5. What does the Mate tool achieve in an assembly?

a) Aligns axes of cylindrical components
b) Makes two faces or planes coincident
c) Creates a cam-follower relationship
d) Rotates components freely

Q6. Which assembly constraint aligns two cylindrical axes?

a) Mate
b) Planar Align
c) Axial Align
d) Insert

Q7. Which tool is used to connect two key points or planes in an assembly?

a) Connect
b) Parallel
c) Tangent
d) Gear

Q8. Which tool is used to move a part along a selected path?

a) Move Component
b) Linear Move
c) Path
d) Drag Component

Q9. What does the Rigid Set tool do?

a) Makes selected components flexible
b) Fixes components with respect to each other
c) Allows components to rotate freely
d) Aligns components to a reference

Q10. What is the function of the Assembly Relationships Manager tool?

a) Automatically applies relationships
b) Displays and allows modification of applied relationships
c) Deletes all relationships in an assembly
d) Locks all components in place

Q11. Which tool is used to copy existing relationships for reapplying them later?

a) Capture Fit
b) FlashFit
c) Move Components
d) Connect

Q12. What is the purpose of the Ground tool in the assembly?

a) To rotate the component
b) To apply a ground relationship to the first part in an assembly
c) To move the component freely
d) To apply a tangent constraint

Q13. Which command is used to make two parts parallel in an assembly?

a) Mate
b) Planar Align
c) Parallel
d) Connect

Q14. What is the default placement method in the assembly environment?

a) Manual placement
b) FlashFit
c) Offset placement
d) Tangent placement

Q15. How can you make a part adjustable when inserting a sub-assembly?

a) By selecting FlashFit
b) By turning on the "Place as Adjustable" toggle button in the Settings rollout
c) By using the Connect tool
d) By using the Drag Component tool

Q16. Which tool is used to apply a cam-follower mate?

a) Connect
b) Cam
c) Mate
d) Planar Align

Q17. What is the use of the Rigid Set tool?

a) To suppress existing constraints
b) To apply axial alignment
c) To make selected components rigid with respect to each other
d) To create a tangent relationship

Q18. How do you make two parts coincident using faces or planes?

a) Use the Mate tool
b) Use the Tangent tool
c) Use the Planar Align tool
d) Use the Connect tool

Q19. Which command allows you to automatically apply assembly relationships between selected components?

a) Assembly Relationship Assistant
b) Assembly Relationships Manager
c) Capture Fit
d) Move on Select

Q20. What does the Assembly Relationships Manager tool allow you to do?

a) Drag and move components
b) Modify or repair assembly relationships
c) Create cam relationships
d) Align components along axes

Q21. What feature ensures that a sub-assembly disperses into individual parts upon placement?

a) FlashFit toggle
b) Disperse after placement toggle button
c) Capture Fit command
d) Path command

Q22. Which option in the Offset rollout allows free lateral movement of a part?

a) Fixed Offset
b) Floating Offset
c) Range
d) Lock rotation

Q23. What parameter is specified to control the rotation of a component?

a) Speed
b) Rotation axis
c) Rotation angle
d) Rotation direction

Q24. Which tool is used to replace an existing part with a standard part from a library?

a) Replace Part
b) Replace Part with New Part
c) Replace Part with Standard Part
d) Replace Part with Copy

Q25. What is required to create a new part in an assembly using the Replace Part with New Part tool?

a) Select and save the part
b) Specify name, template, and location of the new part
c) Open a dialog box and select from library
d) Transfer the part to subassembly

Q26. Which tool allows you to replace a part with another part available on the 3DfindIT portal?

a) Replace Part with Copy
b) Replace Part with Standard Part
c) Replace Part with 3DfindIT Part
d) Replace Part with New Part

Q27. Which tool is used to rotate an under-constrained component in assembly?

a) Linear Motor
b) Rotational Motor
c) Simulate Motor
d) Variable Table Motor

Q28. What must be specified when using a Linear Motor?

a) Axis of rotation
b) Translation limit
c) Angular speed
d) Default motor duration

Q29. What is the purpose of the Simulate Motor tool?

a) To animate motor movement
b) To set rotation limits
c) To replace a motor
d) To create motor patterns

Q30. Which tool creates copies of components along a curve?

a) Mirror Components
b) Duplicate Component
c) Pattern along Curve
d) Clone Component

Q31. Which tool is used to measure dimensions of various 2D elements?

a) Measure Distance
b) Measure Angle
c) Smart Measure
d) Measure Minimum Distance

Q32. What does the Geometry Inspector tool check?

a) Motor limits
b) Errors and small entities in the model
c) Interference between components
d) Assembly statistics

Q33. What does the Assembly Statistics tool display?

a) Physical properties of components
b) Assembly file name, total parts, and subassemblies
c) Interference details
d) Geometry errors

Q34. which Replace Part tool allows the user to modify a new copy without affecting the original?

a) Replace Part with Copy
b) Replace Part with New Part
c) Replace Part with 3DfindIT Part
d) Replace Part with Standard Part

Q35. Which parameter in the Linear Motor tool defines the range of movement?

a) Direction of movement
b) Speed
c) Translation limit
d) Motor duration

Q36. What does the Variable Table Motor tool use to drive a component?

a) Rotation axis
b) Speed and limits
c) A user-created variable table
d) Standard library components

Q37. Which tool provides mass, volume, and other physical properties of an assembly?

a) Properties Manager
b) Properties Tool
c) Geometry Inspector
d) Check Interference

Q38. What does the Check Interference tool analyze?

a) Physical properties of an assembly
b) Errors and small entities in the model
c) Collisions between selected parts
d) Dimensions of parts

Q39. What happens when the Disperse tool is used?

a) The sub-assembly is deleted, and its components are added to the main assembly
b) The selected part is moved to a sub-assembly
c) A new part is created
d) Interference is checked

Q40. Which of the following does the Assembly Statistics tool NOT display?

a) Total number of parts
b) Number of unique subassemblies
c) Physical properties of components
d) Assembly file name

Q41. What is the function of the Clone Component tool?

a) To copy a part along a curve
b) To inspect physical properties
c) To place multiple instances of a component based on geometry recognition
d) To measure distances between components

Q42. What is the purpose of the Inquire Element tool?

a) To measure dimensions of keypoints
b) To obtain information about selected elements
c) To edit material properties
d) To analyze interference

Q43. What information does the Geometry Inspector tool provide?

a) Part dimensions and weights
b) Errors and small entities in the model
c) Interference details
d) Motor simulation properties

Q44. Which tool updates the material properties of components in an assembly?

a) Properties Tool
b) Properties Manager
c) Assembly Statistics
d) Check Interference

Chapter 9

Assembly Design - II

Topics Covered

The major topics covered in this chapter are:

- *Variable and Component Tracker*
- *Update Tools*
- *Reports*
- *Error Assistant, Inter-Part Manager, and Component Structure Editor*
- *Publishing Virtual Components*
- *Engineering Reference Tools*
- *Explode Render Animate Environment*
- *Modes*

INTRODUCTION

In previous chapter, we have discussed about the inserting components in assembly and applying constraints. The constraints are used to arrange the parts in systematic way. Most of the time, we keep some of the components under-constrained so that they can move in predefined motion. In this chapter, we are going to discuss advance assembly tools which allow motion between various components of the assembly. In previous chapter, we have discussed the tools of **Inspect** tab. Now, we will discuss the tools of **Tools** tab in the **Ribbon** and other advanced tools.

TOOLS TAB

The tools in the **Tools** tab are used to manage variables, environments, reports, rendering, and so on. Various tools in this tab are discussed next.

Variables

The **Variables** tool is used to open a spreadsheet that allows you to view, edit, and manage geometric variables and equations. The procedure to use this tool is same as **Peer Variables** tool.

Peer Variables

The **Peer Variables** tool is used to edit variables of any attachment in the Assembly. The steps are given next.

*   Click on the **Peer Variables** tool from the **Variables** group in the **Tools** tab of the **Ribbon**. The **Peer Variables** command box will be displayed; refer to Figure-1.

Figure-1. Peer Variables command box

*   Select the part in assembly whose variables are to be edited or checked. The parameters will be displayed in the dialog box; refer to Figure-2.

Figure-2. Spread Sheet of peer Variable

- Double-click in the empty field of **Name** column and specify desired name of variable. Similarly, double-click in the empty field under **Formula** column next to new variable to create a formula for generating the value. You can use arithmetic operators (+, -, /, *) and logic rules (like greater than, less than, and so on) in the formula. Various logical rules are given next.

Character	Meaning	Where Used	Variable Type
(	Greater Than	Beginning Only	Limit
)	Less Than	End Only	Limit
[	Greater Than or Equal To	Beginning Only	Limit
]	Less Than or Equal To	End Only	Limit
{}	Encloses a Discrete List	Use both as a set	Discrete List
;	Separates values	Between values in a limit or discrete list	Limit and Discrete List

Examples of using logic rules:

1. To define a variable that must be greater than 15 and less than 20, type the following in the Range cell: (15;20)
2. To define a variable that must be greater than or equal to 8 and less than or equal to 15, type the following: [8;15]
3. To define a variable that must be greater than or equal to 9 and less than 20, type the following: [9;20)
4. To define a variable that must be limited to the following list of values: 3, 10, 15, and 20, type the following: {3;10;15;20}

- You can also use the **Formula** button to create formula for a variable. Click in the field where you want to specify formula and then click on the **Formula** button from the dialog box. The Step 1 page of **Function Wizard** dialog box will be displayed; refer to Figure-3. Select desired function from the list and click on the **Next** button. The Step 2 page will be displayed and you will be asked to specify value of formula; refer to Figure-4. Specify desired value in the edit box and click on the **Finish** button.

Figure-4. Step 2 of Function Wizard dialog box

Figure-3. Function Wizard dialog box

- Click on the **Variable Rule Editor** button from the dialog box to edit a parameter. The **Variable Rule Editor** dialog box will be displayed; refer to Figure-5. Select the **Value based on** check box to specify the value of variable based on formula or copied value. Select the **Limit value to** check box to specify upper and lower limits of value in the field or range of values. After setting desired parameters in the dialog box, click on the **OK** button.

Figure-5. Variable Rule Editor dialog box

- Click on the **Refresh** button to update the selected value.
- Click on the **Print** button to print all the parameters of table.
- Click on the **Update All** button to update all the parameters of table in the model.
- Click on the **Activate all** button to activate all parent documents.
- Click on the **Close** button to exit the dialog box.

Similarly, you can use the **Variables** tool of **Variables** group in **Tools** tab of the **Ribbon**.

COMPONENT TRACKER

The **Component Tracker** tool is used to track if there is any change in the components of assembly. The **Component Tracker** dialog box will be displayed; refer to Figure-6. If all the components are updated in assembly then a tick mark will be displayed. Click on the **Close** button to exit the tool. Meaning of various marks are given next.

One or more documents have geometric changes that are not yet reflected at this level. An update is required.

Document is up to date in memory. May need to be updated to disk.

Document has changes and is causing higher level documents to be out of date.

Document is not found.

Document has changes in memory that have not been saved.

Figure-6. Component Tracker dialog box

UPDATE TOOLS

The tools in the **Update** drop-down of **Update** group are used to update components, documents, and relationships in assembly. Click on the **Update Active Level** tool from the **Update** drop-down in the **Tools** tab of the **Ribbon** to update components of current model's active assembly. Click on the **Update All Open Documents** tool from the drop-down to update all the documents currently open in Solid Edge. Click on the **Update Relationships** tool from the **Update** drop-down in the **Ribbon** to update all the relationships applied in the assembly. Toggle on the **Automatic Update** tool from the drop-down to automatically update model and other parameters while working.

LIMITED UPDATE AND LIMITED SAVE

The **Limited Update** and **Limited Save** tools are active only when these options are selected in the **Assembly** page of the **Solid Edge Options** dialog box; refer to Figure-7.

Figure-7. Limited update options

Toggle 'ON' the **Limited Update** and **Limited Save** buttons to let Solid Edge decide when to update and save the model while using minimal resources and keeping balance in performance.

REPORTS

The **Reports** tool is used to generate common assembly reports like Bill of materials, part list, and so on. The Bill of materials is used to list the components of the assembly in the form of a table. The procedure to use this tool is discussed next.

- Click on the **Reports** tool from the **Assistants** group in the **Tools** tab of the **Ribbon**. The **Report** dialog box will be displayed; refer to Figure-8.

Figure-8. Report dialog box

- Select desired radio button (Bill of materials in our case) from the dialog box and click on the **OK** button. The Bill of Materials spread sheet will be displayed with description of assembly parts; refer to Figure-9.

Figure-9. Bill of Material

- Click on the **Save As** button to save the bill of materials in the form of a text file. Click on the **Print** button from the dialog box to get hard copy of the bill of materials.
- Using the **Copy** button, you can copy parameters of the bill of materials.
- Click on the **New Report** button from the dialog box to generate different report.
- Click on the **Close** button from the dialog box to exit.

ERRORS

The **Errors** tool is used to track the errors in parts and assembly that have failed to recompute properly. Using this tool makes easier to fix the errors in parts and assembly. On clicking this tool, the **Error Assistant** dialog box will be displayed;

refer to Figure-10. Select the component which has error and click on the **Zoom** button from the dialog box. The model will be displayed highlighted in the drawing area. Select the error object and click on the **Edit Definition** button to modify the part.

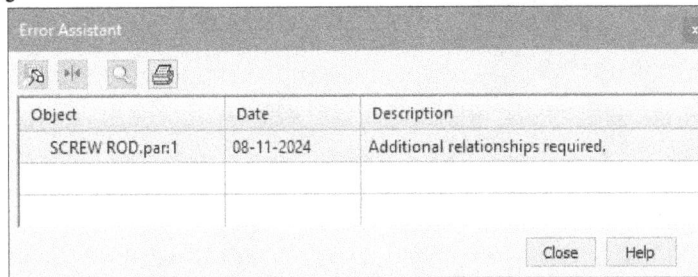

Figure-10. Error Assistant dialog box

INTER-PART MANAGER

The **Inter-Part Manager** tool is used to check the components of model in the form of children or parent hierarchy. Click on the **Inter-Part Manager** tool from the **Assistants** group in the **Tools** tab of the **Ribbon**. The **Inter-Part Links** dialog box will be displayed; refer to Figure-11. Select desired radio button from the dialog box and click on the **Update All** button. The updated list of components and inter-part links will be displayed. Click on the **Close** button to exit from dialog box.

Figure-11. Inter-Part Links dialog box

COMPONENT STRUCTURE EDITOR

The **Component Structure Editor** tool is used to add virtual components in the assembly structure. Virtual components are used to represent the objects which are not in assembly but you want them in bill of materials and other reports. The procedure to use this tool is given next.

- Click on the **Component Structure Editor** tool from the **Assistants** group of the **Tools** tab in the **Ribbon**. The **Virtual Component Structure Editor** dialog box will be displayed; refer to Figure-12.
- Select desired radio button from the dialog box to define what type of object you want to add in the structure.
- Specify desired name of object in the **Name** edit box and click on the **Add Virtual Component** button from the dialog box. The object will be added in the assembly.
- You can also add a real component in the assembly by selecting the **Add as real component** check box and then dragging the component from the left box to right box.
- After setting desired parameters, click on the **OK** button. The component will be added in the assembly.

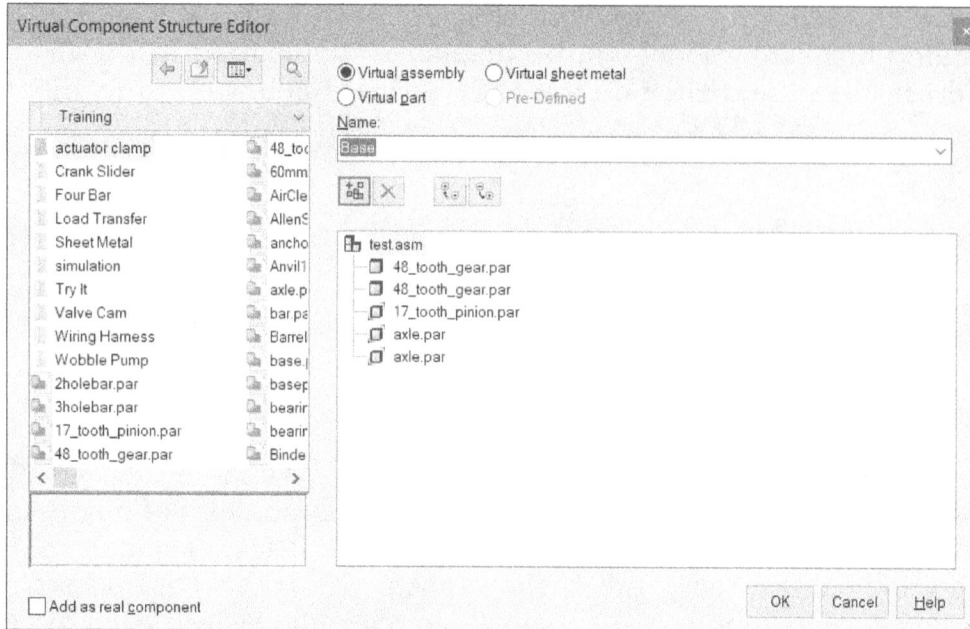

Figure-12. Virtual Component Structure Editor dialog box

PUBLISH INTERNAL COMPONENTS

The **Publish Internal Components** tool is used to create files on disk for internal components and maintain their links to top level assembly. The procedure to use this tool is discussed next.

* Click on the **Publish Internal Components** tool from the **Publish Virtual** drop-down in the **Assistants** group of **Tools** tab in the **Ribbon**. The **Publish Internal Components** dialog box will be displayed; refer to Figure-13.

Figure-13. Publish Internal Components dialog box

* Click on the **Edit file location(s)** button to edit the location of file where the published files are saved.
* Click on the **Edit template(s)** button to edit the template used to create the Solid Edge files.
* Click on the **Reset file location(s)** button to restore the file location to the default.
* Click on the **Reset template(s)** button to restore the template to the default.

- Click on the **Reset material(s)** button to restore the material to the default.
- Click on the **Publish** button from the dialog box. The internal components will be published.

PUBLISHING VIRTUAL COMPONENTS

The **Publish Virtual Components** tool is used to publish the list of virtual components. The procedure to use this tool is given next.

- Click on the **Publish Virtual Components** tool from the **Publish Virtual** drop-down in the **Assistants** group of **Tools** tab in the **Ribbon**. The **Publish Virtual Components** dialog box will be displayed; refer to Figure-14.

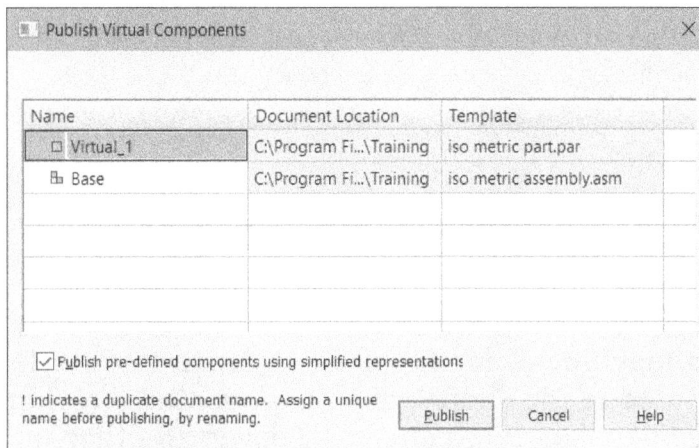

Figure-14. Publish Virtual Components dialog box

- Select the **Publish pre-defined components using simplified representations** check box to publish components in simplified format.
- Click on the **Publish** button from the dialog box. The virtual components will be added in the **Path Finder**.

PUBLISHING TERRAIN MODELS

The **Publish Terrain Models** tool is used to publish virtual objects used to define terrain of environment. The terrain model is imported in the form of an STL file generated by a GIS (Geographic Information System) software. The procedure to import and publish a terrain model is given next.

- Import the STL file of terrain in a new part file and save it in the default location of assembly.
- Create a virtual component using the new created terrain part using the **Component Structure Editor** tool.
- Click on the **Publish Terrain Models** tool from the **Publish Virtual** drop-down in the **Assistants** tab of **Ribbon** and generate the model as discussed for **Publish Virtual Components** tool.

UPDATE STRUCTURE

The **Update Structure** tool in **Assistants** group of **Tools** tab in **Ribbon** is used to update the model after modifications have been performed on the model.

ENGINEERING REFERENCE

The tools in **Engineering Reference** drop-down are used to design a standard part using engineering reference; refer to Figure-15. Note that you need to first save the assembly model before using the tools in this drop-down. Various tools of this drop-down are discussed next.

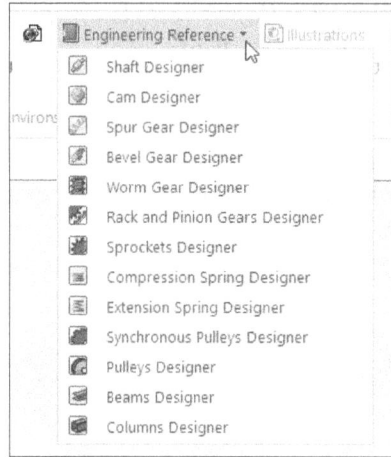

Figure-15. Engineering Reference drop down

Shaft Designer

A shaft is a rotating machine element used to transmit power from one location to other. Power is delivered to the shaft by some tangential force and resultant torque generated in the shaft makes the transfer of power to other linked elements. The standard size of transmission shafts are given as:

    25 mm to 60 mm diameter with 5 mm steps
    60 mm to 110 mm diameter with 10 mm steps
    110 mm to 140 mm diameter with 15 mm steps
    140 mm to 500 mm diameter with 20 mm steps

Standard lengths of shafts are 5 m, 6 m, and 7 m.

While designing a shaft, we need to take care of following stresses:

1. Shear stresses due to the transmission of torque (i.e. due to torsional load).
2. Bending stresses (tensile or compressive) due to the forces acting upon machine elements like gears, pulleys etc. as well as due to the weight of the shaft itself.
3. Stresses due to combined torsional and bending loads.

According to American Society of Mechanical Engineers (ASME) code for the design of transmission shafts, the maximum permissible working stresses in tension or compression may be taken as,
(a) 112 MPa for shafts without allowance for keyways.
(b) 84 MPa for shafts with allowance for keyways.

The maximum permissible shear stress may be taken as
(a) 56 MPa for shafts without allowance for key ways.
(b) 42 MPa for shafts with allowance for keyways.

The **Shaft Designer** tool is used to create a standard shaft based on specified engineering parameters. The procedure to use this tool is given next.

- Click on the **Shaft Designer** tool from the **Engineering Reference** drop-down in the **Environs** group of **Tools** tab in the **Ribbon**. The **Solid Edge Shaft Designer** dialog box will be displayed; refer to Figure-16.

Figure-16. Solid Edge Shaft Designer dialog box

- There are three sections of the dialog box to define parameters of shaft in the **Design Parameters** tab of dialog box; **Shaft Parameters**, **Supports**, and **Loads**. These options are discussed next.

Shaft Parameters

The options in the **Shaft Parameters** section of the dialog box are used to add or remove sections of the shaft. These options are discussed next.

- The **Section** drop-down in the **Shaft Parameters** section is used to select the section of shaft to be modified.
- Select the **Left** or **Right** radio button to define start point of the shaft.
- Click on the **Add <--** button to add a new section before selected section. Click on the **Add-->** button to add a new section after selected section in **Section** drop-down.
- Select desired option from the **Section Type** drop-down to specify what type of shaft section you want to create. Select the **Simple** option from the drop-down to create a simple shaft of specified diameter and length. Select the **Ring Groove** option to create a shaft with ring groove at specified distance from starting of section. Select the **Keyway** option from the drop-down to create a shaft section with cut create for key installation. Select the **Locknut Groove** option from the drop-down to create grooves for installing locknut. Select the **Cone** option from the drop-down to create the shaft section in the form of a cone which has different diameters at start and end point of section. Figure-17 shows different shapes of shaft sections.

Figure-17. Shaft section shapes

• After selecting desired type of shaft, specify the related parameters in the table.

Support Parameters

The options in the **Supports** section are used to how shaft will be supported in the assembly for load distribution. The options in this section are discussed next.

• Select desired number of support whose position is to be modified from the **Support Number** drop-down. There are two supports applied to shaft by default.
• Select desired option from the **Element Number** drop-down to define at which section of shaft will the support be applied.
• Specify desired value in the **Distance** edit box to define the distance of support from starting location of the section.
• Select the **Axially Rigid** check box to make the shaft rigidly held at intersection.

Load Parameters

The options in the **Loads** section are used to specify number of loads, location of load, and type of loads applied on the shaft. These options are discussed next.

• Click on the **Add** button from the **Loads** section of dialog box to add a new load on the shaft. A new load will be applied and preview will be displayed in the preview area of the dialog box.
• Select desired option from the **Load Number** edit box to define which load you want to modify.
• Select desired option from the next drop-down to specify on which section of shaft, the load will be applied.
• Click in the **Distance** edit box and specify the distance from start of section for applying load.
• Specify desired values of radial force, bending moment, axial force, and torque in the respective edit boxes of the **Loads** section.
• Click on the **Material** button from the dialog box and select desired material from the **Material Values** dialog box displayed; refer to Figure-18. Click on the **OK** button from the dialog box to apply the material.

Figure-18. Material Values dialog box

- After setting desired parameters, click on the **Calculate** button from the **Solid Edge Shaft Designer** dialog box. The **Calculated Results** tab will be displayed in the dialog box with a graph showing distribution of load on shaft; refer to Figure-19.

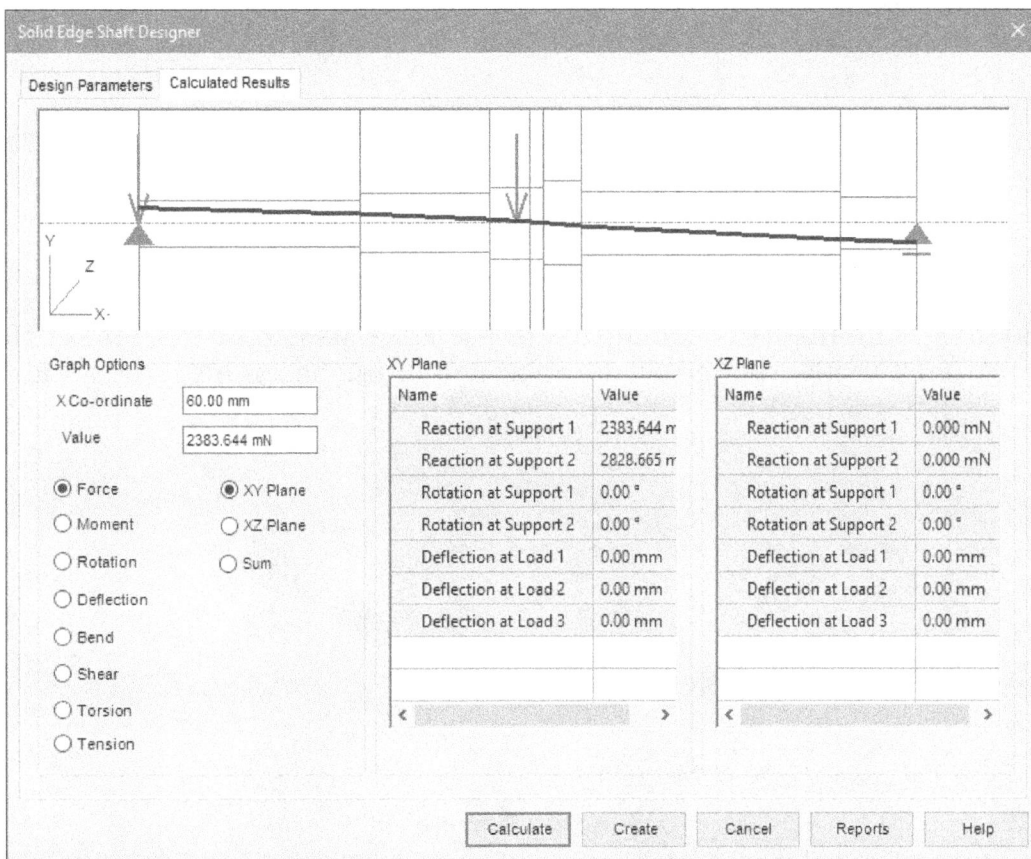

Figure-19. Calculated Results tab

- Select desired radio buttons from the **Graph Options** area to check results.
- Click on the **Reports** button from the dialog box to generate a report file based on results.
- Click on the **Create** button from the dialog box. The **Save As** dialog box will be displayed; refer to Figure-20.

Figure-20. Save As dialog box

- Specify desired name of the shaft file in the **File name** edit box and click on the **Save** button. The file will be saved and added in the assembly; refer to Figure-21.

Figure-21. Shaft created

Finding Torque

Most of the time in real world problems, you will get the load in Watt and RPM like, a shaft need to transfer 20KW load at 200 RPM. In such cases, you need to convert this value into torque. The formula is given next.

$$T = \frac{P \times 60}{2\pi N}$$

Here, T is torque, P is the power to be transferred, and N is the RPM.

After getting the torque value, you can apply it on the model to test the failure of shaft.

Example, A line shaft rotating at 200 r.p.m. is to transmit 20 kW. The shaft may be assumed to be made of mild steel with an allowable shear stress of 42 MPa. Determine the diameter of the shaft, neglecting the bending moment on the shaft.

Finding Bending Moment

Sometimes, you will get the problems like a load of 50kN is working at a distance of 100 mm outside the wheelbase. In such cases, you need to find out the bending moment by using the formula,

$$M = W \times L$$

Here, M is the bending moment, W is the load, and L is the distance from wheelbase.

Example, A pair of wheels of a railway wagon carries a load of 50 kN on each axle box, acting at a distance of 100 mm outside the wheel base. The gauge of the rails is 1.4 m. Find the diameter of the axle between the wheels, if the stress is not to exceed 100 MPa.

Cam Designer

The transformation of one of the simple motions, such as rotation, into any other motions is often conveniently accomplished by means of a cam mechanism. A cam mechanism usually consists of two moving elements, the cam and the follower, mounted on a fixed frame. Cam devices are versatile, and almost any arbitrarily-specified motion can be obtained. In some instances, they offer the simplest and most compact way to transform motions. In Plate cam or disk cam, the follower moves in a plane perpendicular to the axis of rotation of the camshaft. A translating or a swing arm follower must be constrained to maintain contact with the cam profile. Figure-22 shows the nomenclature of a disc cam.

Figure-22. Disc cam nomenclature

The **Cam Designer** tool is used to create standard cam based on specified engineering parameters. The procedure to use this tool is given next.

• Click on the **Cam Designer** tool from the **Engineering Reference** drop-down in the **Environs** group of **Tools** tab in the **Ribbon**. The **Solid Edge Cam Designer** dialog box will be displayed; refer to Figure-23.

Figure-23. Solid Edge Cam Designer dialog box

- Specify the cam parameters and load parameters in respective sections of the dialog box as discussed earlier.
- After setting desired parameters, click on the **Calculate** button to check design and loading graph of Cam in calculated result; refer to Figure-24.

Figure-24. Calculated Result Cam

- Once you find the cam suitable for design, click on the **Create** button from the dialog box. The **Save As** dialog box will be displayed and you will be asked to specify the location for saving model file of Cam. Specify desired name of file and click on the **Save** button. The model file of cam will be created; refer to Figure-25.

Figure-25. Cam Designer

Spur Gear Designer

For any kind of gear design, there are a few requirements of designers. In designing a gear drive, following data is usually given :

1. The power to be transmitted,
2. The speed of the driving gear,
3. The speed of the driven gear or the velocity ratio, and
4. The centre distance.

The following requirements must be met in the design of a gear drive :

(a) The gear teeth should have sufficient strength so that they will not fail under static loading or dynamic loading during normal running conditions.
(b) The gear teeth should have wear characteristics so that their life is satisfactory.
(c) The use of space and material should be economical.
(d) The alignment of the gears and deflections of the shafts must be considered because they effect on the performance of the gears.
(e) The lubrication of the gears must be satisfactory.

The different modes of failure of gear teeth and their possible remedies to avoid the failure, are as follows :

1. **Bending failure**. Every gear tooth acts as a cantilever. If the total repetitive dynamic load acting on the gear tooth is greater than the beam strength of the gear tooth, then the gear tooth will fail in bending, i.e. the gear tooth will break. In order to avoid such failure, the module and face width of the gear is adjusted so that the beam strength is greater than the dynamic load.

2. **Pitting**. It is the surface fatigue failure which occurs due to many repetition of Hertz contact stresses. The failure occurs when the surface contact stresses are higher than the endurance limit of the material. The failure starts with the formation of pits which continue to grow resulting in the rupture of the tooth surface. In order to avoid the pitting, the dynamic load between the gear tooth should be less than the wear strength of the gear tooth.

3. **Scoring**. The excessive heat is generated when there is an excessive surface pressure, high speed or supply of lubricant fails. It is a stick-slip phenomenon in which alternate shearing and welding takes place rapidly at high spots. This type of failure can be avoided by properly designing the parameters such as speed, pressure and proper flow of the lubricant, so that the temperature at the rubbing faces is within the permissible limits.

4. **Abrasive wear**. The foreign particles in the lubricants such as dirt, dust or burr enter between the tooth and damage the form of tooth. This type of failure can be avoided by providing filters for the lubricating oil or by using high viscosity lubricant oil which enables the formation of thicker oil film and hence permits easy passage of such particles without damaging the gear surface.

5. **Corrosive wear**. The corrosion of the tooth surfaces is mainly caused due to the presence of corrosive elements such as additives present in the lubricating oils. In order to avoid this type of wear, proper anti-corrosive additives should be used.

Designing Spur Gear

In order to design spur gears, the following procedure may be followed :
The design tangential tooth load is obtained from the power transmitted and the pitch line velocity by using the following relation :

$$W_T = \frac{P}{v} \times C_S$$

where W_T = Permissible tangential tooth load in newtons,
             P = Power transmitted in watts,
             v = Pitch line velocity in m / s = (πDN)/60
             D = Pitch circle diameter in metres,
We know that circular pitch,

             p_c = π D / T = π m
             D = m.T
Thus, the pitch line velocity may also be obtained by using the following relation, i.e.

$$v = \frac{\pi D.N}{60} = \frac{\pi m.T.N}{60} = \frac{p_c.T.N}{60}$$

where,   m = Module in metres, and
             T = Number of teeth.

The **Spur Gear Designer** tool is used to design standard spur gear based on specified geometry parameters and loads. The procedure to use this tool is given next.

- Click on the **Spur Gear Designer** tool from the **Engineering Reference** drop-down in the **Environs** group of **Tools** tab in the **Ribbon**. The **Solid Edge Spur Gear Designer** dialog box will be displayed. Specify the design parameter of spur gear in dialog box; refer to Figure-26.

Figure-26. Spur Gear Designer dialog box

- You can also specify the material of spur gear by clicking on **Material** button as discussed earlier.
- Click on the **Options** button from the dialog box to define the type of spur gear. The **Design Parameters - Input Conditions** dialog box will be displayed; refer to Figure-27. Select desired radio button from the **Type of Gearing** area to define whether you want to create an internal gear or external gear. Select desired radio button from the **Output Geometrical Parameters** area to specify what type of parameters you want to output for gear. Similarly, select desired radio buttons to define the input conditions and click on the **OK** button.

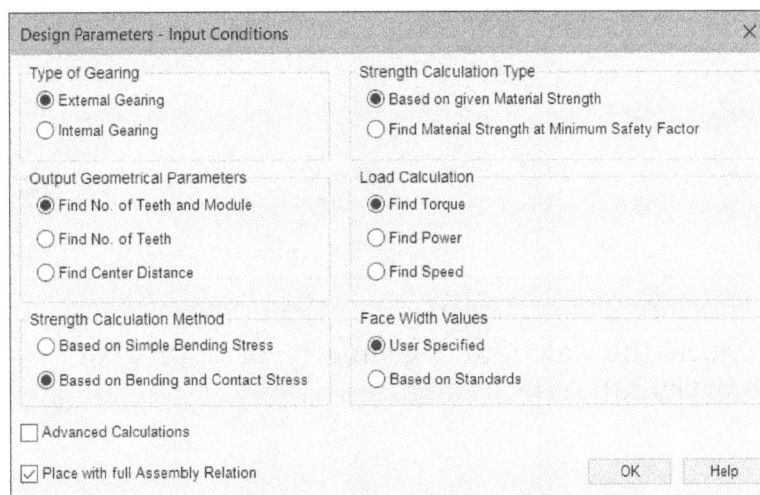

Figure-27. Design Parameters-Input Conditions dialog box

- After setting parameters, click on the **Calculate** button from the dialog box. The **Finalize Output Results** dialog box will be displayed; refer to Figure-28.
- Specify desired parameters in the dialog box and click on the **Apply** button. The **Correction Methods** dialog box will be displayed; refer to Figure-29.

Figure-28. Finalize Output Results dialog box

Figure-29. Correction Methods dialog box

- Select desired option from the **Correction Methods** drop-down to apply size correction of gears and click on the **Apply** button. The **Calculated Results** tab will be displayed in the dialog box; refer to Figure-30.

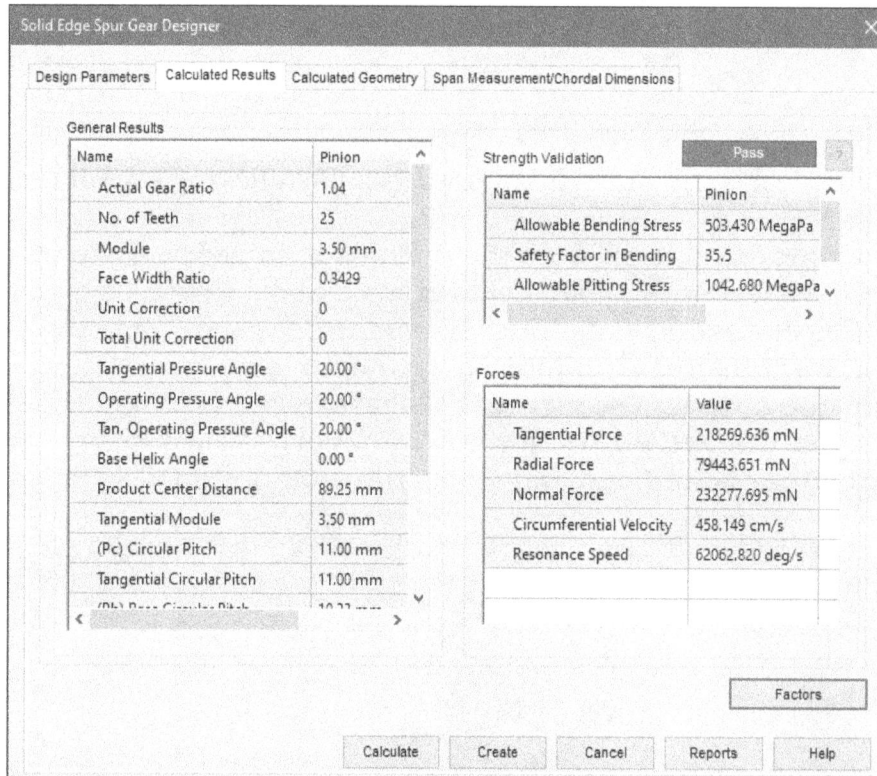

Figure-30. Calculated Results tab of Solid Edge Spur Gear Designer

- You can also check the calculated geometry of spur gear, by clicking on the **Calculated Geometry** tab; refer to Figure-31.

Figure-31. Calculated Geometry tab

- Click on the **Span Measurement/Chordal Dimensions** tab in the dialog box to check chordal dimensions.
- Click on the **Create** button from the dialog box. The **Save As** dialog box will be displayed. Specify the name of gear model and click on the **Save** button. The spur gears will be created; refer to Figure-32.

Figure-32. Spur gears created

Bevel Gear Designer

The bevel gears are used for transmitting power at a constant velocity ratio between two shafts whose axes intersect at a certain angle. The pitch surfaces for the bevel gear are frustums of cones; refer to Figure-33.

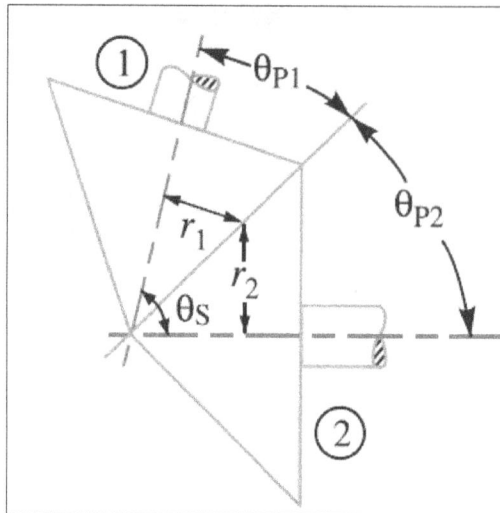

Figure-33. Bevel gear arrangement

Here, r_1 and r_2 are radii of gear 1 and gear 2, respectively.
Θ_{P1} and Θ_{P2} are cone angle for gears
and, Θ_S is shaft angle.

The **Bevel Gear Designer** tool is used to create bevel gears based on specified engineering parameters. The procedure to use this tool is given next.

• Click on the **Bevel Gear Designer** tool from the **Engineering Reference** drop-down in the **Environs** group of **Tools** tab in the **Ribbon**. The **Solid Edge Bevel Gear Designer** dialog box will be displayed; refer to Figure-34.

Figure-34. Design parameters of Bevel gear

- Specify the design parameters and material as discussed earlier and click on the **Calculate** button. The **Calculated Results** tab of dialog box will be displayed; refer to Figure-35.

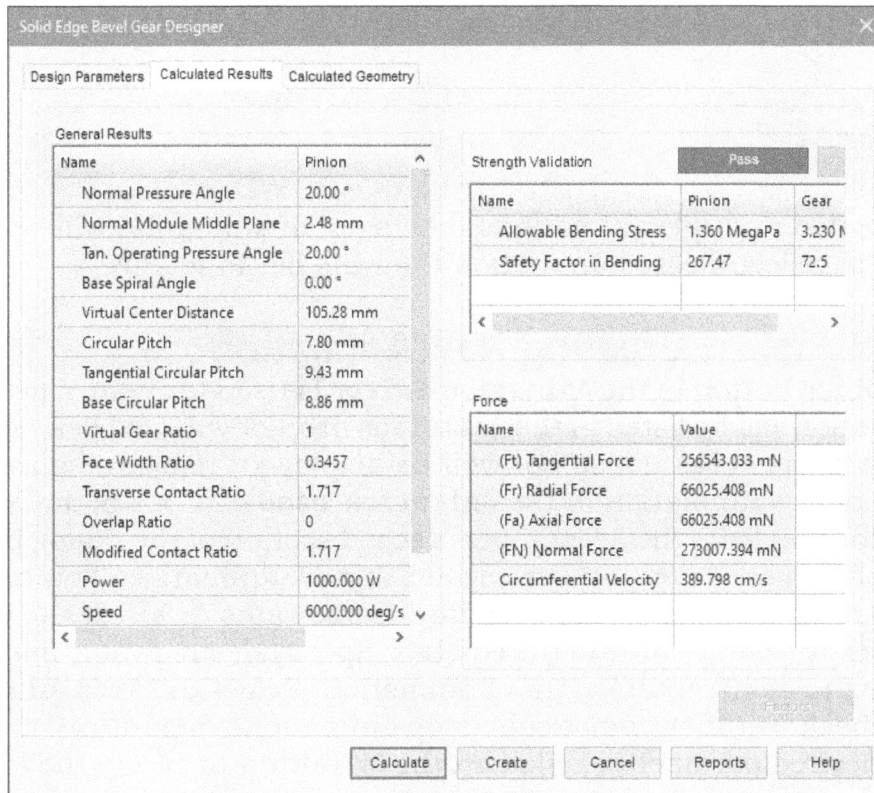

Figure-35. Calculated Results tab for Bevel Gear

Rest of the tools are same as we discussed earlier.

EXPLODE RENDER ANIMATE ENVIRONMENT

The **ERA** stands for **Explode- Render- Animate** environment. Click on the **ERA** tool from the **Environs** group of **Tools** tab in the **Ribbon**. The **Explode - Render - Animate** environment will be displayed and ERA tools will be available in the **Ribbon**; refer to Figure-36.

Figure-36. ERA Ribbon

ANIMATION EDITOR

The **Animation Editor** tool is used to play and produce the animation of assembly. On clicking this tool, the **Animation Editor** window will be displayed at the bottom in the application window; refer to Figure-37. Various tools and options of the **Animation Editor** are discussed next.

Figure-37. Animation Editor

Animation Drop-down

If you have created multiple animations then a list of animation will be displayed in this drop-down. Select desired animation from the list to modify it.

Creating New Animation

The **New Animation** button in the **Animation Editor** is used to create a new animation. The new animation will be listed in the **Animation** drop-down. On clicking this button, the **Animation Properties** dialog box will be displayed; refer to Figure-38. Specify desired name of new animation in the **Animation name** edit box of dialog box. Select desired radio button from the **Frames per second** section of the dialog box to specify the number of frames generated per second in new animation. You can select the **NTSC** and **PAL** radio buttons to use standard frame rates or select the **Custom** radio button and specify desired value in the edit box. Specify desired value in the **Animation length** edit box to define total length of animation. Select the **Stop playback after last event** check box to automatically stop animation when duration has passed. After setting desired parameters, click on the **OK** button to create the animation.

Figure-38. Animation Properties dialog box

Saving Animation

Click on the **Save Animation** button from the **Animation Editor** to save the modifications in animation.

Deleting Animation

Click on the **Delete Animation** button from the **Animation Editor** to delete the animation selected in the **Animation** drop-down. A confirmation box will be displayed. Select the **Yes** button from the confirmation box to delete the animation.

Animation Properties

The **Animation Properties** button in **Animation Editor** is used to modify the properties of the selected animation. On clicking this button, the **Animation Properties** dialog box is displayed. The options in this dialog box have been discussed earlier.

Save as Movie

The **Save as Movie** button in the **Animation Editor** is used to save the current animation in movie clip format (WMV, AVI, and BIP). On clicking this button, the **Save As Movie** dialog box will be displayed; refer to Figure-39.

Figure-39. Save As Movie dialog box

Specify desired name for animation video in the **File name** edit box. Specify desired format for video file in the **Save as type** drop-down. Click on the **Options** button from the dialog box to specify parameters of video. The **Save As Movie Options** dialog box will be displayed; refer to Figure-40.

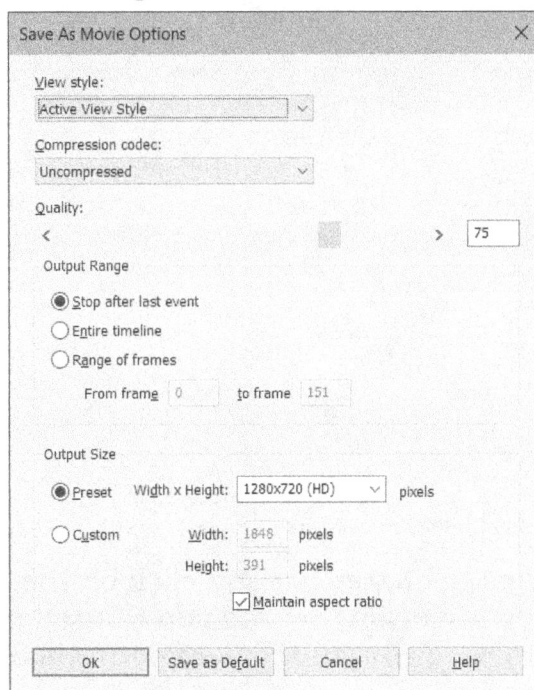

Figure-40. Save As Movie Options dialog box

Specify desired parameters like output size, time line, codec in the dialog box and click on the **OK** button. After setting all the parameters, click on the **Save** button from the dialog box to save the animation video file.

Creating Camera Path and Enabling Camera Movement

The **Camera Path** button in the **Animation Editor** is used to enable or disable the camera movement in animation. After enabling camera movement in animation, now we will create camera path. The procedure to create camera path is given next.

- Right-click on the **Camera** option from the **Animation PathFinder**; refer to Figure-41 and select the **Edit Definition** option from the shortcut menu. The **Camera Path Wizard** dialog box will be displayed; refer to Figure-42.

Figure-41. Edit Definition option for Camera

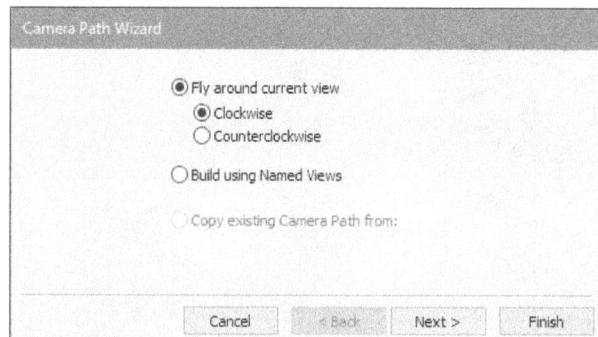

Figure-42. Camera Path Wizard dialog box

- Select desired option from the dialog box like if you want to fly around the model using **Clockwise** or **Counterclockwise** radio button. If you want to use named views for camera rotation then select the **Build using Named Views** radio button. If you want to copy camera path from an existing camera path then select the **Copy existing Camera Path from** radio button and then select desired path from the drop-down. After selecting desired radio button (in our case, it is **Build using Named Views** radio button) and click on the **Next** button. The options in the dialog box will be displayed as shown in Figure-43.

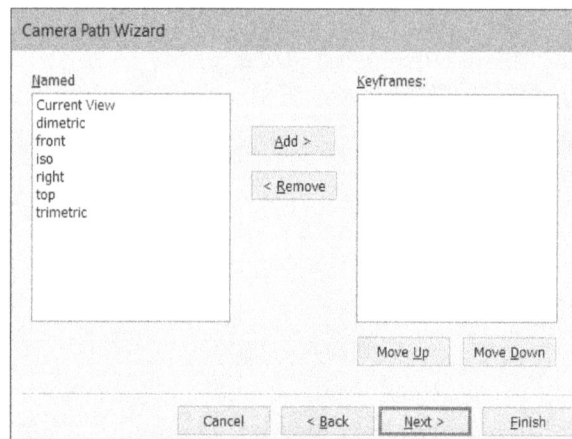

Figure-43. Views for camera path

- Select desired plane from the **Named** list and click on the **Add >** button. The plane will be added in the keyframe. One by one add desired planes and then click on the **Next >** button. Preview page of the camera path will be displayed in the dialog box.

- Click on the **Preview** button to check preview of camera path. Click on the **Play** button to check the camera path.
- Click on the **Finish** button from the dialog box to create the camera path.

SHOW CAMERA PATH

The **Show Camera Path** button from the **Animation PathFinder** to display path of camera movement during animation.

WORKING WITH ANIMATION

The buttons at the top in the **Animation Editor** are used to play, stop, restart, and switch frames in the animation.

CREATING KEYSHOT ANIMATION

Once you have checked the animation, click on the **KeyShot Animate** button from the **Animation Editor** box if you want to create animation using the KeyShot software. The procedure to use this button is given next.

- Click on the **KeyShot Animate** button from the **Animation Editor**. The **KeyShot Options** dialog box will be displayed; refer to Figure-44.

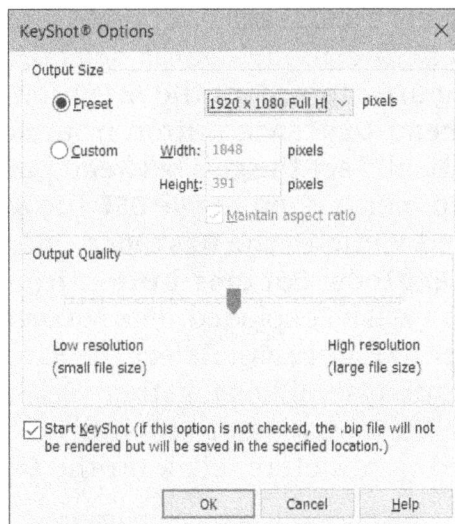

Figure-44. KeyShot Options dialog box

- Select the **Preset** radio button and then select desired resolution for animation video from the drop-down. You can also specify custom resolution of animation video by selecting the **Custom** radio button.
- Set desired output quality using the **Output Quality** slider and click on the **OK** button. The **KeyShot For Solid Edge** application window will be displayed with rendering of model.
- Click on the **Animation** button at the bottom in the application window to check animation and save the animation using the **Save** tool from **File** menu.
- Close the application by **Close** button at the top right corner.

KEYSHOT RENDER

The **KeyShot Render** tool is used to render an assembly or part by applying realistic material and environment. On clicking this tool from the **KeyShot** group of **Home** tab in **Ribbon**, the KeyShot application will be displayed. We will discuss about the options of this application later in this book.

AUTO EXPLODE

The **Auto Explode** tool is used to explode the assembly automatically. The procedure to use this tool is given next.

*   Click on the **Auto Explode** tool from the **Explode** group in the **Home** tab of the **Ribbon**. The **Auto Explode** command box will be displayed; refer to Figure-45.

Figure-45. Auto Explode command box

*   Select desired option from the **Select** drop-down. If you want to explode a subassembly then select the **Subassembly** option from the drop-down and select desired subassembly from the model. After selecting assembly/subassembly, right-click in the drawing area to accept the selection.
*   Select the **Automatic Spread Distance** button from **Settings Step** rollout of the command box to automatically set the gap between parts. If you want to manually specify gap between components then toggle **OFF** the **Automatic Spread Distance** button and specify desired value in the **Distance** edit box.
*   Click on the **Automatic Explode Options** button from **Settings Step** rollout to specify how subassemblies will be exploded. The **Automatic Explode Options** dialog box will be displayed; refer to Figure-46. Select the **Bind all subassemblies** check box to keep the exploded subassemblies together. Select desired radio button from the **Explode Technique** area to explode parts individually or using subassembly level. After setting desired parameters, click on the **OK** button.

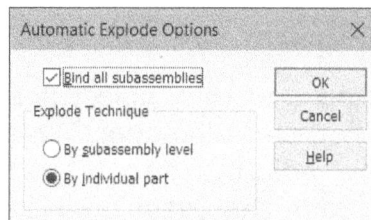

Figure-46. Automatic Explode Options dialog box

*   Click on the **Explode** button from the command box to apply automatic explode view of assembly. The exploded view of assembly will be displayed; refer to Figure-47.

Figure-47. Auto Exploded view

- Click on the **Finish** button from the command box and press **ESC** to exit.

EXPLODE

The **Explode** tool is used to explode the assembly manually. The procedure to use this tool is given next.

- Click on the **Explode** button from the **Explode** group in the **Home** tab of **Ribbon**. The **Explode** command box will displayed; refer to Figure-48.

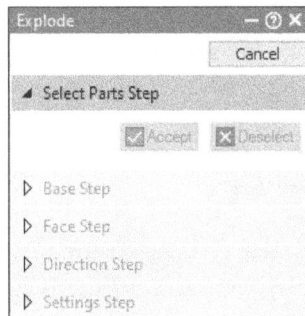

Figure-48. Explode command box

- Select the part you want to move and right-click in the drawing area. You will be asked to select a base part.
- Select desired base part to use as reference for explode. You will be asked to select a face of base part.
- Select desired face of base part. You will be asked to specify direction in which you want to move exploded part.
- Click to specify the direction and specify desired value in the **Distance** edit box. After specifying value, click on the **Explode** button from the command box. The selected part will move at specified distance; refer to Figure-49. Click on the **Finish** button to complete exploding the component. Repeat the procedure to explode other components and then press **ESC** to exit the tool.

Figure-49. Explode manually

VISUAL EXPLODE

The **Visual Explode** tool is used to separate and display individual components of an assembly, making it easier to see how the parts fit together and relate to one another. The procedure to use this tool is discussed next.

*   Click on the **Visual Explode** tool from the **Explode** group in the **Home** tab of **Ribbon**. The **Explode** command box will displayed; refer to Figure-50.

Figure-50. Visual explode command box

*   You can explode the components using the slider in the explode rollout.
*   Click on the **Explode Stepwise** button to control how far the parts move apart using the slider at the top.
*   Click on the **Save Configuration** button to save current exploded state as a new configuration, the **New Configuration** dialog box will be displayed; refer to Figure-51.

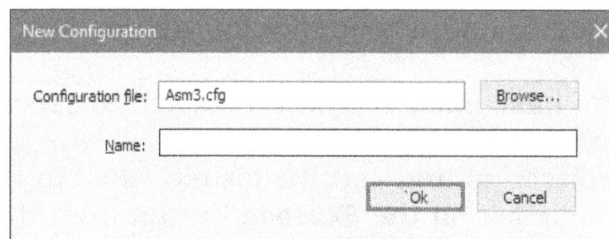

Figure-51. New configuration dialog box

*   Click on the **Browse** button if you want to use pre-existing configuration file.
*
*   Click in the **Name** edit box and specify desired name of the new configuration file.
*   After specifying desired parameters, click on the **OK** button from the dialog box.

- Click on the **Close** button from the command box. The visual explode will be created; refer to Figure-52.

Figure-52. Visual explode created

REPOSITION

The **Reposition** tool is used to reorder an exploded part. The procedure to use this tool is given next.

- Click on the **Reposition** tool from the **Modify** group in the **Ribbon**. The **Reposition** command box will be displayed; refer to Figure-53.

Figure-53. Reposition command box

- Select the part to be repositioned. You will be asked to select an earlier exploded component to position the new component.
- Select desired component. You will be asked to specify the direction of movement.
- Click in desired direction. The component will move accordingly; refer to Figure-54. Press **ESC** to exit the tool.

Figure-54. Reposition view

REMOVING COMPONENT FROM EXPLODED STATE

The **Remove** tool in **Modify** group of **Ribbon** is used to remove a component from exploded state and return it to assembled position. To do so, select the component to be removed from the exploded state and click on the **Remove** tool from **Modify** group in **Ribbon**. The component will return to original assembled position and will be hidden.

COLLAPSING EXPLODED COMPONENT

The **Collapse** tool is used to return selected exploded component to its assembled position. To do so, select the component to be collapsed and click on the **Collapse** tool from the **Modify** group in the **Home** tab of **Ribbon**. The component will return to assembled position.

The **Drag Component** and **Move on Select** tools work in the same way as discussed earlier.

UNEXPLODE

The **Unexplode** tool is used to return the components of assembly to their native assembled position. The procedure to use this tool is given next.

- Click on the **Unexplode** tool from the **Modify** group of **Home** tab in the **Ribbon**. The **Solid Edge** message box will displayed as shown in Figure-55. You are asked to either save the exploded view file or the data will erase. If you do not want to save the exploded view then click on **Yes** button to continue. The unexploded view of model will be displayed; refer to Figure-56. Click on the **No** button and save the exploded view as a new configuration if you do not want to see the message box next time.

Figure-55. Solid Edge message box

Figure-56. Unexplode

BINDING SUB-ASSEMBLY

The **Bind** tool in **Modify** group is used to keep all the components of a sub-assembly together. To use this tool, select the sub-assembly from the **Path Finder** that you want to bind together and click on the **Bind** tool from the **Modify** group of **Home** tab in the **Ribbon**. The sub-assembly components will be exploded as a single unit.

UNBINDING SUB-ASSEMBLY

The **Unbind** tool is used to make sub-assemblies explode when exploded of assembly is generated. To do so, select the sub-assembly to be unbound from the **Path Finder** and click on the **Unbind** tool from the **Modify** group in the **Home** tab of the **Ribbon**.

FLOW LINES

The tools in the **Flow Lines** group are used to create and display annotation lines showing how components are assembled in exploded view. Various tools in this group are discussed next.

Drop

The **Drop** tool is used to create an individual entry in the **Explode PathFinder** for each flow line created automatically while exploding; refer to Figure-57. You can select a flow line to see its path and what its connects to. It converts the event flow lines in the exploded assembly into annotation flow lines. You can use annotation flow lines of exploded views in draft documents.

Figure-57. Explode path finder

Draw

The **Draw** button is used to draw annotation flow lines for components of exploded view. You can use annotation flow lines to create an exploded view of assembly model on a drawing sheet. The procedure to use this tool is given next.

- Click on the **Draw** tool from the **Flow Lines** group in the **Home** tab of **Ribbon**. The **Draw** command box will be displayed; refer to Figure-58.

Figure-58. Draw command box

- Select desired start key point and end key point to define flow line.
- Specify desired value of length of the segments which terminate at ports in the **Port segment length** edit box. Preview of the line will be displayed; refer to Figure-59.

Figure-59. Preview of flow lines

- Click on the **Next** and **Previous** buttons to switch between variations of flow lines between same points.
- Click on the **Change Orientation** button from command box and select desired face to orient middle section of flow line.
- Click on the **Finish** button to create flow line and press **ESC** to exit the tool.

Modify

The **Modify** tool is used to modify flow lines. Modification an event flow line or annotation flow line between two exploded parts in an assembly. You can:

- Change the shape of an event flow line.
- Split annotation flow line into multiple segments.
- Change the location of the entire flow line.
- Edit the length of an end segment or the position of a joggle segment.

The procedure to use this tool is given next.

- Click on the **Modify** tool from the **Flow Lines** group of the **Home** tab in the **Ribbon**. The **Modify** command box will be displayed; refer to Figure-60.

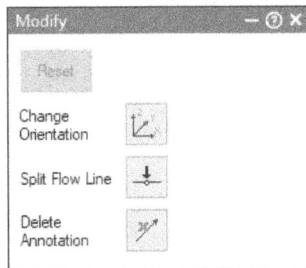

Figure-60. Modify command box

- To change orientation, select the **Change Orientation** button from the command box and select the face/plane which you want to use for orientation and then select the flow line to be oriented. After selecting flow line, right-click in the drawing area.
- To split a flow line, select the **Split Flow Line** button from the command box and then click at desired location on flow line to split it. Right-click in drawing area to exit the button.
- To delete an annotation flow line, select the **Delete Annotation** button from the command box and select desired flow line to be deleted. Right-click in the drawing area to delete the flow line.
- Press **ESC** to exit the tool.

Select the **Flow Lines** and **Flow Line Terminators** buttons to display flow lines and flow line terminators, respectively.

CONFIGURATIONS

Configurations are variations of assembly. A display configuration captures the display status of the parts, assemblies, assembly sketches, weld beads, and reference planes in an assembly. The tools to create and manage configurations are available in the **Configurations** group of the **Home** tab in **Ribbon**; refer to Figure-61. Various tools in this group are discussed next.

Figure-61. Configurations group

Saving and Retrieving Display Configurations

The **Display Configurations** tool is used to save the current displaying configurations of assembly components and retrieve them later. The procedure to use this tool is given next.

*   After performing desired modifications of model like exploded view, changed position of assembly components, and so on; click on the **Display Configurations** tool from the **Configurations** group of **Home** tab in the **Ribbon**. The **Display Configurations** dialog box will be displayed; refer to Figure-62.

Figure-62. Display Configurations dialog box

*   Click on the **New** button from the dialog box. The **New Configuration** dialog box will be displayed.
*   Specify desired name in the edit box and click on the **OK** button. A new configuration will be created.
*   Select desired configuration and click on the **Rename** button to change its name.
*   Select the configuration and click on the **Update** button after making changes in the model to update the configuration.
*   If you want to delete a configuration then select it from dialog box and click on the **Delete** button.
*   Click on the **Close** button from the dialog box to exit the tool.

You can apply a configuration by selecting it from the **Current Configuration** drop-down in the **Configurations** group of the **Ribbon**; refer to Figure-63.

Figure-63. Current Configuration drop-down

Configuration Manager

The **Configuration Manager** tool is used to display/hide components in an assembly configuration. The procedure to use this tool is given next.

*   Click on the **Configuration Manager** tool from the **Configurations** group of **Home** tab in the **Ribbon**. The **Configuration Manager** dialog box will be displayed; refer to Figure-64.

Figure-64. Configuration Manager dialog box

*   Select or clear check boxes from the configurations to add or remove respective components from them.
*   Click on the **Save** button from the dialog box to save the configuration.

Configuration Options

The **Configuration Options** tool is used to specify parameters for configurations. The procedure to use this tool is given next.

*   Click on the **Configuration Options** tool from the **Configurations** group in the **Home** tab of **Ribbon**. The **Display Configuration Options** dialog box will be displayed; refer to Figure-65.

Figure-65. Display Configuration Options dialog box

- Select the **Fit view after apply** check box to automatically fit view when a display configuration is applied.
- Select the **List exploded configuration in modeling mode** check box to display exploded view configurations in the modeling mode after you exit ERA environment. After selecting the check box, select desired radio button from the When section to specify when the configuration will be displayed.
- Select the **Use single default configuration** check box to a single default configuration rather than using different configurations for different users of software.
- Similarly, set the other options in the dialog box and click on the **OK** button.

Copying Components to Display Configuration

The **Copy Current Display** tool is used to copy selected component to a different display configuration. The procedure to use this tool is given next.

- Select the component you want to copy from the **Explode PathFinder** and click on the **Copy Current Display** tool from the **Configurations** group of **Home** tab in the **Ribbon**. The **Copy Current Display** dialog box will be displayed; refer to Figure-66.

Figure-66. Copy Current Display dialog box

- Select desired check boxes from the dialog box to specify the configurations in which the component will be copied.
- Click on the **OK** button to apply the changes.

Taking Snapshot and Restoring

The **Take Snapshot** tool is used to keep a record of shown and hidden components in the current configuration of assembly. After applying the show/hide settings to components, click on the **Take Snapshot** tool from the **Configurations** group of the **Home** tab in the **Ribbon**. A snapshot of settings will be saved.

Open another configuration of assembly with different show/hide settings and change the settings of components. Click on the **Restore Snapshot** button to apply the settings earlier saved using **Take Snapshot** tool.

Unloading Hidden Parts

The **Unload Hidden Parts** tool is used to remove hidden parts from the temporary memory of assembly. This tool is available in the **Configurations** group of **Home** tab in the **Ribbon**.

MODES

The tools in the **Modes** group are used to active different assembly modes for getting better performance. Select the **Large Assembly Performance** toggle button to increase the performance of manipulating a large assembly at the cost of better visualization. Select the **Limited Update** toggle button to reduce the frequency of automatic model updation when you have made changes in the assembly. Select the **Limited Save** toggle button to reduce the load on system when you save the assembly. When this button is **ON**, only the directly modified parts will be saved with assembly.

PRACTICE

Generate exploded views of all the assemblies discussed earlier in Practical and Practices of Chapter 8.

SELF ASSESSMENT

Q1. What is the purpose of constraints in assembly?

a) To create a new model
b To arrange parts systematically
c) To edit variables
d) To design shafts

Q2. The Peer Variables tool is used to?

a) Create new assemblies
b) Edit variables of attachments in assembly
c) Check for geometric errors
d) Add virtual components

Q3. Which logic rule is used to define a variable greater than or equal to 8 and less than or equal to 15?

a) (8;15)
b) [8;15]
c) {8;15}
d) [9;20)

Q4. What does the 'Component Tracker' tool do?

a) Adds virtual components
b) Tracks changes in assembly components
c) Updates the assembly model
d) Publishes terrain models

Q5. Which of the following stress values is permissible for shafts without keyways under ASME code?

a) 84 MPa (Tension)
b) 42 MPa (Shear)
c) 112 MPa (Tension)
d) 56 MPa (Shear)

Q6. Which section of the Shaft Designer tool specifies the number and type of loads applied?

a) Shaft Parameters
b) Supports
c) Loads
d) Reports

Q7. What does the 'Errors' tool assist with in Solid Edge?

a) Adding virtual components
b) Checking and fixing recompute failures in parts or assemblies
c) Publishing terrain models
d) Editing variables

Q8. Which parameter is NOT included in the Loads section of Shaft Designer?

a) Radial Force
b) Axial Force
c) Torque
d) Material Weight

Q9. Which of the following options describes virtual components?

a) Real components added to the assembly
b) Objects not in the assembly but included in reports like Bill of Materials
c) Components used exclusively for rendering
d) Components that are always constrained

Q10. Which tool is used to add virtual components to the assembly structure?

a) Publish Internal Components
b) Inter-Part Manager
c) Component Structure Editor
d) Errors

Q11. What type of file is typically imported when using the 'Publish Terrain Models' tool?

a) STL file
b) TXT file
c) CSV file
d) DXF file

Q12. What must be done before using tools in the Engineering Reference drop-down?

a) Enable Automatic Update
b) Save the assembly model
c) Add virtual components
d) Run the Errors tool

Q13. What is the standard diameter range of transmission shafts between 60 mm and 110 mm?

a) 5 mm steps
b) 10 mm steps
c) 15 mm steps
d) 20 mm steps

Q14. Which tool allows users to check the hierarchy of components in the form of parent and child links?

a) Publish Internal Components
b) Component Structure Editor
c) Inter-Part Manager
d) Component Tracker

Q15. What is the purpose of the 'Formula' button in the Peer Variables tool?

a) To create formulas using logic rules
b) To reset the model parameters
c) To track changes in variables
d) To generate a Bill of Materials

Q16. Which of the following describes 'Limited Update and Limited Save'?

a) Saves the assembly in multiple formats
b) Updates only when necessary to conserve resources
c) Publishes internal components
d) Enables automatic error tracking

Q17. In shaft design, which of the following stresses occurs due to torsional loads?

a) Bending stress
b) Shear stress
c) Compressive stress
d) Axial stress

Q18. What is the permissible shear stress for shafts with keyways, according to ASME code?

a) 42 MPa
b) 56 MPa
c) 84 MPa
d) 112 MPa

Q19. What is the main function of the 'Error Assistant' dialog box?

a) Highlight errors in parts or assembly
b) Save error logs to a text file
c) Create a hierarchy of components
d) Design shafts based on stress values

Q20. What is the primary function of a cam mechanism?

a) To generate power
b) To transform one motion into another
c) To increase speed
d) To maintain balance

Q21. Which type of cam allows the follower to move in a plane perpendicular to the camshaft axis?

a) Plate cam or disk cam
b) Cylindrical cam
c) Translating cam
d) Swing arm cam

Q22. Which of the following is NOT a requirement in gear design?

a) Sufficient tooth strength
b) Effective use of space and material
c) Maximum number of teeth
d) Proper alignment of gears

Q23. What is the primary cause of pitting in gear teeth?

a) Poor lubrication
b) High-speed operation
c) Excessive heat generation
d) Hertz contact stresses

Q24. How can abrasive wear in gears be minimized?

a) Reducing tooth size
b) Using anti-corrosive additives
c) Installing filters for lubricants
d) Increasing gear speed

Q25. Which parameter determines the pitch line velocity of a spur gear?

a) Module and number of teeth
b) Face width and dynamic load
c) Lubricant viscosity
d) Shaft angle

Q26. Bevel gears are primarily used to transmit power between shafts that are:

a) Parallel
b) Intersecting at an angle
c) Perpendicular
d) Rotating in the same direction

Q27. What shape is the pitch surface of a bevel gear?

a) Cylindrical
b) Conical frustum
c) Flat disk
d) Elliptical

Q28. What does ERA stand for?

a) Enhanced Rendering and Animation
b) Explode, Render, Animate
c) Engineering Reference Application
d) Environment for Rendering and Assembly

Q29. What elements are required for a cam mechanism to function?

a) Cam, follower, and lubricant
b) Cam, follower, and a fixed frame
c) Camshaft, gears, and a follower
d) Shaft, disk, and load

Q30. Which button in the Cam Designer dialog box is used to verify the design and loading graph?

a) Save
b) Create
c) Calculate
d) Load

Q31. What must be done to ensure that a translating or swing arm follower maintains contact with the cam profile?

a) Increase the rotation speed
b) Constrain it appropriately
c) Adjust the camshaft diameter
d) Use a larger cam

Q32. What determines the tangential tooth load in spur gear design?

a) Pitch circle diameter and module
b) Power transmitted and pitch line velocity
c) Shaft deflection and alignment
d) Dynamic load and material strength

Q33. Which type of failure is caused by repetitive dynamic loads exceeding the gear tooth's beam strength?

a) Pitting
b) Bending failure
c) Scoring
d) Corrosive wear

Q34. What is the purpose of using high-viscosity lubricant oil in gear systems?

a) To prevent overheating
b) To reduce dynamic load
c) To form a thicker oil film and allow particle passage
d) To increase gear speed

Q35. Which dialog box allows you to specify whether a spur gear will be internal or external?

a) Design Parameters - Input Conditions dialog box
b) Finalize Output Results dialog box
c) Save As dialog box
d) Correction Methods dialog box

Q36. What is the cone angle of a bevel gear directly related to?

a) Shaft speed
b) Shaft angle
c) Pitch circle diameter
d) Gear material

Q37. What must be specified in the Bevel Gear Designer dialog box before calculating results?

a) Lubricant viscosity
b) Design parameters and material
c) Shaft alignment details
d) Number of teeth only

Q38. Which frame rate options are available when creating a new animation?

a) FPS only
b) NTSC, PAL, and Custom
c) Low, Medium, and High
d) Static and dynamic

Q39. What type of video formats can be selected when saving an animation as a movie?

a) JPEG, PNG, TIFF
b) WMV, AVI, BIP
c) PDF, DOC, XLS
d) GIF, MP4, MOV

Q40. Which tool is used to enable or disable camera movement in animation?

a) Animation PathFinder
b) Camera Path
c) Camera Wizard
d) KeyShot Options

Q41. How can you preview the created camera path?

a) By clicking the Play button in the Preview page.
b) By selecting the Path Preview option from the menu.
c) By enabling Camera Path in the Animation Editor.
d) By clicking on the Finish button directly.

Q42. Which button is used to create animations using the KeyShot software?

a) Animation Editor
b) KeyShot Animate
c) KeyShot Render
d) Auto Explode

Q43. What option should you select in the KeyShot Options dialog to specify custom video resolution?

a) Preset
b) Custom
c) Resolution Slider
d) Output Quality

Q44. What does the "Auto Explode" tool do?

a) Explodes components manually.
b) Automatically sets gaps between parts in the assembly.
c) Creates flow lines for exploded components.
d) Removes exploded components from the assembly.

Q45. Which option in the Auto Explode dialog keeps subassemblies together during explosion?

a) Automatic Spread Distance
b) Explode Technique
c) Bind all subassemblies
d) Collapse Subassemblies

Q46. Which tool is used to manually move parts during an explosion process?

a) Explode
b) Auto Explode
c) Reposition
d) Unexplode

Q47. What is the purpose of the "Split Flow Line" option in the Modify tool?

a) To delete annotation flow lines.
b) To divide a flow line into multiple segments.
c) To change the orientation of flow lines.
d) To bind all flow lines together.

Q48. Which configuration tool ensures that a single default configuration is used across all users?

a) Copy Current Display
b) Configuration Manager
c) Configuration Options
d) Display Configurations

Q49. How can you save the show/hide settings of assembly components?

a) By clicking the Unexplode button.
b) By using the Take Snapshot tool.
c) By using the Unload Hidden Parts tool.
d) By selecting the Copy Current Display tool.

Q50. What does the Bind tool do in an exploded assembly?

a) Explodes subassemblies as individual components.
b) Keeps all components of a subassembly together.
c) Returns components to their native position.
d) Draws annotation flow lines for subassemblies.

Q51. What does the Show Camera Path button do?

a) Enables or disables camera movement in the animation.
b) Displays the path of camera movement during animation.
c) Edits the camera path definition.
d) Rotates the model for camera movement.

Q52. What is the purpose of the KeyShot Render tool?

a) To adjust animation resolution.
b) To render assemblies or parts with realistic materials and environments.
c) To automatically create animations from keyframes.
d) To configure flow lines in an exploded view.

Q53. What does the Automatic Spread Distance button do in the Auto Explode tool?

a) Removes gaps between parts in an exploded assembly.
b) Manually adjusts distances between components.
c) Automatically sets gaps between components.
d) Binds subassemblies in an exploded state.

Q54. What happens when the Remove tool is used on an exploded component?

a) The component moves to a different location in the assembly.
b) The component returns to its assembled position and becomes hidden.
c) The exploded component is bound with other components.
d) The exploded component is permanently deleted.

Q55. What does the Drop tool do in the Flow Lines group?

a) Deletes existing flow lines.
b) Splits flow lines into multiple segments.
c) Creates individual entries for flow lines in the Explode PathFinder.
d) Automatically adjusts the orientation of flow lines.

Q56. How can you modify the shape of an event flow line?

a) Use the Draw tool to redraw it.
b) Select the Modify tool and adjust its orientation or position.
c) Use the Collapse tool to reset its shape.
d) Adjust it using the Configuration Manager.

Q57. Which tool copies display settings to another configuration?

a) Configuration Manager
b) Take Snapshot
c) Copy Current Display
d) Display Configurations

Q58. What is the purpose of the Unload Hidden Parts tool?

a) Saves hidden parts for future use.
b) Returns hidden parts to the assembly.
c) Removes hidden parts from temporary memory.
d) Hides selected parts from the assembly.

Q59. What does the Limited Update toggle button do?

a) Reduces frequency of automatic updates to the assembly model.
b) Improves visualization quality of large assemblies.
c) Saves only modified parts in the assembly.
d) Automatically updates all components of the assembly.

Q60. Which toggle button is used to save only directly modified parts in an assembly?

a) Limited Save
b) Large Assembly Performance
c) Limited Update
d) Configuration Options

FOR STUDENT NOTES

Chapter 10

Assembly - III

Topics Covered

The major topics covered in this chapter are:

- *Introduction to XpressRoute*
- *Introduction to Electrical Routing*
- *Introduction to Frame Design*
- *Publishing Model to HTML Web Page*

INTRODUCTION TO XPRESROUTE

The tools in the Xpresroute environment are used to create tube and pipe routes in the assembly. To activate this environment, click on the **Xpresroute** tool from the **Environs** group in the **Tools** tab of **Ribbon**. The Xpresroute environment will be displayed; refer to Figure-1.

Figure-1. XpresRoute environment

CREATING PATH FOR ROUTE

The **Routing Path** tool is used to automatically create path for routing pipes and tubes. The procedure to use this tool is given next.

* Click on the **Routing Path** tool from the **3D Draw** group in the **Home** tab of **Ribbon**. The **Routing Path** command box will be displayed; refer to Figure-2 and you will be asked to specify start point of route path.

Figure-2. Routing Path command box

* Select desired color/style of path from **Style** drop-down in the command box. The colors displayed in drop-down are based on material.
* Select desired points (centers of circles) to specify start point and end point of route. Preview of route path will be displayed; refer to Figure-3.

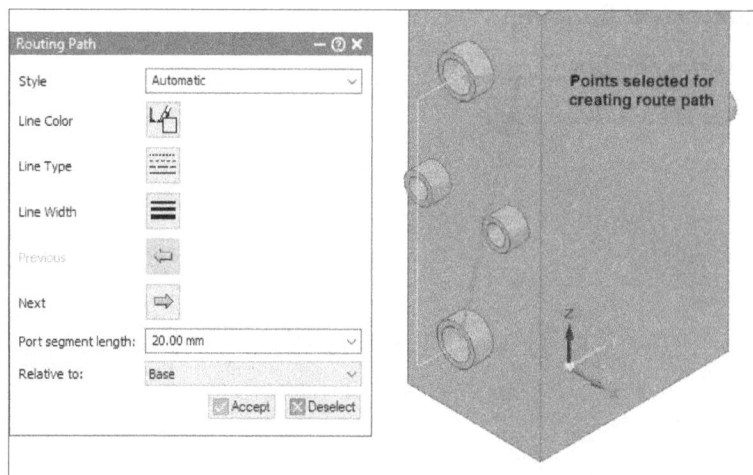

Figure-3. Preview of route path

- Specify desired value in the **Port segment length** edit box to specify length of route at the start and end points of route path. Press **TAB** to check the preview of route path.
- Click on the **Next** and **Previous** buttons to switch between different shapes of path curves.
- Click on the **Accept** button from the command box or right-click in the drawing area to create the route path; refer to Figure-4.

Figure-4. Route path created

- Press **ESC** to exit the tool.

CREATING 3D LINE OF ROUTE PATH

The **3D Line** tool is used to create a 3D Line of route path. The procedure to use this tool is given next.

- Click on the **3D Line** tool from the **3D Draw** group in the **Home** tab of **Ribbon**. You will be asked to select an existing 3D sketch or start a new 3D sketch. To start a new 3D sketch, right-click in the drawing area. The **3D Line** command box will be displayed with tips; refer to Figure-5.

Figure-5. 3D Line command box

- Click on desired location to specify start point of 3D Line. The other end of 3D Line will get attached to cursor; refer to Figure-6.

Figure-6. Specifying end point of line segment

• Click at desired location or specify length and angle in edit boxes of the command box to specify end point.

CREATING 3D ARCS

You can create 3D arcs by using two tools; 3D Arc by 3 Points and Tangent 3D Arc. These tools are discussed next.

Creating 3D Arc by 3 Points

The **3D Arc by 3 Points** tool is used to create segment in the form of an arc. The procedure to use this tool is given next.

• Click on the **3D Arc by 3 Points** tool from the **3D Draw** group in the **Home** tab of **Ribbon**. The **3D Arc by 3 Points** command box will be displayed; refer to Figure-7.

Figure-7. 3D Arc by 3 Points command box

• Click to specify the start point and end point of arc segment. You will be asked to specify radius of arc segment.
• Specify desired value of radius in the **Radius** edit box of command box and press **ENTER**. You will be asked to specify desired side to create the arc.
• Click on desired side. The arc segment will be created.
• Press **ESC** to exit the tool.

Creating Tangent 3D Arc

The **Tangent 3D Arc** tool is used to create tangent arc to a 3D path curve. The procedure to use this tool is given next.

• Click on the **Tangent 3D Arc** tool from the **3D Draw** group in the **Home** tab of **Ribbon**. You will be asked to select a 3D sketch or start a new 3D sketch.

- Select desired 3D sketch and create the tangent arc by specifying start and end points; refer to Figure-8.

Figure-8. Creating 3D tangent arc

Similarly, you can use other 3D sketching tools.

Note: You can move any other sketch entities create in route 3D sketch by selecting the entity and then dragging its key points.

CREATING ROUTE USING PATHS

The **Route** tool in **3D Draw** group is used to create route using a path segment and then continuing to next selected openings. The procedure to use this tool is given next.

- Click on the **Route** tool from the **3D Draw** group in the **Home** tab of **Ribbon**. You will be asked to select 3D curve path to be used for creating the route.
- Select desired path segment to be continued. You will be asked to select next circular opening.
- Select desired openings one by one to continue route path. After specifying desired path, right-click in the drawing area to create the path.

CREATING TUBE ALONG PATH

The **Tube** tool is used to create tube of desired material along selected route path. The procedure to use this tool is given next.

- Click on the **Tube** tool from the **Tubing** group in the **Home** tab of **Ribbon**. The **Tube Options** dialog box will be displayed; refer to Figure-9.

Figure-9. Tube Options dialog box

- Specify desired template, file location, and other properties for tube in the **File properties** area of the dialog box.
- Select desired material for tube from the **Material** drop-down in dialog box.
- Clear the **Use default value** check boxes and specify desired values for bend radius, outer diameter, minimum flat, wall thickness etc. in the **Tube properties** area of the dialog box.
- After setting desired parameters, click on the **OK** button. The **Tube** command box will be displayed; refer to Figure-10.

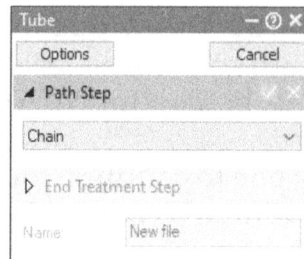

Figure-10. Tube command box

- If you want to change the tube options again then click on the **Options** button from the command box.
- Select the path curves for creating pipe and specify desired name for pipe parts in the **Name** edit box of the command box. The **Preview** button will be displayed in the command box.
- Click on the **Preview** button to check preview of pipe.
- Click on the **Finish** button from the command box to create the tube. The **Tube Options** dialog box will be displayed again for next pipe segment. Specify the parameters for next segment and create the pipes or press **ESC** to exit the tool.

CREATING TRIM TUBE

The **Trim Tube** tool is used to trim the side or corner of a tube, or to miter the ends of connected tubes, according to the sketch used to create the tube connections. The procedure to use this tool is discussed next.

*   Click on the **Trim Tube** tool from the **Tubing** group in the **Home** tab of **Ribbon**. The **Assembly Feature Options** dialog box will be displayed; refer to Figure-11.

Figure-11. Assembly feature option dialog box

*   Select **Create Assembly features** radio button to indicate that the entire feature will be created and stored within the assembly document.
*   Select **Create Assembly-Driven Part feature** radio button to specify that only the reference plane and profile will be stored in the assembly document.
*   After specifying desired parameters, click on the **OK** button from the dialog box. The **Trim Tube** command box will be displayed; refer to Figure-12.

Figure-12. Trim tube command box

*   Select desired primary tube in the graphics window. You will be asked to select the secondary tube.
*   Select desired secondary tube from the graphics window then click on the **OK** button from the command box. The ends of the connected tube will be mitered; refer to Figure-13.

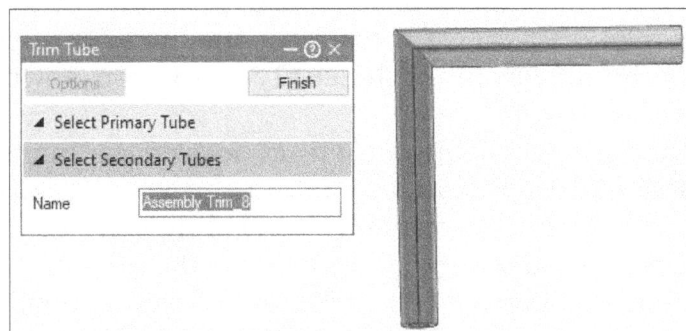

Figure-13. Trim tube created

*   Click on the **Finish** button to exit the command box.

CREATING PIPING ROUTE

The **Piping Route** tool is used to create pipes and connectors for piping system while following the route path. Note that to use this tool, you need to install Standard Parts and Standard Parts Administrator application separately using installation media provided by Siemens. The procedure to use this tool is given next.

- After creating the route path, click on the **Piping Route** tool from the **Tubing** group in the **Home** tab of **Ribbon**. The **Piping Options** dialog box will be displayed; refer to Figure-14.

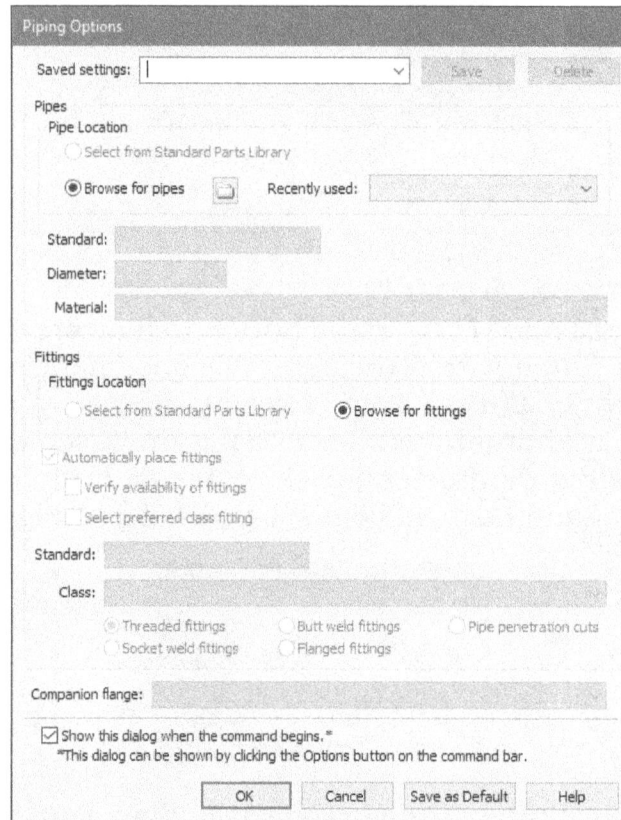

Figure-14. Piping Options dialog box

- Select desired option from the **Pipe Location** area of the dialog box to define source part files for creating pipe. By default, the **Select from Standard Parts Library** radio button is selected so you can use the standard parts for piping using the library of Solid Edge. Select the **Browse for pipes** radio button and click on the Browse button next to it if you want to select the part file of pipe manually. You can also select desired pipe from the **Recently used** drop-down if you have earlier created pipes in the system.
- If you are using standard parts library then set desired parameters in the **Standard**, **Diameter**, and **Material** drop-downs to define shape and size of pipes.
- Similarly, you can set desired parameters in **Fittings** area to define pipe fittings.
- After setting desired parameters, click on the **OK** button. The **Piping Route** command box will be displayed; refer to Figure-15 and you will be asked to select path segment.

Figure-15. Piping Route command box

- Select desired 3D path and right-click in the drawing area. The nodes which need pipe fittings will be highlighted on path; refer to Figure-16.

Figure-16. Nodes which need pipe fittings

- Select desired node on which you want to apply pipe fitting. The **Standard Parts** dialog box will be displayed; refer to Figure-17.

Figure-17. Standard Parts dialog box

- Select desired pipe fitting from desired category and click on the **Select** button. Preview of the fitting will be displayed.
- After applying all desired fittings, click on the **Preview** button to check preview of piping.
- Click on the **Finish** button to create the piping route and press **ESC** to exit the tool.

To exit the XpresRoute environment, click on the **Close XpresRoute** tool from the **Close** group in the **Home** tab of **Ribbon**.

INTRODUCTION TO ELECTRICAL ROUTING

The tools in the Electrical Routing environment are used to create wiring harness and cable for electrical system of component. To activate the **Electrical Routing** environment, click on the **Electrical Routing** tool from the **Environs** group in the **Tools** tab of **Ribbon**. The **Electrical Routing** environment will be displayed; refer to Figure-18.

Figure-18. Electrical Routing environment

CREATING PATHS

The tools in the **Paths** group are used to create paths for electrical routing. These tools work in the same way as discussed for XpresRoute. Create desired 3D path for electrical route.

CREATING WIRES

The **Wire** tool is used to generate wire using the selected 3D path. The procedure to use this tool is given next.

- Click on the **Wire** tool from the **Electrical Routing** group in the **Home** tab of the **Ribbon**. The **Wire** command box will be displayed; refer to Figure-19.

Figure-19. Wire command box

- Create the path for electrical wire as discussed. If you have earlier created path for wire then click on the **Use Existing Path** button from the command box. You will be asked to select 3D path.
- Select desired path segment and right-click in the drawing area. The command box will be displayed as shown in Figure-20.

Figure-20. Command box with Wire Type options

- Select desired wire type from the **Material** drop-down in the command box. After selecting the wire type, click on the **Properties** button from command box if you want to modify wire properties. The **Properties - Wire** dialog box will be displayed; refer to Figure-21.

Figure-21. Properties-Wire dialog box

- Specify desired parameters in the dialog box to modify properties of wire and click on the **OK** button. The **Wire** command box will be displayed again.
- Click on the **Preview** button from the command box to check preview of wire.
- Specify desired name of wire in the **Name** edit box of command box and click on **Finish** button to create the wire. Press **ESC** to exit the tool.

CREATING CABLE

The **Cable** tool is used to combine two or more wires for making a cable. The procedure to use this tool is given next.

- Click on the **Cable** tool from the **Electrical Routing** group in the **Home** tab of **Ribbon**. The **Cable** command box will be displayed; refer to Figure-22.

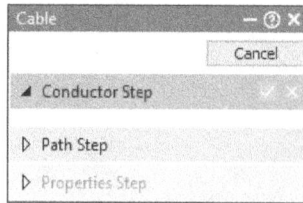

Figure-22. Cable command box

- Click at desired point of wire to define start point of cable and right-click. You will be asked to select a path for cable.
- Select desired path curve and right-click in the drawing area. The option to select cable type will be displayed in the command box.
- Select desired cable type and click on the **Finish** button to create the cable.

Similarly, you can use the **Bundle** and **Splice** tools. Click on the **Close Electrical Routing** tool from the **Close** group in the **Home** tab of **Ribbon**.

INTRODUCTION TO FRAME DESIGN

The Frame Design environment is used to create steel frame members along selected paths and edges. To activate the Frame Design environment, click on the **Frame** tool from the **Environs** group in the **Tools** tab of **Ribbon**. The tools to create frame design will be displayed; refer to Figure-23. Most of the tools in this environment are same as discussed for XpresRoute and Electrical Routing. Here, we will discuss the Frame tool used for creating frame members.

Figure-23. Tools for frame design

CREATING FRAME MEMBERS

The **Frame** tool is used to insert frame members along selected paths and edges. The procedure to use this tool is given next.

- Click on the **Frame** tool from the **Frame** group in the **Home** tab of **Ribbon**. The **Frame Options** dialog box will be displayed; refer to Figure-24.
- Select desired radio buttons from the **Preferred Frame Orientation** area of the dialog box to define orientation of frame members.
- Set desired options in the **Corner Treatment Options** area of the dialog box to define how corners will be created at intersections of two frame members.
- Click on the **Options** button from the dialog box to check advanced options of the frame; refer to Figure-25.

Figure-24. Frame Options dialog box

Figure-25. Advanced options

- Set desired check boxes in the dialog box to define the placement of frame members and click on the **OK** button. The **Frame** command box will be displayed; refer to Figure-26.

Figure-26. Frame command box

- Click on the **Select Cross Section Component** button from the command box to select desired frame member. The **Frame** dialog box will be displayed; refer to Figure-27.

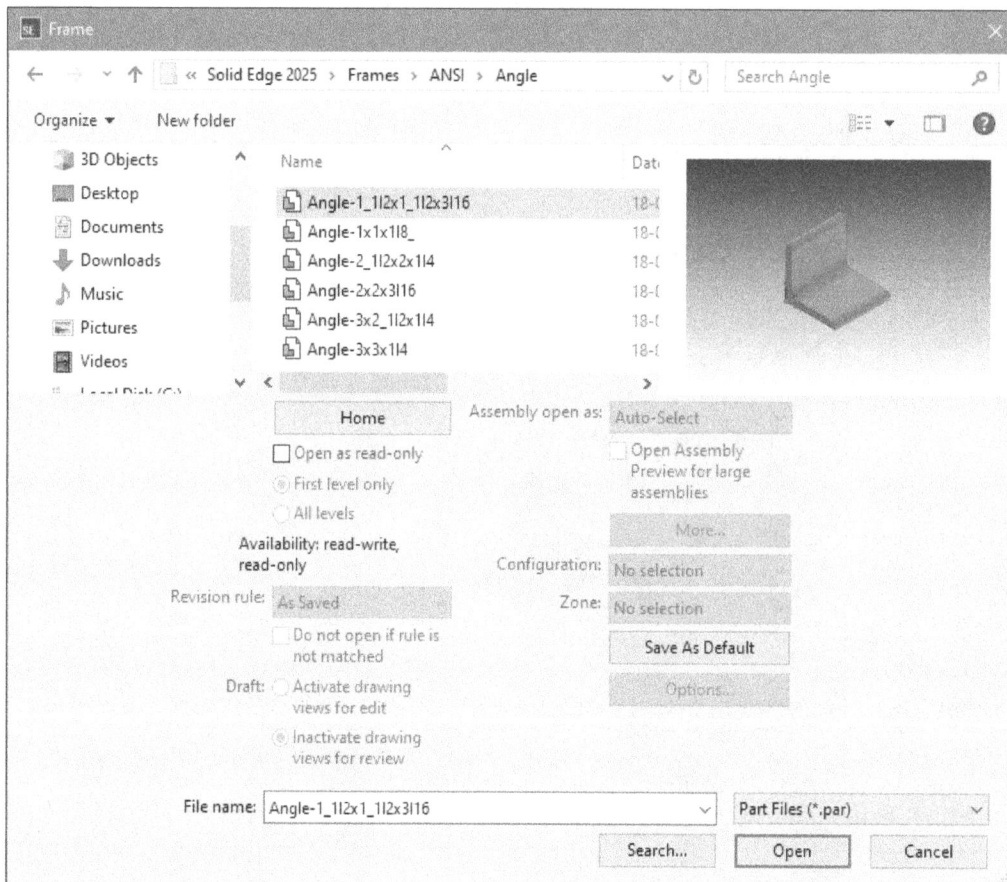

Figure-27. Frame dialog box

- Select desired frame member part file from the default library folder (C:\Program Files\Siemens\Solid Edge 2026\Frames in our case) and click on the **Open** button. You will be asked to select the edges or path curves from the model.
- Select desired entities and right-click in the drawing area. Preview of the frame will be displayed; refer to Figure-28.

Figure-28. Creating frame

- Click on the **Finish** button from the command box to create the frame. Press **ESC** to exit the tool.

APPLYING FRAME END CAP

The **Frame End Cap** tool is used to close end of frame with specified thickness and cap type. The procedure to use this tool is given next.

- Click on the **Frame End Cap** tool from the **Frame** group in the **Home** tab of **Ribbon**. The **End Cap Options** dialog box will be displayed; refer to Figure-29.

Figure-29. End Cap Options dialog box

- Select desired material type from the **Material** drop-down and define thickness of end cap plate in the **Thickness** edit box.
- Select the **Inward** or **Outward** radio button to create end cap inside or outside of frame member, respectively. Select the **Offset** check box if you want to decrease the size of end cap by specified value near end point of frame member.
- Select the **Inset** check box to specify distance from end point of frame member at which cap will be placed. This check box is available for Inward cap type only.
- Select the **Apply Corner Treatment** check box to chamfer or fillet at the edges of end cap.
- If you want to create end cap based on frame member's cross-section then select the **Create end caps based on frame cross-section** check box.

- Select the end face(s) of frame member at which you want to apply end cap(s) and click on the **Apply** button. The end caps will be created; refer to Figure-30.

Figure-30. End caps created

- Click on the **OK** button to exit the dialog box.

CREATING GUSSET PLATE

The **Gusset Plate** tool is used to add structural rib to faces of frame members for increasing strength of frame. The procedure to use this tool is given next.

- Click on the **Gusset Plate** tool from the **Frame** group in the **Home** tab of **Ribbon**. The **Gusset Plate Options** dialog box will be displayed; refer to Figure-31.

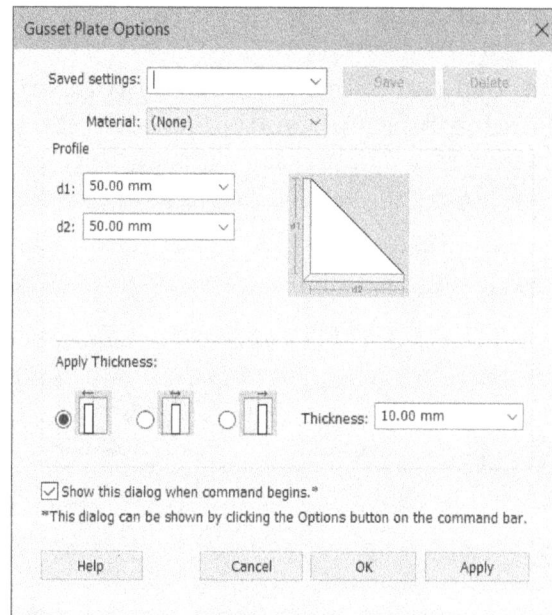

Figure-31. Gusset Plate Options dialog box

- Specify desired length of gusset in **d1** and **d2** edit boxes.
- Set the other parameters as discussed earlier and click on the **OK** button. The **Gusset Plate** command box will be displayed; refer to Figure-32.

Figure-32. Gusset Plate command box

- Click on two intersecting faces and right-click in the drawing area, preview of gusset will be displayed; refer to Figure-33.

Figure-33. Preview of gusset plate

- Select desired option from the **Alignment** drop-down to define location of gusset along the edge.
- Specify desired value in **Offset** edit box and right-click in the drawing area. The gusset plate will be created. Press **ESC** to exit the tool.

TRIMMING/EXTENDING FRAME MEMBER

The **Trim/Extend** tool is used to remove intersecting portion of frame members or extend a frame member up to specified point. The procedure to use this tool is given next.

- Click on the **Trim/Extend** tool from the **Frame** group in the **Home** tab of the **Ribbon**. The **Trim/Extend** command box will be displayed; refer to Figure-34.

Figure-34. Trim Extend command box

- Select the face of frame member to be trimmed/extended and right-click in the drawing area. You will be asked to select plane/face/body to which you want to trim or extend the frame member.
- Select desired face/plane/body and right-click to perform trimming; refer to Figure-35.

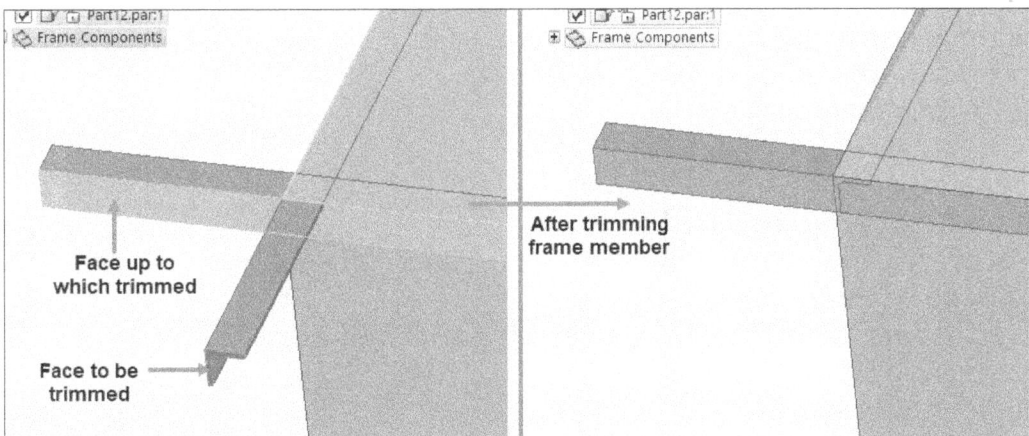

Figure-35. Trimming frame member

- Click on the **Finish** button from command box to complete operation and press **ESC** to exit the tool.

Click on the **Close Frame** tool from the **Close** group in the **Home** tab of **Ribbon** to exit the environment.

SELF ASSESSMENT

Q1. Which tool is used to create paths for routing pipes and tubes?

a) 3D Line
b) Tube
c) Routing Path
d) Frame

Q2. What should you press to preview the route path?

a) Enter
b) TAB
c) ESC
d) Accept

Q3. What is the purpose of the Tube tool?

a) To create a path for electrical wires
b) To create a tube along a selected route path
c) To trim frame members
d) To add gusset plates

Q4. Which parameter can be adjusted in the Tube Options dialog box?

a) Wall thickness
b) Corner treatment
c) Wire type
d) Web page title

Q5. What is the purpose of the Wire tool?

a) To create 3D arcs
b) To generate wires using a 3D path
c) To trim frame members
d) To combine wires into a cable

Q6. What is the Frame tool used for?

a) To combine wires into a cable
b) To insert frame members along selected paths and edges
c) To create tangent arcs
d) To create gusset plates

Q7. Which tool is used to add structural ribs for strength?

a) Trim/Extend
b) Frame End Cap
c) Gusset Plate
d) Publish

Q8. What does the Accept button do when creating a route path?

a) Exits the Routing Path tool
b) Finalizes the route path creation
c) Switches between path curves
d) Displays the Routing Path command box

Q9. What is required to use the Piping Route tool?

a) The Tube tool must be activated
b) Standard Parts and Standard Parts Administrator must be installed
c) A wire must be created
d) A frame must be designed

Q10. Which area in the Piping Options dialog box is used to define the source part files?

a) Standard
b) Pipe Location
c) Fittings
d) Material

Q11. Which option is available for the Inward cap type when applying a Frame End Cap?

a) Material
b) Offset
c) Corner treatment
d) Thickness

Q12. What does the Gusset Plate tool do?

a) Adds a structural rib for frame strength
b) Trims the intersecting portion of frame members
c) Inserts frame members along paths
d) Creates end caps on frame members

Q13. What is the purpose of the Publish tool in the Environs group?

a) To save a frame design
b) To generate an HTML file of the active model
c) To combine wires into a cable
d) To create frame members along selected paths

Q14. Which group in the Tools tab is used to activate the XpresRoute environment?

a) 3D Draw
b) Close
c) Environs
d) Frame

Q15. What is the purpose of the Port segment length edit box?

a) To define the radius of a 3D arc
b) To specify the length of the route path at start and end points
c) To define the material of the path
d) To switch between path shapes

Q16. What is highlighted after selecting a 3D path in the Piping Route tool?

a) The Material drop-down menu
b) Nodes that require pipe fittings
c) The Pipe Location area
d) The Preview button

Q17. What does the Standard Parts dialog box allow you to do?

a) Specify the end point of a path
b) Select desired pipe fittings
c) Define the length of a gusset plate
d) Trim frame members

Q18. Which tool is used to create multiple wires combined into a single cable?

a) Wire tool
b) Cable tool
c) Splice tool
d) Electrical Routing tool

Q19. What is the purpose of the Corner Treatment Options in the Frame Options dialog box?

a) To specify the material of frame members
b) To define how corners are created at intersections
c) To trim intersecting portions of frame members
d) To apply end caps to frame members

Q20. What does the Offset check box do in the Frame End Cap tool?

a) Decreases the size of the end cap near the frame end
b) Creates an outward cap type
c) Changes the frame orientation
d) Defines placement of the frame member

Q21. What does the Alignment drop-down in the Gusset Plate tool define?

a) The material type of the plate
b) The size of the frame member
c) The location of the gusset along the edge
d) The preview type of the gusset

Q22. What is the purpose of the Trim/Extend tool?

a) To create arcs tangent to a path
b) To remove or extend portions of frame members
c) To create wiring paths for electrical routing
d) To apply end caps to frames

FOR STUDENT NOTES

FOR STUDENT NOTES

FOR STUDENT NOTES

Chapter 11

Drawing

Topics Covered

The major topics covered in this chapter are:

- *Introduction*
- *Starting Drawing Environment*
- *Placing Views*
- *Creating Section View using Cutting Plane*
- *Creating Parts List*
- *Creating Hole Table and Bend Table*
- *Dimensioning Tools*

INTRODUCTION

Drawing is the engineering representation of a model on the paper. For manufacturing a model in real world, we need some means by which we can tell the manufacturer what to manufacture. For this purpose, we create drawings from the models. These drawings have information like dimensions, material, tolerances, objective, precautions and so on. In Solid Edge, we create drawings by using the Drawing environment. There are two methods to start drawing environment; creating drawing from assembly/part environment and create a new drawing. The procedure to start drawing environment is given next.

STARTING DRAWING ENVIRONMENT

There are two methods to start drawing environment in Solid Edge. These methods are discussed next.

Starting a New Drawing

The procedure to use **New** tool for starting a new drawing is given next.

- Click on the **New** tool from the **New** cascading menu of the **Application** menu. The **New** dialog box will be displayed as discussed earlier.
- Select desired **.dft** template from the dialog box and click on the **OK** button from the dialog box. The drawing environment will be displayed without a model selected for placing views.

Starting a Drawing using Model or Assembly

The procedure to start drawing environment using a model/assembly is given next.

- Open the part file or assembly file in Solid Edge for which you want to generate the views.
- Click on the **Drawing of Active Model** option from the **New** drop-down in the **Quick Access Toolbar**; refer to Figure-1. The **Create Drawing** dialog box will be displayed; refer to Figure-2.

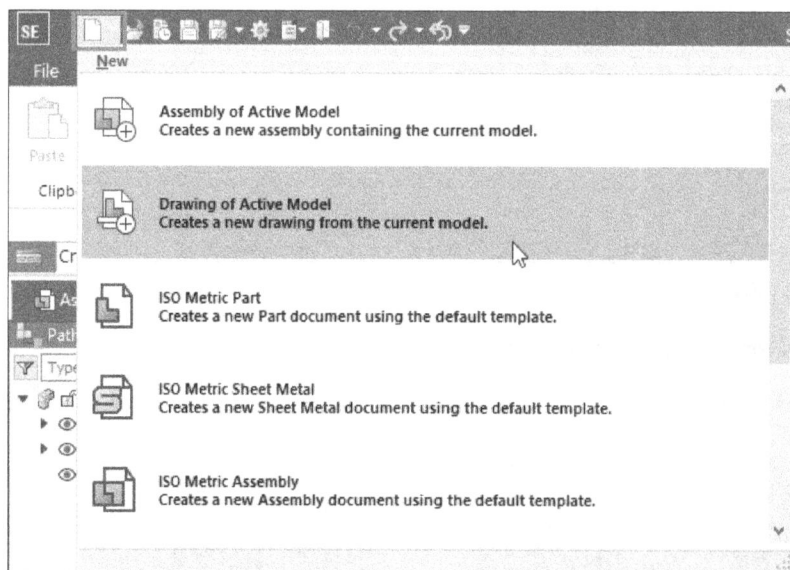

Figure-1. Drawing of Active Model tool

Figure-2. Create Drawing dialog box

- Click on the **Browse** button from the dialog box and select desired template for creating drawing.
- Select the **Run Drawing View Creation Wizard** check box if you want to create primary views using Drawing View Creation Wizard.
- After setting parameters, click on the **OK** button. If you have not selected the **Run Drawing View Creation Wizard** check box then three standard views will be placed automatically in the drawing. If you have selected the **Run Drawing View Creation Wizard** check box then **View Wizard** command box will be displayed and you will be asked to place the primary view. Click at desired locations to place the views and press **ESC** to exit the tool. You will learn more about the **View Wizard** command box in next topic.

PLACING VIEWS WITH VIEW WIZARD

The **View Wizard** tool is used to select the model for creating and placing the drawing views. The procedure to use this tool is given next.

- Click on the **View Wizard** tool from the **Drawing Views** group in the **Home** tab of **Ribbon**. The **Select Attachment** dialog box will be displayed; refer to Figure-3.

Figure-3. Select Attachment dialog box

- Click on the **Browse** button from the dialog box to select desired model. The **Select Model** dialog box will be displayed.
- Select desired model and click on the **Open** button. The front view of model will get attached to cursor and the **View Wizard** command box will be displayed; refer to Figure-4.

Figure-4. View Wizard command box

- Select desired option from the **Style** drop-down in the command box to define view placement style and unit for view.
- Click in the **Caption** edit box and specify desired text or symbol to be displayed below the view.
- Click on the **Show Caption** button from the command box. A list of options will be displayed; refer to Figure-5. Select desired options in the list to include them with view.

Figure-5. Show Captions list

- Click on the **Saved Settings** button from the command box to save current settings of the command box for using them later. The **Drawing View Wizard** dialog box will be displayed; refer to Figure-6. Specify desired name for the setting in the **Saved Settings** edit box and click on the **Save** button. After that click on the **OK** button from the dialog box.

Figure-6. Drawing View Wizard dialog box

- Click on the **Drawing View Wizard Options** button from the command box. The **Drawing View Wizard** dialog box will be displayed; refer to Figure-7.

Figure-7. Drawing View Wizard dialog box

- Select the **Designed part** radio button from the **Part and Sheet Metal Drawing View** section to use designed model as it looks while creating. Select the **Simplified part** radio button to use simplified version of model with lesser details for creating view. Select the **Flat pattern** radio button from the section to use flat pattern of sheet metal part while creating the view.
- If you have applied PMI annotations to the model then select desired PMI view created in the Part/Assembly environment from the **Congifuration or PMI model view** drop-down. The check boxes below the drop-down will become active. Select the **Include PMI dimensions from model views** check box to automatically import PMI dimension in the view. Similarly, select the **Include PMI dimensions from model views** check box to include the annotations specified in PMI.
- Select the **Show tube centerlines** check box to display center lines in the view.
- Similarly, you can set the other options in the dialog box for displaying various entities in the view.
- After setting desired parameters, click on the **OK** button from the dialog box.
- Click on the **Drawing View Layout** button from the command box to set desired standard view of model. The **Drawing View Wizard** dialog box will be displayed as shown in Figure-8.

Figure-8. Drawing View Wizard dialog box

- Select desired primary view from the **Primary View** list box and select desired buttons to create multiple views. After setting parameters, click on the **OK** button. The preview of selected views will be displayed attached to cursor; refer to Figure-9.

Figure-9. Multiple views attached to cursor

- Click on the **View Orientation** button from the command box to select the primary view to be used for projection; refer to Figure-10 and select desired view option.

Figure-10. View Orientation options

- Click on the **Best Fit** button from the command box to automatically fit all the views in the sheet.
- Click on the **Set View Scale** button from the command box to use specified scale value for size of views.
- Click on the **Set Sheet Scale** button from the command box to match scale of views with scale of sheet.
- Click on the **Model Display Settings** button to specify display settings of the model in the views. The **Drawing View Properties** dialog box will be displayed; refer to Figure-11. Set desired parameters and click on the **OK** button.

Figure-11. Drawing View Properties dialog box

- Click on the **Shading Options** button to define how 3D models will be displayed in the view; refer to Figure-12. Select desired option from the list to specify shading style.

Figure-12. Shading Options

- Select the **Use Model Colors** button from the command box to use colors specified for model in the view.
- After setting desired parameters, click at desired location in the sheet. The views will be created; refer to Figure-13.

Figure-13. Views created

UPDATING VIEWS

The **Update Views** tool in the **Drawing Views** group is used to update views in the drawing based on changes applied in the model.

CREATING PRINCIPAL VIEWS

The **Principal** tool in the **Drawing Views** group is used to create principal views based on selected view in the drawing. The procedure to use this tool is given next.

- Click on the **Principal** tool from the **Drawing Views** group in the **Home** tab of the **Ribbon**. The **Principal** command box will be displayed; refer to Figure-14 and you will be asked to select the view to be used for creating projected principal views.

Figure-14. Principal command box

- Select desired base view from the drawing. The projected principal view will get attached to cursor.
- Click at desired locations. The views will be created based on their angle from base view.
- Press **ESC** to exit the tool.

CREATING AUXILIARY VIEW

The auxiliary view is a projected view created by making selected edge of view parallel to screen. The procedure to create auxiliary view is given next.

- Click on the **Auxiliary** tool from the **Drawing Views** group in the **Home** tab of **Ribbon**. The **Auxiliary** command box will be displayed; refer to Figure-15.

Figure-15. Auxiliary command box

- Select the **Parallel** or **Perpendicular** button from the command box to create a view parallel or perpendicular to selected edge of model.
- Select two key points of model or select desired edge of model. The view will get attached to the cursor; refer to Figure-16.

Figure-16. Creating auxiliary view

- Click at desired location to place the view. The auxiliary view will be created.

CREATING DETAIL VIEW

The **Detail** tool is used to create a scaled up view of local region of the model. The procedure to use this tool is given next.

- Click on the **Detail** tool from the **Drawing Views** group in the **Home** tab of the **Ribbon**. The **Detail** command box will be displayed; refer to Figure-17.

Figure-17. Detail command box

- By default, the **Circular Detail View** button is used to create a circular boundary around the feature to create detailed circular view. If you want to use sketched curves for defining boundary of detail view then click on the **Define Profile** button from the command box and click on the view to be used for creating detail view. The sketching environment will be displayed. Create desired closed loop sketch and click on the **Close Detail Envelope** button from the **Close** group of the **Ribbon**. The detail view will get attached to cursor.
- Click at desired location to place the view.

CREATING BROKEN VIEW

The **Broken** tool is used to represent very long objects in the drawing by breaking them at specific span. The procedure to use this tool is given next.

- Select desired view to be used for broken view and click on the **Broken** tool from the **Drawing Views** group in the **Home** tab of the **Ribbon**. You will be asked to specify location of first break line.
- Select desired option from the **Style** drop-down to define style of break lines.
- Click on the **Vertical Break** or **Horizontal Break** button from the command box to define direction of break lines.
- Click on the **Break Line Type** button from the command box and select desired option from the list displayed to define type of break lines.
- Specify desired value of distance between two break lines in the **Break gap** edit box of the command box.
- Similarly, set the other parameters in the command box.
- Click at desired location to specify first break line and then click to specify location of end break line; refer to Figure-18.

Figure-18. Break view lines

- Click on the **Finish** button from the command box to create the view.
- Press **ESC** to exit the tool.

CREATING SECTION VIEW CUTTING PLANE

The **Cutting Plane** tool is used to create cutting plane for generating section view of the model. The procedure to use this tool is given next.

- Click on the **Cutting Plane** tool from the **Drawing Views** group in the **Home** tab of **Ribbon**. The **Cutting Plane** command box will be displayed; refer to Figure-19 and you will be asked to select the view for which you want to create the cutting plane.

Figure-19. Cutting Plane command box

- Select desired view from the drawing. You will be asked to draw shape of cutting plane.
- Draw the shape of cutting plane using line or arc. Note that the sketch for cutting plane should be a linear curve which is not self-intersecting; refer to Figure-20.

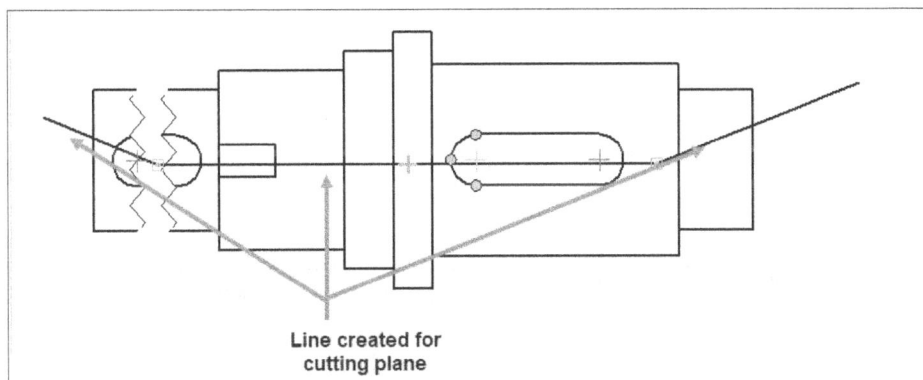

Figure-20. Creating lines for cutting plane

- After creating sketch, click on the **Close Cutting Plane** tool from the **Close** group in the **Ribbon**. The preview lines/curves for cutting plane will be displayed; refer to Figure-21.

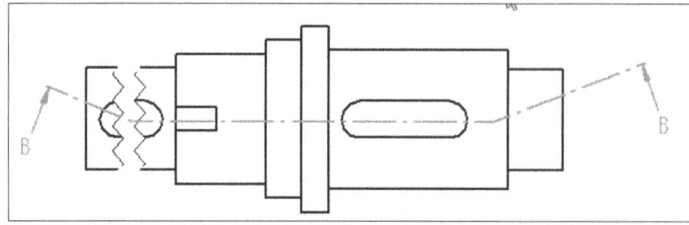

Figure-21. Cutting plane lines

- Click above or below the line to define direction of cutting plane. The plane will be created.

CREATING SECTION USING CUTTING PLANE LINE

The **Section** tool is used to create section plane using cutting plane line. The procedure to use this tool is given next.

- Click on the **Section** tool from the **Drawing Views** group in the **Home** tab of the **Ribbon**. The **Section** command box will be displayed.
- Select desired cutting plane line. The section view will get attached to cursor and the **Section** command box will be displayed; refer to Figure-22.

Figure-22. Section command box and section view

- Set desired values in the **Fill**, **Angle**, and **Spacing** edit boxes.
- Select the **Section Only** button from the command box to display section only and do not display hidden edges.
- Click at desired location to place the section view after setting desired parameters. The section view will be created. Press **ESC** to exit the tool.

CREATING BROKEN-OUT DRAWING VIEW

The **Broken-Out** tool is used to create broken out view for checking inside region of part. The procedure to use this tool is given next.

- Click on the **Broken-Out** tool from the **Drawing Views** group in the **Home** tab of **Ribbon**. You will be asked to select the drawing view to be used for creating broken-out view.
- Select desired drawing view. You will be asked to create sketch for broken out section.

- Create desired closed loop sketch for section and click on the **Close Broken Out Section** button from the **Close** group in the **Ribbon**. The **Broken-Out** command box will be displayed and you will be asked to specify depth of broken-out section view.
- Specify desired value in the **Depth** edit box and press **ENTER**. You will be asked to select the view which will be sectioned.
- Select desired drawing view. The broken out section view will be displayed; refer to Figure-23. Press **ESC** to exit the tool.

Figure-23. Broken-out section view

CREATING PARTS LIST

The **Parts List** tool is used to create a list of parts with balloons. The procedure to use this tool is given next.

- Click on the **Parts List** tool from the **Tables** group in the **Home** tab of the **Ribbon**. The **Parts List** command box will be displayed and you will be asked to select a drawing view.
- Select desired view from the drawing. The table of parts list will get attached to cursor.
- Click at desired location to place the table. The balloons will be applied to parts with parts list; refer to Figure-24.

Figure-24. Parts list with balloons

CREATING HOLE TABLE

The **Hole Table** tool is used to create a table of holes in the drawing. The procedure to use this tool is given next.

- Click on the **Hole Table** tool from the **Table** group in the **Home** tab of **Ribbon**. The **Hole Table** command box will be displayed and you will be asked to specify location of origin point.
- Click at desired point to specify x and y location of origin for table; refer to Figure-25. You will be asked to select the view which contains holes.

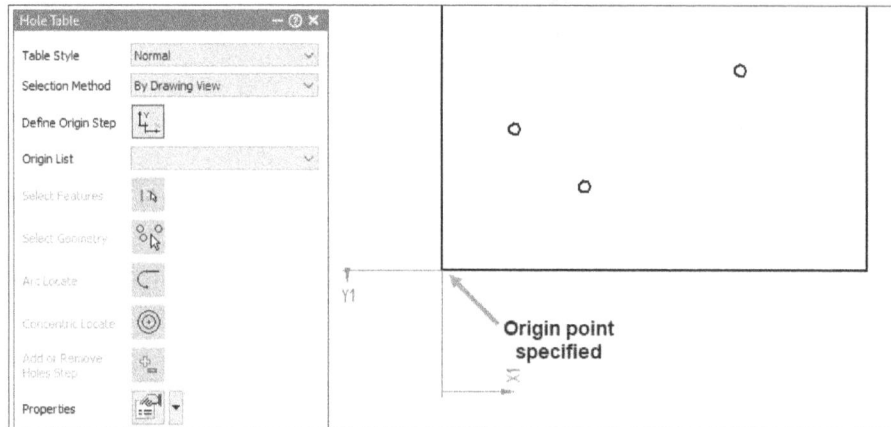

Figure-25. Origin point specified

- Select desired view with holes. Note that the holes should have been created by **Hole** tool in modeling environment. After selecting view, right-click in the drawing to accept the selection. The table will get attached to cursor.
- Click at desired location to place the table. The holes will be annotated with hole table; refer to Figure-26.

Hole	X	Y	Size
1.1	40.67 mm	120.39 mm	3.17 mm
1.2	39.63 mm	22.68 mm	3.17 mm
1.3	65.77 mm	82.3 mm	3.17 mm
1.4	99.61 mm	102.22 mm	3.17 mm
1.5	82.79 mm	54.07 mm	3.17 mm
1.6	73.22 mm	138.07 mm	3.17 mm
1.7	24.36 mm	115.65 mm	3.17 mm
1.8	20.09 mm	38.24 mm	3.17 mm

Figure-26. Holes with table

- Press **ESC** to exit the tool.

CREATING BEND TABLE

The **Bend Table** tool is used to create table of bends in the sheetmetal part. Note that there should be a sheetmetal part with bends in the drawing to use this tool. The procedure to use this tool is given next.

- Click on the **Bend Table** tool from the **Tables** group in the **Home** tab of the **Ribbon**. The **Bend Table** command box will be displayed and you will be asked to select the drawing view which contains flat pattern of sheet metal part.
- Select desired drawing view. The bend table will get attached to cursor.
- Select the **Auto-Callout** button from the command box to automatically add callouts to bends in the drawing view.

- Select the **Place Table** button from the command box to create bend table and place it in drawing sheet.
- Click at desired location in the drawing to place the table. The table will be created; refer to Figure-27.

Sequence	Feature	Radius	Angle	Direction	Included Angle
1	Bend 1	0.002 mm	180.00 deg	Up	0.00 deg
2	Bend 2	8.000 mm	25.18 deg	Down	154.82 deg
3	Bend 3	1.000 mm	90.00 deg	Up	90.00 deg

Figure-27. Bend table

- Press **ESC** to exit the tool.

Similarly, you can use the other tools in the **Tables** group to generate respective tables.

DIMENSIONING TOOLS

Most of the tools used for creating dimensions have been discussed earlier in chapters related to modeling and PMI. We will now discuss rest of the tools for dimensioning and annotation.

Creating Symmetric Diameter Dimension

The **Symmetric Diameter** tool is used to create a diameter dimension which is double of the distance between two selected points to create diameter dimension. The procedure to use this tool is given next.

- Click on the **Symmetric Diameter** tool from the **Dimension** group in the **Home** tab of **Ribbon**. The **Symmetric Diameter** command box will be displayed and you will be asked to select the points.
- One by one click on the two points to be dimensioned. The dimension will get attached to cursor; refer to Figure-28.

Figure-28. Diameter dimension

- Click at desired location to place the dimension and press **ESC** to exit the tool.

Applying Chamfer Dimension

The **Chamfer Dimension** tool is used to dimension chamfers in the drawing. The procedure to apply chamfer dimension is given next.

- Click on the **Chamfer Dimension** tool from the **Dimension** group in the **Home** tab of the **Ribbon**. The **Chamfer Dimension** command box will be displayed and you will be asked to select a chamfer for dimensioning.
- Select desired option from the **Orientation** drop-down in the command box to specify how dimension will be placed. Select the **Along Axis** option from the drop-down to create dimension aligned to selected chamfer edge. Select the **Callout Perpendicular** option from the drop-down to create chamfer dimension callout perpendicular to selected edge. Select the **Callout Parallel** to place the dimension parallel to selected edge.
- Select the chamfer line to be dimensioned and then select the base line to be used for aligning dimension. The dimension get attached to cursor; refer to Figure-29.

Figure-29. Chamfer dimension created

- Click at desired location to place the dimension.

Retrieving Dimension

The **Retrieve Dimensions** tool is used to copy PMI dimensions and annotations of the model in to a drawing view. The procedure to use this tool is given next.

- Click on the **Retrieve Dimensions** tool from the **Dimension** group in the **Home** tab of the **Ribbon**. The **Retrieve Dimensions** command box will be displayed; refer to Figure-30.

Figure-30. Retrieve Dimensions command box

- Select desired buttons from the command box to define which type of dimensions and annotations will be retrieved in the drawing views.
- Select desired drawing views to copy dimensions and press **ESC** to exit the tool.

Line Up Text

The **Line Up Text** tool is used to align dimension and annotation texts. The procedure to use this tool is given next.

- Click on the **Line Up Text** tool from the **Dimension** group in the **Home** tab of the **Ribbon**. The **Line Up Text** command box will be displayed; refer to Figure-31.

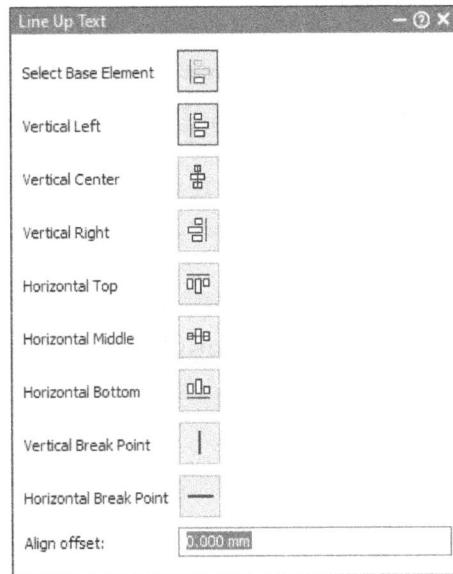

Figure-31. Line Up Text command bar

- Select desired button from the command box to define alignment style and then select the dimensions & annotations to align them.
- Press **ESC** to exit the tool.

Copying Attributes

The **Copy Attributes** tool is used to copy attributes from one annotation to other. The procedure to use this tool is given next.

- Click on the **Copy Attributes** tool from the **Dimension** group in the **Home** tab of the **Ribbon**. The **Copy Attributes** command box will be displayed; refer to Figure-32.

Figure-32. Copy Attributes command box

- Select desired option from the drop-down to define which attributes are to be copied.
- Select desired dimension/annotation you want to copy from the drawing and then select the other dimensions to which you want to copy the attributes.
- After setting parameters, press **ESC** to exit the tool.

Removing Dimensions from Alignment Set

Click on the **Remove from Alignment Set** tool from the **Dimension** group in the **Home** tab of **Ribbon** to remove selected dimensions from alignment set. After removing a dimension from alignment set, you can individually move the selected dimension.

Stacking Dimensions

The **Arrange Dimensions** tool in **Dimension** group of **Ribbon** is used to stack dimensions one over the other at fixed distance between them. To use this tool, select it from the

Dimension group in the **Home** tab of the **Ribbon**. The **Arrange Dimensions** command box will be displayed. Specify desired gap multiplier in the command box and then select the dimensions to be arranged.

Similarly, the **Update Retrieved Dimensions** tool in **Dimension** group of **Ribbon** is used to update the changes made in PMI of model in drawing.

The tools in the **Annotation** group have been discussed earlier in the book.

PRACTICAL 1

Create engineering drawing of model as shown in Figure-33. The procedure to create this model has been discussed earlier in Chapter 7 of this book.

Figure-33. Practical 1

Starting A Drawing File

• Start Solid Edge using Start menu or desktop icon, if not started yet.
• Click on the **ISO Metric Draft** tool from the **New** cascading menu of the **Application** menu. The drawing creation interface will be displayed with a new draft document opened.

Placing Views

• Click on the **View Wizard** tool from the **Drawing Views** group in the **Home** tab of **Ribbon**. The Select Model dialog box will be displayed.
• Select the model file of PMI Practical in Chapter 7 of resource kit. The Front view of model will get attached to cursor.
• Make sure the scale value in **Scale value** edit box of command box is specified as **4** and then click at the right side in the drawing area to place the view; refer to Figure-34.

Figure-34. Placing front view

- Move the cursor to left side of model. The right view of model will get attached to cursor.
- Click on the left side of front view to place the view; refer to Figure-35 and then press **ESC** to exit the tool.

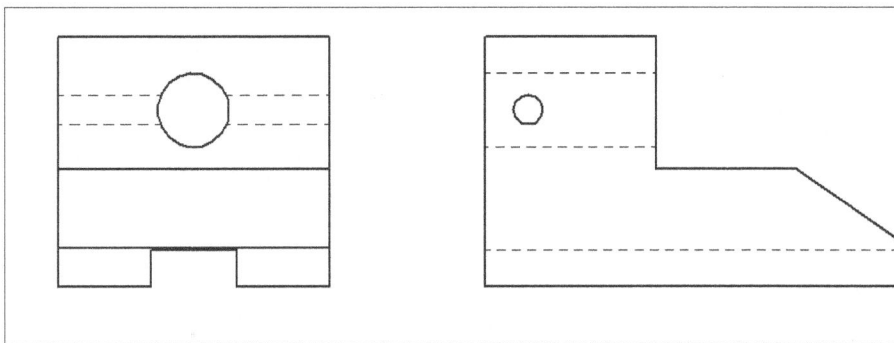

Figure-35. Views placed in drawing

Retrieving and Applying Dimensions

- Click on the **Retrieve Dimensions** tool from the **Dimension** group in the **Home** tab of **Ribbon**. The **Retrieve Dimensions** command box will be displayed.
- Select the two views one by one. All the dimensions from PMI data will be retrieved in the drawing views. Press **ESC** to exit the tool.
- Move the dimensions and annotations to desired locations. Delete the extra dimensions and generate a neat drawing as shown in Figure-36.

Figure-36. Drawing after applying dimensions

PRACTICE 1

Create the model and generate drawing as shown in Figure-37.

Figure-37. Production drawing for practice 1

PRACTICE 2

Create the model and drawing per the views shown in Figure-38.

Figure-38. Practice 2

SELF ASSESSMENT

Q1. Which of the following is NOT a method to start a drawing environment in Solid Edge?

(A) Creating a new drawing
(B) Drawing from assembly/part environment
(C) Direct printing
(D) Using the Drawing View Creation Wizard

Q2. What happens when the "Run Drawing View Creation Wizard" checkbox is selected?

(A) Views are automatically placed
(B) The wizard assists in placing primary views
(C) Only front views are displayed
(D) Models are converted into templates

Q3. Which tool is used to create and place drawing views in Solid Edge?

(A) Update Views
(B) View Wizard
(C) Model Display Settings
(D) Principal

Q4. What does the "Set View Scale" button do?

(A) Matches the scale of the view to the sheet
(B) Adjusts the size of the view
(C) Resizes the entire sheet
(D) Defines the shading style

Q5. Which view is created by making a selected edge parallel to the screen?

(A) Detail View
(B) Auxiliary View
(C) Broken View
(D) Section View

Q6. What is the purpose of a Broken View?

(A) To show intricate details
(B) To highlight specific regions
(C) To represent long objects by breaking them
(D) To create a sectional view

Q7. What must be done first to create a Section View?

(A) Create a cutting plane
(B) Select a shading style
(C) Set the view orientation
(D) Use the Broken-Out tool

Q8. Which tool is used to align dimension and annotation texts?

(A) Copy Attributes
(B) Line Up Text
(C) Symmetric Diameter
(D) Retrieve Dimensions

Q9. The Parts List tool is used to create:

(A) A table of dimensions
(B) A table of holes
(C) A list of parts with balloons
(D) An assembly outline

Q10. What is the Hole Table tool used for?

(A) To annotate bend lines
(B) To create a list of holes in the drawing
(C) To align text annotations
(D) To define chamfer dimensions

Q11. Which tool is used to create a scaled-up view of a specific region?

(A) Auxiliary Tool
(B) Broken-Out Tool
(C) Detail Tool
(D) Chamfer Tool

Q12. What is the function of the Retrieve Dimensions tool?

(A) To copy model annotations into a drawing
(B) To adjust alignment of text
(C) To scale views
(D) To define cutting planes

Q13. Which tool allows the creation of a table for bends in a sheet metal part?

(A) Parts List
(B) Hole Table
(C) Bend Table
(D) Model Display Settings

Q14. Which tool is used to create primary views in the drawing environment?

(A) Update Views
(B) View Wizard
(C) Cutting Plane
(D) Hole Table

Q15. Which view placement option ensures that views fit well on the drawing sheet?

(A) Set Sheet Scale
(B) Best Fit
(C) View Orientation
(D) Model Display Settings

Q16. What type of sketch is required for creating a cutting plane?

(A) Closed loop
(B) Open curve
(C) Linear and non-self-intersecting
(D) Circular boundary

Q17. What does the Symmetric Diameter tool calculate?

(A) Radius of a circle
(B) Diameter as double the distance between two points
(C) Perimeter of a circle
(D) Width of a rectangle

Q18. What is the purpose of the "Retrieve Dimensions" tool?

(A) To create new dimensions
(B) To align dimensions
(C) To copy PMI dimensions to the drawing
(D) To delete dimensions

Q19. What is required for using the Bend Table tool?

(A) Holes in the model
(B) A flat pattern of a sheet metal part
(C) PMI annotations
(D) A cutting plane

Q20. Which of the following statements about section views is correct?

(A) Section views are used to enlarge small details of a model.
(B) Section views represent the interior details of a model.
(C) Section views always require auxiliary views.
(D) Section views do not require cutting planes.

Q21. The "Principal" tool in the View Wizard is used to create:

(A) Auxiliary views
(B) Primary views like front, top, and side
(C) Enlarged detailed views
(D) Cutting planes

Q22. The "Update Views" button in the drawing environment is used to:

(A) Edit view scales
(B) Reflect changes made to the 3D model
(C) Delete views
(D) Align drawing annotations

Q23. What is the purpose of the "Broken-Out" section view?

(A) To break the drawing into multiple sheets
(B) To display interior details in a localized area
(C) To align dimensions automatically
(D) To add exploded views

Q24. Which annotation tool adds reference information such as tolerances?

(A) Dimensions
(B) Balloon
(C) Feature Control Frame
(D) Parts List
Q25. What does the "Shading Options" tool affect?

(A) Annotation visibility
(B) Model display style in the drawing view
(C) Sheet background color
(D) View alignment

Q26. How can a sheet size be changed after a drawing has been created?

(A) By using the "Sheet Setup" command
(B) By adjusting the view properties
(C) By scaling the drawing manually
(D) It cannot be changed once created

Q27. What does the "Break View" tool achieve in Solid Edge drawings?

(A) Highlights individual parts in an assembly
(B) Removes unnecessary details to focus on specific areas
(C) Shortens the representation of long components
(D) Splits the drawing into multiple sheets

Q28. How can you change the font style of annotations in Solid Edge drawings?

(A) Modify the "Annotation Style" settings
(B) Edit the default template properties
(C) Use the "Retrieve Dimensions" command
(D) Adjust the "Line Style" settings

Chapter 12

Sheetmetal Design

Topics Covered

The major topics covered in this chapter are:

- *Starting Sheetmetal Design*
- *Creating Tab Feature*
- *Creating Different types of Flanges*
- *Creating Hem, Dimple, and Louver Features*
- *Creating Drawn Cutouts, Bead, Gusset, Cross Brake, Emboss, and Etch Features*
- *Modifying Bend Corners*
- *Bending, Unbending, and Rebending*
- *Creating Jog features,*
- *Sheetmetal Modification Tools*

INTRODUCTION

Sheet metal work is an important aspect of Mechanical engineering. Many parts around us are manufactured via sheetmetal processes. For example, car body, vents in houses, Air-conditioner ducts, spoon, metal bowls, and so on. The sheetmetal parts generally have thickness ranging from fraction of millimeter to 12.5 millimeters i.e. up to half inch. Like welding and machining, sheetmetal also has its own processes like, bending, punching, stamping, spinning, rolling, and so on. In this chapter, we will discuss about the tools available in Solid Edge related to sheetmetal designing. Note that we will be working in Ordered modeling environment in this chapter. You can use the same tools in Synchronous modeling environment as well.

STARTING SHEETMETAL PART DOCUMENT

In Solid Edge, Sheetmetal parts are designed in a different environment. The procedure to create sheetmetal part document is given next.

- Click on the **New** tool from the **New** cascading menu in the **Application** menu. The **New** dialog box will be displayed.
- Select desired .psm template from the dialog box (**iso metric sheet metal.psm** in our case) and click on the **OK** button. The Sheetmetal interface will be displayed with **Ordered** environment selected by default in a dialog asking you to **Continue with Synchronous** or **Continue with Ordered** environment; refer to Figure-1.

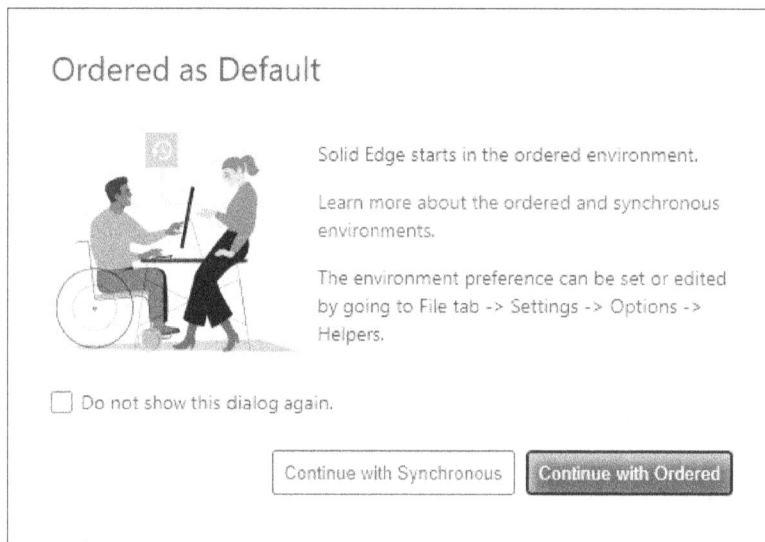

Figure-1. Dialog asking to select Synchronous or Ordered environment

- Select **Continue with Ordered** button in the dialog or if dialog not appears, select **Ordered** radio button from the **Model** group in the **Tools** tab of the **Ribbon**. The interface will be displayed in **Ordered** environment; refer to Figure-2.

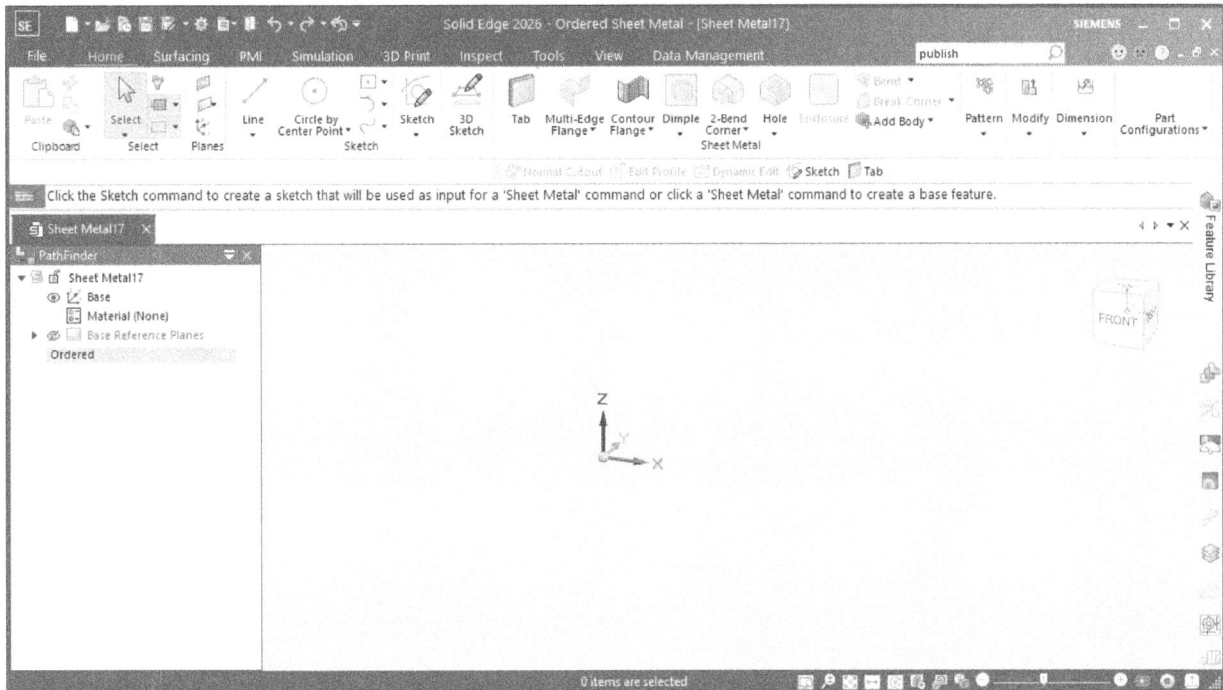

Figure-2. Sheet metal environment interface

CREATING TAB FEATURE

Tab is a flat plate of sheet metal generally used as base for flanges and other features. The procedure to create a tab feature is given next.

- Create a closed loop sketch to be used for defining boundary of tab feature.
- Click on the **Tab** tool from the **Sheet Metal** group in the **Home** tab of **Ribbon**. The **Tab** command box will be displayed; refer to Figure-3 and you will be asked to select a sketching plane.

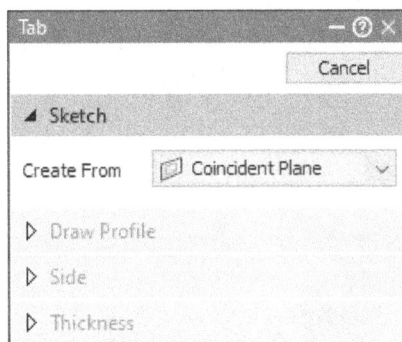

Figure-3. Tab command box

- Select desired sketch from the graphics window and then click on the **OK** button from the **Sketch** rollout. If you have not created sketch earlier then select a plane/ face and create the sketch.
- Now, click on the either side of the tab preview to specify the direction of tab thickness.
- Expand the **Thickness** node from dialog box and specify desired thickness in the **Thickness** edit box; refer to Figure-4.

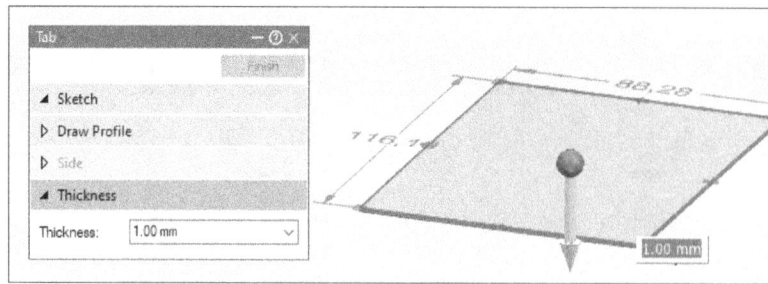

Figure-4. Internal Face Loops option

- Click on the screen at desired side of sketch, the tab feature will be created; refer to Figure-5.

Figure-5. Creating the tab

- Click in the **Name** edit box and specify desired name of the tab. Click on the **Finish** button to exit the dialog box.

CREATING FLANGE FEATURE

The **Flange** feature is used to create sheet metal walls on selected edges of tab. The procedure to create the feature is given next.

- Click on the **Flange** tool in **Multi-Edge Flange** drop-down from the **Sheet Metal** group in the **Home** tab of **Ribbon**. The **Flange** command box will be displayed; refer to Figure-6.

Figure-6. Flange command box

- Click on the **Options** button from the command box. The **Flange Options** dialog box will be displayed; refer to Figure-7. Clear the **Use default value** check box to specify desired bend radius value in the **Bend radius** edit box. Select the **Bend relief**, **Extend relief**, and **Corner relief** check boxes to apply relief cuts at bends and corners of the flange sheet.

Figure-7. Flange Options dialog box

- Specify desired values in the edit box and click on the **OK** button to apply to flange.
- Select the **Material Inside** button from **Material Placement** area in the command box to create flange inside the selected edge. Similarly, select the **Material Outside** and **Bend Outside** buttons to create flange wall and bend of flange outside selected edge.
- Click on desired edge to create the flange. The flange will get attached to cursor; refer to Figure-8.

Figure-8. Flange attached to cursor

- Select desired button from the command box to define width of flange. Select the **Full Width** option from **Flange Placement** drop-down in the command box to create flange with width equal to length of selected edge. Select the **Centered** option from the drop-down to create flange of specified width at center of selected edge. Select the **At End** option from the drop-down to create flange of specified width at selected end point and then select the end point from where you want to start the flange; refer to Figure-9. Select the **From Both Ends** option from the drop-down to create flange at specified distance from the end points. Select the **From End** option from the drop-down to create flange at specified distance from the selected end point.
- Specify desired distance value in **Distance** edit box to define length of flange and press **ENTER**. Preview of the flange will be displayed; refer to Figure-10.

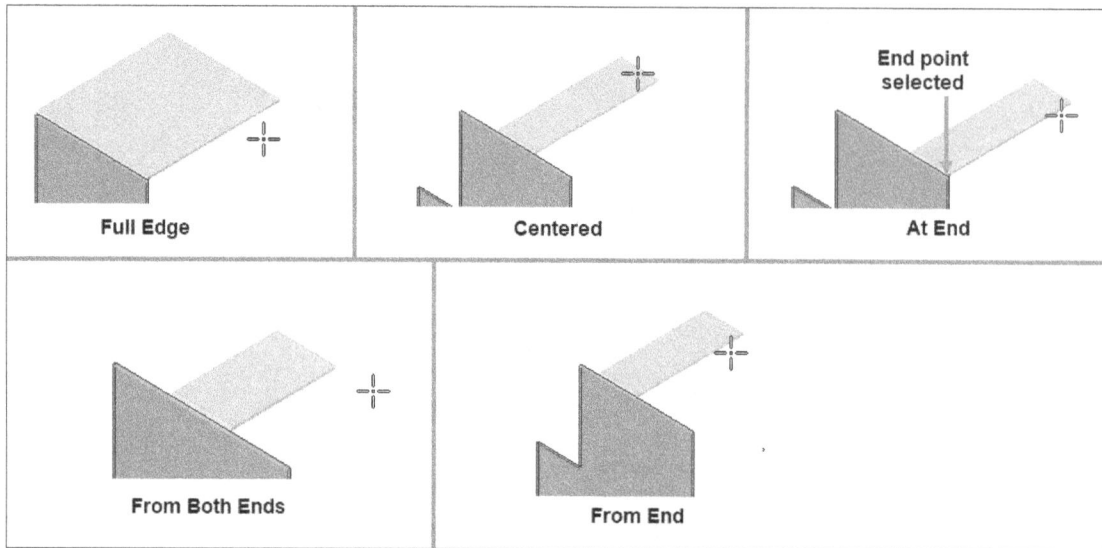

Figure-9. Specifying width of flange

Figure-10. Sketch for flange

*   Double-click on desired dimensions to modify the flange and click on the **Finish** button. The flange will be created. Press **ESC** to exit the tool.

CREATING MULTI-EDGE FLANGE

The **Multi-Edge Flange** tool is used to create flanges using multiple edges of the sheet metal part. The procedure to use this tool is given next.

*   Click on the **Multi-Edge Flange** tool from the **Sheet Metal** group in the **Home** tab of the **Ribbon**. The **Multi-Edge Flange** command box will be displayed similar to **Flange** command box.
*   Click on the **Options** button from the command box. The **Multi-Edge Flange Options** dialog box will be displayed; refer to Figure-11.

Figure-11. Multi-Edge Flange Options dialog box

- Select the **Miter** check box from the **Miters and Corners** tab to create miter cuts by joining flanges of multiple edges. Set the other parameters as discussed earlier and click on the **OK** button from the dialog box.
- Select the edges of the model on which you want to apply flange and press **ENTER**. Preview of flange will be displayed; refer to Figure-12.

Figure-12. Preview of multi-edge flange

- Set desired angle for flanges in the **Angle** edit box of command box.
- Select the **Trim** toggle button from the command box to trim intersecting flanges; refer to Figure-13.

Figure-13. Trimmed flanges

- Set the other parameters as discussed earlier and click on desired side of sheet for creating flange.
- Click on the **Finish** button from the command box to create flange. Press **ESC** to exit the tool.

CREATING CONTOUR FLANGE

The **Contour Flange** tool is used to create flange at desired edge with specified shape. The procedure to use this tool is given next.

- Click on the **Contour Flange** tool from the **Contour Flange** drop-down in the **Sheet Metal** group of the **Home** tab in the **Ribbon**. The **Contour Flange** command box will be displayed; refer to Figure-14.

Figure-14. Contour Flange command box

- Select desired edge to be used for creating flange. You will be asked to specify location of plane for creating sketch of flange shape.
- Click at desired location to specify location of plane and create desired sketch for flange; refer to Figure-15.

Figure-15. Sketch created for contour flange

- After creating sketch, click on the **Close Sketch** button from the **Close** group in the **Home** tab of the **Ribbon**. The flange will get attached to cursor; refer to Figure-16.

Figure-16. Contour flange attached to cursor

- Select the **Finite Extent** button to specify distance up to which the flange will be created. Select the **To End** button to create flange up to end point of selected edge. Select the **Chain** button from the command box to create the flanges along multiple edges of base feature in a chain; refer to Figure-17.

Figure-17. Contour flanges using chain selection

- Click on the **OK** button from the command box to create the flange and click on the **Finish** button. Press **ESC** to exit the tool.

CREATING LOFTED FLANGE

The **Lofted Flange** tool is used to create flange using sketched sections. The procedure to use this tool is given next.

- Click on the **Lofted Flange** tool from the **Contour Flange** drop-down in the **Sheet Metal** group of the **Home** tab in the **Ribbon**. The **Lofted Flange** command box will be displayed; refer to Figure-18.

Figure-18. Lofted Flange command box

- Select desired plane/face to create the first sketch of lofted flange and create desired open sketch; refer to Figure-19. You will be asked to specify the start point of section.

Figure-19. Sketch 1 for lofted section

- Select desired vertex of sketch and right-click in the drawing area. You will be asked to select a plane for creating second section of sketch.
- Select the **Parallel Plane** or other option from the **Create-From Options** drop-down in the command box and select desired sketching plane. You will be asked to create sketch for second section.
- Create desired open sketch and click on the **Close Sketch** button from **Ribbon**. You will be asked to select starting vertex of second section.
- Select desired vertex and right-click in the drawing area. You will be asked to specify direction of applying thickness to feature.
- Click at desired side of profile. The preview of flange will be displayed; refer to Figure-20.

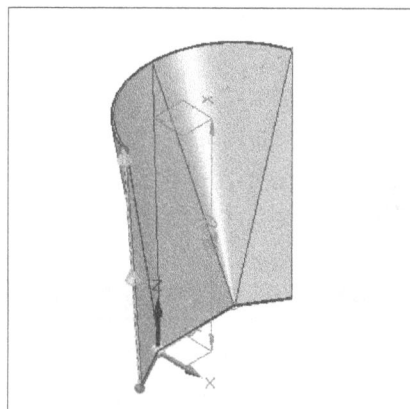
Figure-20. Preview of lofted flange feature

- Click on the **Finish** button from the command box to create the feature and press **ESC** to exit the tool.

CREATING HEM FEATURE

The **Hem** tool is used to create hem bend in the sheet metal at selected edges. The procedure to use this tool is given next.

- Click on the **Hem** tool from the **Contour Flange** drop-down in the **Sheet Metal** group in the **Home** tab of **Ribbon**. The **Hem** command box will be displayed; refer to Figure-21.

Figure-21. Hem command box

- Click on the **Options** button from the command box. The **Hem Options** dialog box will be displayed; refer to Figure-22.

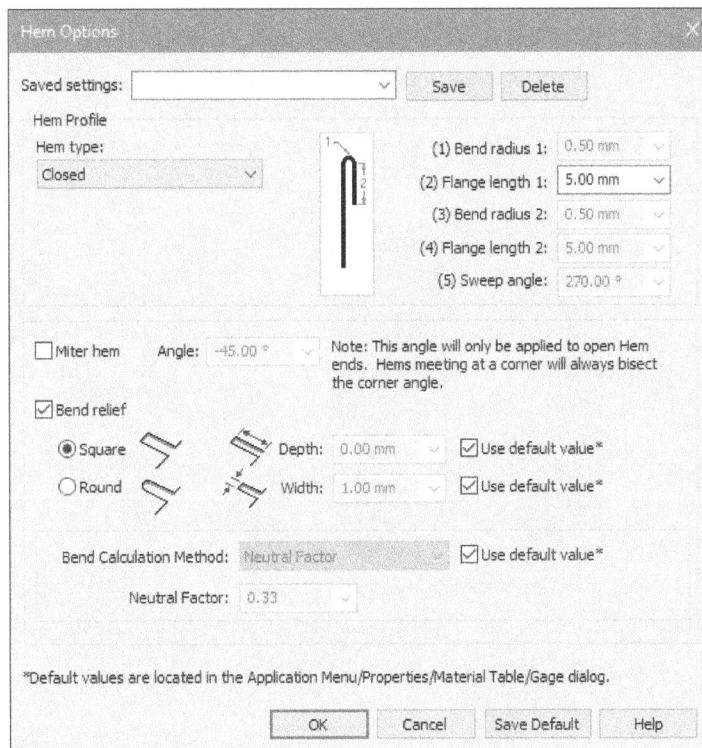

Figure-22. Hem Options dialog box

- Select desired option from the **Hem type** drop-down in the **Hem Profile** section and set desired parameters in the dialog box.
- Select the **Miter hem** check box to apply miter cut at the end sides of the hem feature. Click on the **OK** button from the dialog box to apply parameters.

- Select desired edge to apply the feature.
- Click on the **Material Inside**, **Material Outside**, or **Bend Outside** button to define location of hem.
- After setting parameters, click on the **OK** button and then **Finish** button from the command box to create the hem feature.
- Press **ESC** to exit the tool.

CREATING DIMPLE FEATURE

The **Dimple** tool is used to create dent in sheet metal with specified boundary curves. The procedure to use this tool is given next.

- Click on the **Dimple** tool from the **Dimple** drop-down in the **Sheet Metal** group of the **Home** tab in the **Ribbon**. The **Dimple** command box will be displayed; refer to Figure-23 and you will be asked to select a planar face of sheet metal.

Figure-23. Dimple command box

- Select desired face of sheet metal. The sketching environment will be displayed and you will be asked to create a closed loop sketch of dimple.
- Create the sketch as desired for dimple; refer to Figure-24. After creating sketch, click on the **Close Sketch** button from the command box. You will be asked to specify depth of dimple.

Figure-24. Sketch created for dimple

- Click at desired depth level to specify depth of dimple; refer to Figure-25. You can double-click on the depth dimension to change it.

Figure-25. Specifying depth of dimple

- Click on the **Finish** button from the command box and then press **ESC** to exit the tool.

CREATING LOUVER FEATURE

The **Louver** tool is used to create louver cuts in the sheet. The procedure to use this tool is given next.

- Click on the **Louver** tool from the **Dimple** drop-down in the **Sheet Metal** group of the **Home** tab in the **Ribbon**. The **Louver** command box will be displayed; refer to Figure-26 and you will be asked to select a face of sheet metal.

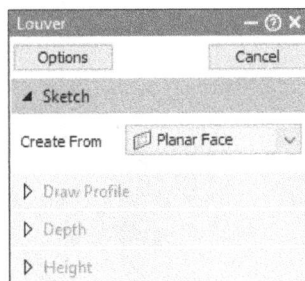

Figure-26. Louver command box

- Click on the **Options** button from the command box. The **Louver Options** dialog box will be displayed; refer to Figure-27.

Figure-27. Louver Options dialog box

- Select desired radio button from the dialog box to define what type of louver will be created. Select the **Include rounding** check box to specify radius at louver edges. After specifying parameters, click on the **OK** button.
- Click on the face of sheet metal to create sketch for louver. The sketching environment will be displayed.

- Create a line for louver at desired location and click on the **Close Sketch** tool from the **Ribbon**. You will be asked to specify depth of louver; refer to Figure-28.

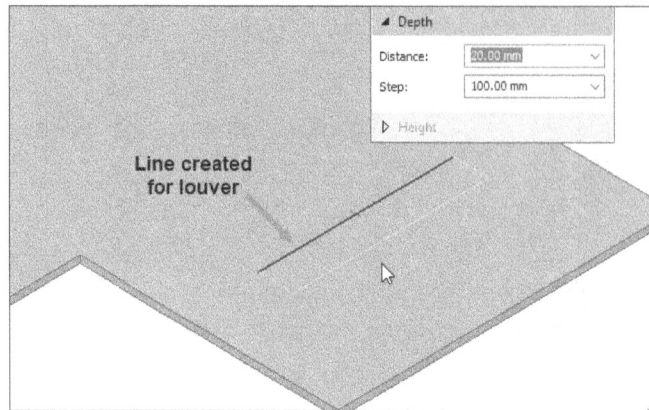

Figure-28. Specifying depth of louver

- Click at desired location to specify depth of louver or specify desired depth value in the **Distance** edit box of **Depth** rollout.
- Specify desired value in the **Step** edit box to set the increment by which the value will change.
- Select the **Offset Dimension** or **Full Dimension** button from **Height** rollout to specify height style of louver and then specify desired height value for louver in the **Distance** edit box. The preview of louver will be created; refer to Figure-29.

Figure-29. Preview of louver

- Click on the **Finish** button to create the louver. Press **ESC** to exit the tool.

CREATING DRAWN CUTOUT

The **Drawn Cutout** tool is used to cut material from the sheet using closed sketch boundary and use the boundary to draw flange walls. The procedure to use this tool is given next.

- Click on the **Drawn Cutout** tool from the **Dimple** drop-down in the **Sheet Metal** group of the **Home** tab in the **Ribbon**. The **Drawn Cutout** command box will be displayed; refer to Figure-30.

Figure-30. Drawn Cutout command box

- Click on the **Options** button from the command box. The **Drawn Cutout Options** dialog box will be displayed; refer to Figure-31. Set desired parameters in the dialog box like taper angle, round radius at edges, corner radius, and so on. After setting desired parameters, click on the **OK** button.

Figure-31. Drawn Cutout Options dialog box

- Click at desired face of sheet metal. The sketching environment will be displayed.
- Create desired closed loop sketch for cutout and click on the **Close Sketch** tool from the **Ribbon**. You will be asked to specify height/depth of cutout.
- Click at desired level to create the cutout. The preview of cutout will be displayed; refer to Figure-32.

Figure-32. Preview of cutout

- Select the **Profile Represents Die** or **Profile Represents Punch** button from the command box to define how boundaries will be used for creating cutout.
- After setting desired parameters, click on the **Finish** button from the command box. The cutout will be created. Press **ESC** to exit the tool.

CREATING BEAD FEATURE

The **Bead** tool is used to create dent in sheet metal using line curves as reference. The procedure to use this tool is given next.

- Click on the **Bead** tool from the **Dimple** drop-down in the **Sheet Metal** group of the **Home** tab in the **Ribbon**. The **Bead** command box will be displayed as shown in Figure-33 and you will be asked to select a plane for creating base sketch.

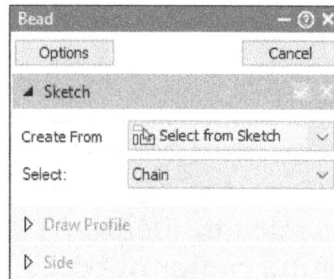

Figure-33. Bead command box

- Click on the **Options** button from the command box. The **Bead Options** dialog box will be displayed; refer to Figure-34.

Figure-34. Bead Options dialog box

- Select desired radio button from the **Cross Section** area of the dialog box to define shape of bead and specify the parameters of beads in the edit boxes of this area.
- Select desired option from the **End Conditions** area to define how ends of beads will be created.
- Set the other parameters as desired in the dialog box and click on the **OK** button.
- Click on the face of sheet metal on which you want to create bead. The sketching environment will be displayed.
- Create desired open line sketch (note that lines should be in tangent chain) and click on the **Close Sketch** button. Preview of bead will be displayed; refer to Figure-35.

Figure-35. Preview of bead

• Click at desired side of sheet metal to create the bead and click on the **Finish** button from the command box. Press **ESC** to exit the tool.

CREATING GUSSET FEATURE

The **Gusset** tool is used to create gussets in sheet metal for supporting flange walls. The procedure to use this tool is given next.

• Click on the **Gusset** tool from the **Dimple** drop-down in the **Sheet Metal** group of the **Home** tab in **Ribbon**. The **Gusset** command box will be displayed; refer to Figure-36.

Figure-36. Gusset command box

• Click on the **Gusset Options** button from the command box. The **Gusset Options** dialog box will be displayed; refer to Figure-37.

Figure-37. Gusset Options dialog box

• Set desired profile and thickness of gusset in the dialog box and click on the **OK** button.

- Select desired bend from the model to apply gusset. The gusset will get attached to cursor; refer to Figure-38.

Figure-38. Gusset attached to cursor

- Select the **Single** option from the **Pattern Type** drop-down in the command box to create single gusset at bend.
- Select the **Fit** option from the **Pattern Type** drop-down and specify total number of gussets to be created in the **Count** edit box. Press **TAB** after specifying value to check preview.
- Select the **Fill** option from the **Pattern Type** drop-down to create as many gussets as possible on a bend with specified gap. Specify desired distance in the **Spacing** edit box and click on the **OK** button to create the gussets.
- Select the **Fixed** option from the **Pattern Type** drop-down and specify gap as well as number of instances of gussets in respective edit boxes of the command box.
- After setting desired parameters, click on the **OK** button from the command box to create the gussets.
- Click on the **Finish** button from command box and then press **ESC** to exit the tool.

TAB AND SLOT

The **Tab and Slot** tool is used to create interlocking tab and slot features between two parts in a sheet metal or assembly design. Note that the two parts should be two separate bodies. The procedure to use this tool is discussed next.

- Click on the **Tab and Slot** tool from the from the **Dimple** drop-down in the **Sheet Metal** group of the **Home** tab in the **Ribbon**. The **Select** dialog box will be displayed; refer to Figure-39.

Figure-39. Select dialog box

- Select desired edge of the model from which the tab is to be created. The preview of tab will be displayed and you will be asked to select the slot face.
- Select desired face of the model on which the slot is to be created. The preview of slot face will be displayed; refer to Figure-40.

Figure-40. Displayed the parameters of tab and slot

- Click in the **Number of Tabs** edit box and specify desired value to define how many tabs will be generated along the selected edge.
- Click in the **Tab Length** edit box and specify the length of each tab, determining how far it extends from the tab edge into the slot face.
- Click in the **Tab Spacing** edit box and specify desired value to set the distance between consecutive tab.
- Click in the **Start Offset** and **End Offset** edit box and specify desired value to define the distance from the start and end of the edge.
- Select **Tab and Slot - Fit** button to create a fit-type pattern.
- Select **Tab and Slot - Fill** button to create a fill-type pattern.
- Select **Tab and Slot - Single Slot Gap Value** button to apply a single input value for the slot gap.
- Select **Tab and Slot - Multiple Slot Gap Value** button to apply a multiple slot gap value.
- Click in the **Slot Length Gap** edit box and specify desired value to define the gap between the slot and tab faces along the tab length direction.
- Click in the **Slot Width Gap** edit box and specify desired value to define the gap between slot and tab faces along the tab width direction.
- Click in the **Slot Depth Gap** edit box and specify desired value to define the gap between slot and tab faces along the tab depth direction.
- After specify desired parameters, click on the **RMB** button then the **Extend** rollout will be displayed; refer to Figure-41.

Figure-41. Displaying the extend of tab and slot

- The options of the **Extent** rollout have been discussed earlier. Click **LMB** button. The interlocking tab and slot feature will be created; refer to Figure-42.

Figure-42. Tab and slot created

- Click on the **Finish** button from command box and then press **ESC** to exit the tool.

CREATING CROSS BRAKE

The **Cross Brake** tool is used to create cross-brake feature on the sheet metal using sketch. The procedure to use this tool is given next.

- Create the line sketch for cross brake on the sheet metal; refer to Figure-43.

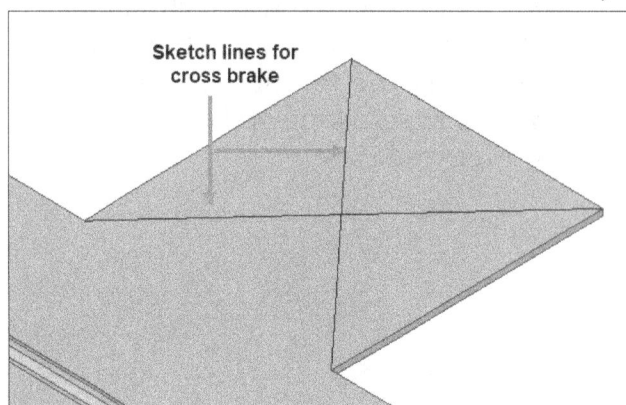

Figure-43. Sketch for cross brake

- Click on the **Cross Brake** tool from the **Dimple** drop-down in the **Sheet Metal** group of the **Home** tab in the **Ribbon**. The **Cross Brake** command box will be displayed; refer to Figure-44.

Figure-44. Cross Brake command box

- Select the face of sheet on which you want to create cross brake and right-click in the drawing area. You will be asked to select the sketch lines.
- Select desired lines and right-click in the drawing area.
- Specify the bend angle in the **Angle** edit box and click to specify direction of cut. Preview of feature will be displayed.
- Click on the **Finish** button from the command box and then press **ESC** to exit the tool.

CREATING ETCH FEATURE

The **Etch** tool is used to create etch feature using selected sketch. The procedure to use this tool is given next.

- Click on the **Etch** tool from the **Dimple** drop-down in the **Sheet Metal** group of the **Ribbon**. The **Etch** command box will be displayed; refer to Figure-45.

Figure-45. Etch command box

- Click on the **Etch Options** dialog box will be displayed; refer to Figure-46. Set desired color, width, and line type for etching. Click on the **OK** button from the dialog box to apply changes.

Figure-46. Etch Options dialog box

- Select desired curves to be used for creating etch feature and click on the **OK** button. The etch feature will be created; refer to Figure-47.

Figure-47. Etch feature created

CREATING EMBOSS FEATURE

The **Emboss** tool is used to emboss a solid body on sheet metal. Note that to use this tool, you will need a solid body along with sheet metal body. The procedure to use this tool is given next.

- Click on the **Emboss** tool from the **Dimple** drop-down in the **Sheet Metal** group of **Home** tab in the **Ribbon**. The **Emboss** command box will be displayed; refer to Figure-48.

Figure-48. Emboss command box

- Select the sheet metal part on which the solid body will be embossed. You will be asked to select the tool body (solid body) to be used for embossing.
- Select desired solid body; refer to Figure-49.

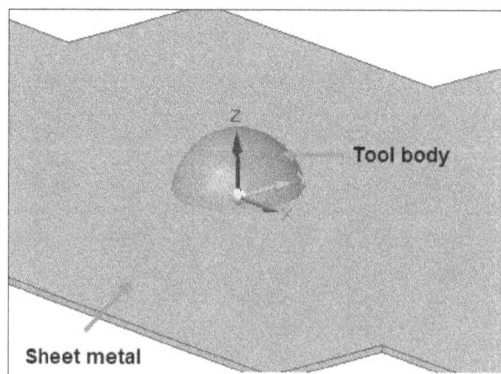

Figure-49. Objects selected for embossing

- Select the **Thicken** button from the command box to create embossed sheet of thickness equal to sheet metal thickness. If this button is not selected then a cut is made equal to boundary of tool body.
- Click on the **Direction** button from the command box to flip direction of embossing.
- Specify desired value in the **Clearance** edit box to generate gap of specified value between tool body and target body.
- You can also press **t** to toggle between thicken On and Off. Similarly, press **f** to flip the direction of embossing.

*   After setting desired parameters, click on the **OK** button from the command box to create the feature. Press **ESC** to exit the tool.

CLOSING 2-BEND CORNER

The **Close 2-Bend Corner** tool is used to close or overlap flange walls at 2-bend corners. The procedure to use this tool is given next.

*   Click on the **Close 2-Bend Corner** tool from the **2-Bend Corner** drop-down in the **Sheet Metal** group of the **Home** tab in the **Ribbon**. The **Close 2-Bend Corner** command box will be displayed; refer to Figure-50.

Figure-50. Close 2-Bend Corner command box

*   Select desired option from the **Treatment** drop-down in the command box and specify respective parameters in the edit boxes of the command box.
*   Select the **Close** button or **Overlap** button to create closed bends or overlapping bends at flange walls. If you have selected the **Overlap** button then specify desired value in **Overlap ratio** edit box to define the extent of two overlapping flange walls. For example if you have specified **0.5** in the edit box then half of thickness of flange walls will be overlapping.
*   After specifying the parameters, select two consecutive bends. The bends will be made closed or overlapping and preview will be displayed; refer to Figure-51.

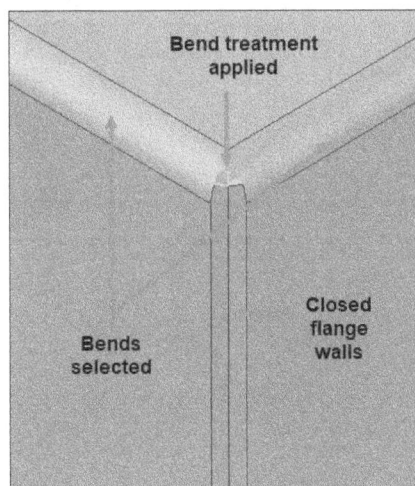

Figure-51. Preview of closed bends

*   After checking preview, click on the **OK** button from command box to create feature.
*   Click on the **Finish** button and then press **ESC** key to exit the tool.

CLOSING 3-BEND CORNERS

The **Close 3-Bend Corner** tool is used to modify corners created at intersection of 3 bends. The procedure to use this tool is given next.

*   Click on the **Close 3-Bend Corner** tool from the **2-Bend Corner** drop-down in the **Sheet Metal** group of **Home** tab in the **Ribbon**. The **Close 3-Bend Corner** command box will be displayed; refer to Figure-52.

Figure-52. Close 3-Bend Corner command box

*   Select desired option from **Treatment** drop-down to define corner condition to be applied and set desired parameters in the edit boxes of the command box.
*   Select two bends of three bends set to be used for modifying the cut; refer to Figure-53.

Figure-53. Bends selected for 3-Bends

*   Click on the **OK** button from the command box to create the feature.
*   Click on the **Finish** button from the command box and press **ESC** to exit the tool.

CREATING RIPPED CORNERS

The **Rip Corner** tool is used to create ripped corners by replacing the bends. The procedure to use this tool is given next.

*   Click on the **Rip Corner** tool from the **2-Bend Corner** drop-down in the **Sheet Metal** group of the **Home** tab in the **Ribbon**. The **Rip Corner** command box will be displayed; refer to Figure-54.

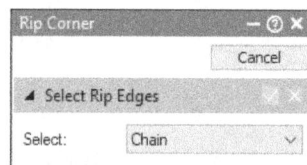

Figure-54. Rip Corner command box

*   Select desired edges to be ripped. Right-click in the drawing area to accept the selection.
*   Click on the **Preview** button from the command box to check preview of ripping.
*   Click on the **Finish** button from the command box to create the feature.

CUT

The **Cut** tool is used to remove material from model. It can create simple hole and other cuts using the close loop sketch sections. The procedure to create a cut is discussed next.

- Click on the **Cut** tool from **Hole** drop-down in the **Sheet Metal** panel of **Home** tab in the **Ribbon**. The **Cut** command box will be displayed; refer to Figure-55.

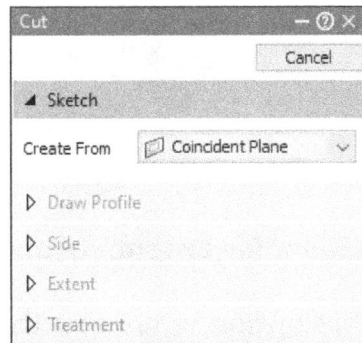

Figure-55. Cut command box

- Select plane or features face, create a sketch which you want to use for removing material. Note that if you are using plane to create sketch then the plane should be parallel to the face through which you want to cut the material.
- After creating closed loop sketch, click on the **Close Sketch** button from **Close** panel in the **Home** tab of **Ribbon**. Preview of cut feature will be displayed with extent of feature attached to cursor and the updated **Cut** command box will be displayed as shown in Figure-56.

Figure-56. Updated Cut command box

- Set desired parameters in the command box. The feature preview will be displayed; refer to Figure-57.

Figure-57. Preview of cut

Various option in the command box for **Extent** rollout are discussed next.

- **Through All Cut** 🔲 :- If this option is chosen then the cut is created through all part. Refer to Figure-58.

Figure-58. Cut Through All

- **Through Next** 🔲 :- If this option is selected then the cut is made through next face of the model; refer to Figure-59.

Figure-59. Cut Through Next

- **Cut - From/To Extent** 🔲 :- Select this option if you want to define the start point and end point of cut feature by selecting faces, planes, or other references.

Note that the selected references should be in parallel planes for defining extents. After selecting the **From/To Extent** button, select the start reference and then end reference. The preview of feature will be displayed; refer to Figure-60.

Figure-60. Cut From Or To Extent

- **Cut - Finite Extent** [icon] :- If this option is selected then you need to define the distance upto which the cut will be created in the **Distance** edit box; refer to Figure-61.

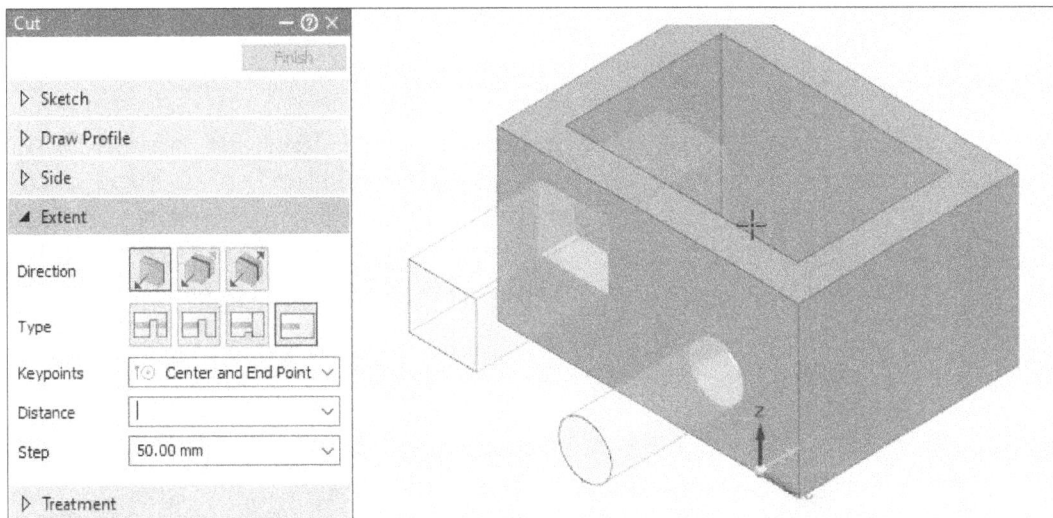

Figure-61. Cut Finite Extent

- Select **Active Body** radio button from **Cut Preview** area to cut only the active body. Select **Selected Bodies** option to allow selection of bodies to be cut by the feature.
- Select **Treatment Parameters Prompt** check box from **Treatment** rollout to prompt treatment step everytime after completion of feature.

Various option in the command box for **Treatment** rollout are discussed next.

- **No Treatment** [icon] :- Select this option to remove the draft or crown treatment from the feature; refer to Figure-62.

Figure-62. No Treatment applied

- **Draft** :- Select this option to add draft to the feature; refer to Figure-63. Specify desired angle value for direction of draft in the **Angle** edit box. Toggle the **Flip Side** button to flip the angle for direction.

Figure-63. Draft applied

- **Crown** :- Select this option to add crown to the feature; refer to Figure-64. Select desired method of defining the crown properties from **Type** drop-down. Specify desired radius, offset value, and takeoff angle of crown in the **Radius(r)**, **Offset(o)**, and **Takeoff(t)** edit boxes, respectively. Toggle the **Flip Side** button to flip the side on which the crown is applied. Toggle the **Flip Curvature** button to flip the side on which the curvature is applied.

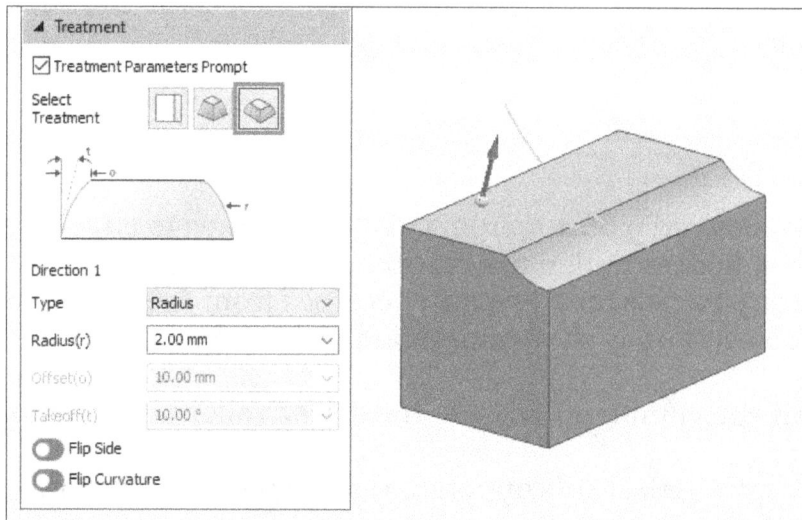

Figure-64. Crown applied

The other options in the command box have been discussed earlier.

NORMAL CUTOUT

The **Normal Cutout** tool is used to remove material from the solid by using sketch projected on solid face. So, you can create cut from non planar body faces by projecting sketch on them. The procedure to use this tool is same as discussed in **Cut** tool.

ENCLOSURE

The **Enclosure** tool is used to enclose existing design body into box or cylinder. The procedure to use this tool have been discussed earlier.

CREATING BEND

The **Bend** tool is used to create bend in sheet metal using specified bend lines. The procedure to use this tool is given next.

*   Click on the **Bend** tool from the **Bend** drop-down in the **Sheet Metal** group of **Home** tab in the **Ribbon**. The **Bend** command box will be displayed; refer to Figure-65.

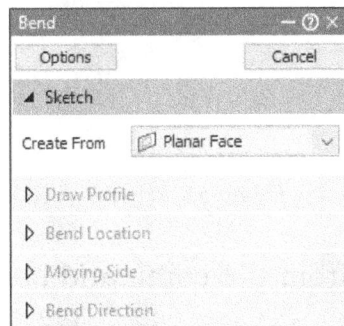

Figure-65. Bend command box

*   Click on the **Options** button from the command box. The **Bend Options** dialog box will be displayed; refer to Figure-66.

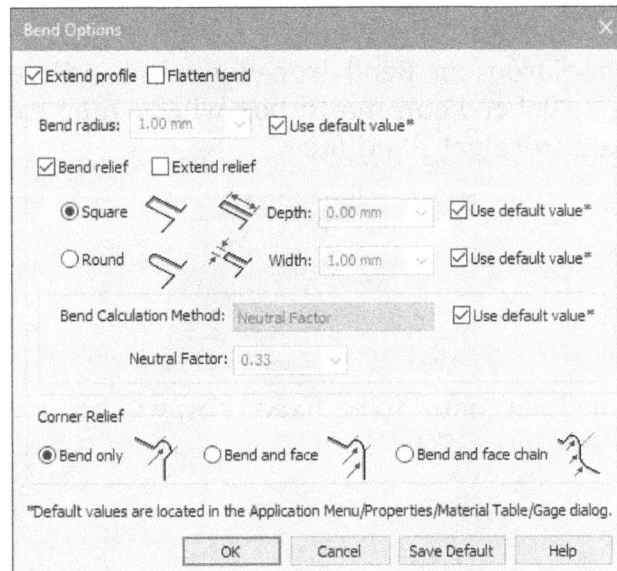

Figure-66. Bend Options dialog box

- Select the **Extend profile** check box from the dialog box to extend the bend line automatically to the edges of sheet.
- Select the **Flatten bend** check box from the dialog box to flatten the bend after creating it.
- Set the other parameters as discussed earlier and click on the **OK** button.
- Select desired planar face of sheet metal to draw the bend line. The sketching environment will be displayed.
- Create a straight line to define bend line; refer to Figure-67 and click on the **Close Sketch** button from **Ribbon**. You will be asked to specify direction of bend.
- Click on the left or right side of bend line to define direction of bend. You will be asked to specify which side of profile will move.
- Click on desired side to define which side will be folded by bend line. You will be asked to specify whether you want to bend the sheet inward or outward.
- Move the cursor to desired side and click to specify the bend direction. Preview of the bend will be displayed; refer to Figure-68.

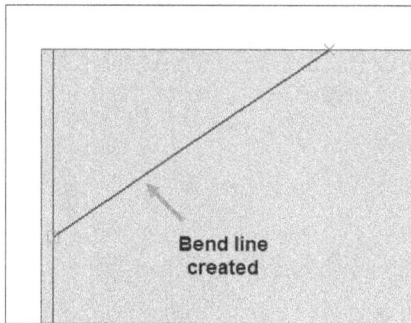

Figure-67. Bend line

Figure-68. Preview of the bend

- Click on the **Finish** button from the command box to create bend and press **ESC** to exit the tool.

UNBENDING SHEET METAL

The **Unbend** tool is used to unbend selected portion of sheet metal. The procedure to use this tool is given next.

- Click on the **Unbend** tool from the **Bend** drop-down in the **Sheet Metal** group of **Home** tab in the **Ribbon**. The **Unbend** command box will be displayed; refer to Figure-69 and you will be asked to select fixed face.

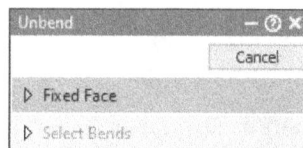

Figure-69. Unbend command box

- Select desired face of sheet metal to be fixed. You will be asked to select the bends to be unbend; refer to Figure-70.

Figure-70. Selecting the fixed face of unbend

- Select desired bends from the model; refer to Figure-71. The preview of unbend feature will be displayed.

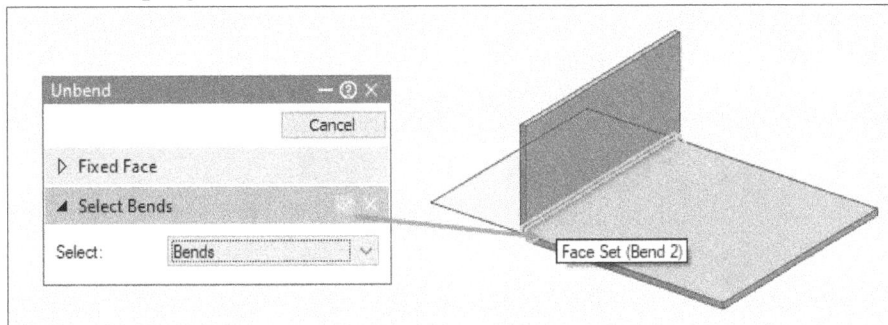

Figure-71. Selecting the bend of unbend

- Select the **All Bends** option from the **Select** drop-down if you want to select all the bends in model for unbend.
- Click on the **OK** button from the **Select Bends** rollout. The unbend will be created and the **Finish** button will be displayed; refer to Figure-72.

Figure-72. Preview of unbend

- Specify desired name in the **Name** edit box and click on the **Finish** button to close the dialog box.

REBENDING

The **Rebend** tool is used to rebend selected unbend features. The procedure to use this tool is given next.

- Click on the **Rebend** tool from the **Bend** drop-down in the **Sheet Metal** group of **Home** tab in the **Ribbon**. The **Rebend** command box will be displayed; refer to Figure-73.

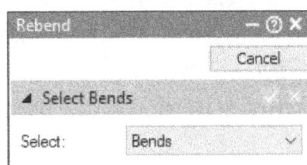

Figure-73. Rebend command box

- Select the **Bends** option from the **Select** drop-down in the command box to select the bends which were unbend earlier.
- Select the **All Bends** option from the drop-down if you want to rebend all the unbends created in the model.
- After selecting bends, click on the **Finish** button from the command box to create features.

CREATING JOG FEATURE

The **Jog** tool is used to jog features in the sheet metal. The procedure to use this tool is given next.

- Click on the **Jog** tool from the **Bend** drop-down in the **Sheet Metal** group of **Home** tab in the **Ribbon**. The **Jog** command box will be displayed; refer to Figure-74.

Figure-74. Jog command box

- Select desired option from the **Create-From** drop-down in the command box to define plane for creating jog line or select a sketch of jog line created on sheetmetal face.
- Select/create the sketch for jog line and then right-click in the drawing area. You will be asked to specify side to be moved by jog.
- Click on desired side of jog line to specify which side will be moved.
- The jogged face will be attached to cursor; refer to Figure-75. Select desired buttons from the command box to define how bends will be applied to jog.

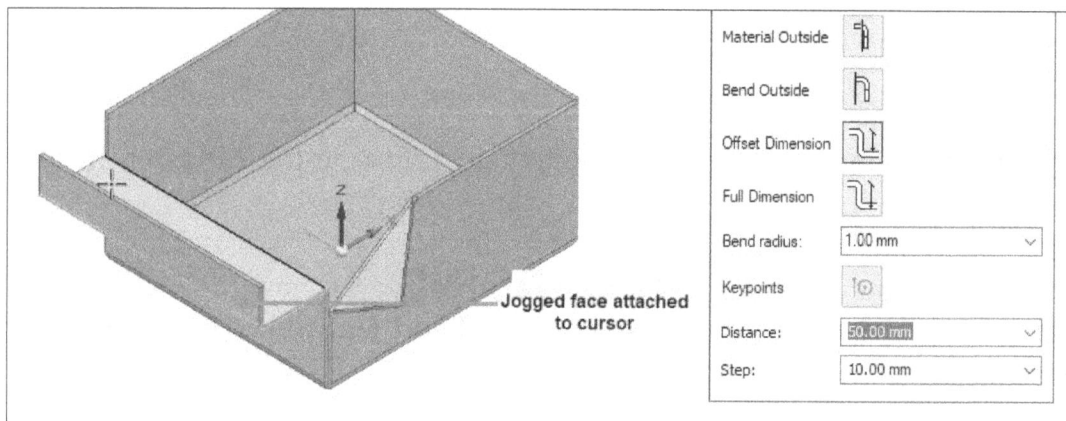

Figure-75. Jogged face attached to cursor

- Specify desired radius of bend in the **Bend radius** edit box.
- Click at desired location to specify length of jog base or click in the **Distance** edit box of command box and enter the value.
- Click on the **Finish** button from the command box to create the feature. Press **ESC** to exit the tool.

APPLYING BEND BULGE RELIEF

The **Bend Bulge Relief** tool is used to provide relief at the bends which cause bulge in the sheet. The procedure to use this tool is given next.

- Click on the **Bend Bulge Relief** tool from the **Bend** drop-down in the **Sheet Metal** group of the **Home** tab in the **Ribbon**. The **Bend Bulge Relief** command box will be displayed; refer to Figure-76.

Figure-76. Bend Bulge Relief command box

- Click on the **Bend Bulge Relief Options** button from the command box. The **Bulge Relief Options** dialog box will be displayed; refer to Figure-77.

Figure-77. Bulge Relief Options dialog box

- Select desired radio button from the dialog box and specify desired parameters. Click on the **OK** button to apply settings.
- Select desired end point of bend which is deformed. The preview of relief will be displayed; refer to Figure-78.

Figure-78. Preview of bulge relief

- Click on the **OK** button from the command box to apply the relief.
- Click on the **Finish** button from the command box and then press **ESC** to exit.

APPLYING BREAK CORNER

The **Break Corner** tool is used to apply round or chamfer to selected edge of sheet metal. The procedure to use this tool is given next.

- Click on the **Break Corner** tool from the **Break Corner** drop-down in the **Sheet Metal** group of **Home** tab in the **Ribbon**. The **Break Corner** command box will be displayed; refer to Figure-79.

Figure-79. Break Corner command box

- Select the **Radius Corner** or **Chamfer Corner** button from the command box to create round or chamfer respectively at the selected corner edge.
- Select desired edge to be modified from the model and specify desired value of radius/setback in the **Break** edit box of command box. Preview of corner break will be displayed; refer to Figure-80.

Figure-80. Preview of corner break

- Click on the **OK** button from the command box to apply corner break. Click on the Finish button to create the feature.
- Press **ESC** to exit the tool.

SHEET METAL MODIFICATION TOOLS

The tools in the **Modify** group are used to modify the shape and size of sheet metal. Various tools in this **Modify** group which have not been discussed are discussed next.

Modifying Bend Angle

The **Bend Angle** tool in **Modify** group is used to modify angle of selected bend. The procedure to use this tool is given next.

- Click on the **Bend Angle** tool from the **Modify** group in the **Home** tab of the **Ribbon**. The **Bend Angle** command box will be displayed; refer to Figure-81.

Figure-81. Bend Angle command box

- Select the bend whose angle is to be modified. You will be asked to select a planar face to be fixed.
- Select desired face to be used for angle reference. The **Angle** edit box will be displayed in the command box.
- Specify desired value in the edit box and press **ENTER**. Preview of the modified bend will be displayed; refer to Figure-82.

Figure-82. Modified bend angle

- Click on the **Finish** button from the command box to apply modification and press **ESC** to exit the tool.

Modifying Bend Radius

The **Bend Radius** tool is used modify radius of bend. The procedure to use this tool is given next.

- Click on the **Bend Radius** tool from the **Modify** group of the **Home** tab in the **Ribbon**. The **Bend Radius** command box will be displayed; refer to Figure-83.

Figure-83. Bend Radius command box

- Select the bends to be modified and press **ENTER**. The New radius edit box will be displayed in the command box.
- Specify desired value in the edit box and press **ENTER**. The **Preview** button will be displayed.
- Click on the **Preview** button to check preview of modified bend. Click on the **Finish** button to apply modifications.
- Press **ESC** to exit the tool.

The other tools in this group have been discussed earlier.

PRACTICAL 1

In this practical, we will create a sheet metal design as shown in Figure-84. The drawing is shown in Figure-85.

Figure-84. Model for Practical 1

Figure-85. Drawing for Practical 1

Starting Sheet Metal Modeling

- Double-click on the Solid Edge icon from the desktop to start Solid Edge if not started already.
- Click on the **New** button from **New** cascading menu of **File** menu. The **New** dialog box will be displayed.
- Select the **Metric** option from left pane in the dialog box and select **metric sheet metal.psm** template from the dialog box; refer to Figure-86.

Figure-86. Template selected

- Click on the **OK** button from the dialog box. A new sheet metal document will open in application window. Continue with Ordered environment if prompted to select otherwise set Ordered environment using options in **Tools** tab of **Ribbon**.

Creating Base Tab

- Click on the **Tab** tool from the **Sheet Metal** group of **Home** tab in the **Ribbon**. The **Tab** command box will be displayed.
- Select the **Coincident Plane** option from the drop-down in the command box and select the Top plane from drawing area. The sketching environment will be displayed.
- Create the sketch and apply dimensions as shown in Figure-87.

Figure-87. Sketch for tab

- Click on the **Close Sketch** button from the **Close** group in the **Home** tab of the **Ribbon**. The **Tab** command box will be displayed again and you will be asked to specify thickness of tab.

- Specify the thickness as **2.5** in the **Thickness** edit box and click above the sketch to specify side on which thickness will be applied. Preview of tab will be displayed; refer to Figure-88.

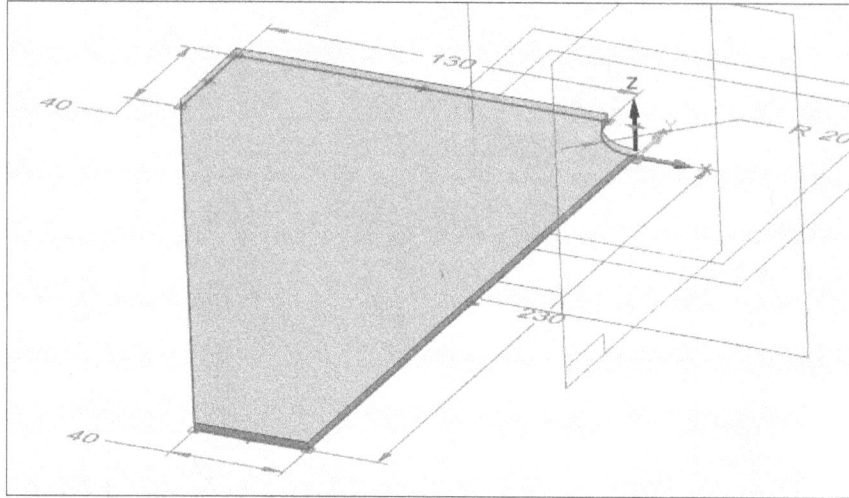

Figure-88. Preview of tab

- Click on the **Finish** button from the command box to create tab feature and exit the tool by pressing **ESC** key.

Creating Flange Walls

- Click on the **Multi-Edge Flange** tool from the **Multi-Edge Flange** drop-down in the **Sheet Metal** group of **Home** tab in the **Ribbon**. The **Multi-Edge Flange** command box will be displayed.
- Select the edge of tab as shown in Figure-89 and then right-click in empty area. You will be asked to specify length of flange.

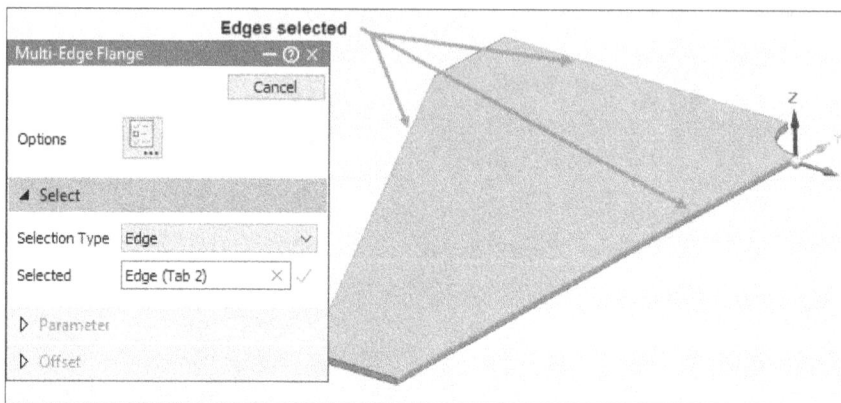

Figure-89. Edges selected for multi-edge flange

- Move the cursor downward and enter the length of flange as **60** in the **Distance** edit box of the command box. Preview of flange will be displayed; refer to Figure-90.

Figure-90. Preview of flange

• Click on the **Finish** button and press **ESC** to exit the tool.

Creating Flanges

• Click on the **Flange** tool from the **Multi-Edge Flange** drop-down in the **Sheet Metal** group of **Home** tab in the **Ribbon**. The **Flange** command box will be displayed.
• Select edge of model as shown in Figure-91 and move cursor outward. One end of flange will be attached to cursor.

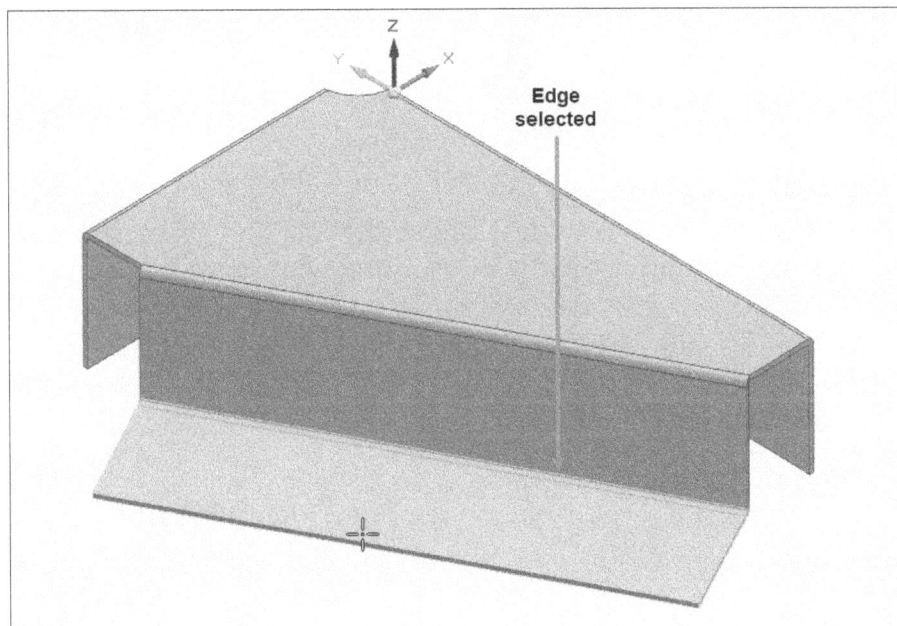

Figure-91. Edge selected

• Click at some distance to define length of flange. Preview of flange will be displayed along with the **Flange** command box.
• Click on the **Profile** node in the **Flange** command box. The sketching environment will become active; refer to Figure-92.

Figure-92. Sketch of flange

- Modify the sketch as shown in Figure-93 using sketching tools. Note that end points of open sketch should be on the same construction line as the original.

Figure-93. Modified sketch

- After creating sketch, click on the **Close Sketch** tool from the **Close** group in the **Home** tab of the **Ribbon**. Preview of modified flange will be displayed; refer to Figure-94.

Figure-94. Preview of modified flange

- Similarly, create flanges on other sides of model; refer to Figure-95 using profile sketches as shown in Figure-96.

Figure-95. After creating flanges

Figure-96. Sketches for flanges

Creating Cuts

- Click on the **Cut** tool from the **Hole** drop-down in the **Sheet Metal** group of **Home** tab in the **Ribbon**. The **Cut** command box will be displayed.
- Select the top face of model to use it as sketching plane and create circles as shown in Figure-97.

Figure-97. Circles created for cut

- Click on the **Close Sketch** tool from the **Close** group of **Ribbon**. You will be asked to specify direction of cut.
- Move the cursor downward so that cut feature passes through the faces of the model; refer to Figure-98 and click in the drawing area to create the cuts.

Figure-98. Preview of cut extent

- Click on the **Finish** button from command box and press **ESC** to exit the tool. Final model will be created; refer to Figure-99.

Figure-99. Final model

PRACTICE 1

Create the sheet metal model as shown in Figure-100. Dimensions are given in Figure-101.

Figure-100. Sheetmetal model

100 — 300

300

100

A

Detail A

10

4X ∅15

20 — 20

250

25

25 — 350 — 300 — 10

150

100

80

Thickness of sheet= 1mm
Bend Radius= 5mm

Figure-101. *Drawing views*

PRACTICE 2

Create the sheet metal model as shown in Figure-102. Dimensions are given in Figure-103.

Figure-102. Practice 2 Model

Figure-103. Practice 2 drawing views

SELF ASSESSMENT

Q1. What is the maximum thickness generally considered for sheet metal parts?

a) 5 mm
b) 12.5 mm
c) 15 mm
d) 20 mm

Q2. Which of the following is NOT a common sheet metal process?

a) Bending
b) Welding
c) Stamping
d) Rolling

Q3. What is the primary function of the Tab tool?

a) To create a flange
b) To create a flat plate as the base feature
c) To create embossments
d) To apply a hem

Q4. Which tool is used to create sheet metal walls on selected edges of a tab?

a) Tab
b) Flange
c) Hem
d) Dimple

Q5. What is the purpose of the Bend relief option in the Flange tool?

a) To create smoother flanges
b) To apply relief cuts at bends
c) To increase the flange width
d) To decrease the flange length

Q6. What does the Miter option do in the Multi-Edge Flange tool?

a) Creates flanges at multiple edges
b) Joins flanges with miter cuts
c) Increases flange length
d) Reduces flange thickness

Q7. Which tool is used to create a dent in the sheet metal?

a) Hem
b) Dimple
c) Louver
d) Bead

Q8. What is the primary purpose of the Gusset tool?

a) To create cross brakes
b) To support flange walls
c) To create embossed features
d) To create dimpled features

Q9. Which feature is used to create raised or solid shapes on sheet metal?

a) Louver
b) Emboss
c) Etch
d) Hem

Q10. What does the Close 2-Bend Corner tool do?

a) Applies embossing to two corners
b) Creates overlapping bends at flange walls
c) Adjusts the thickness of corner flanges
d) Trims intersecting edges at two bends

Q11. What is the purpose of the Etch tool?

a) To create dents
b) To engrave features with a sketch
c) To emboss a solid body
d) To add thickness to sheet metal

Q12. Which feature uses line curves to create a dent or indentation?

a) Dimple
b) Bead
c) Louver
d) Emboss

Q13. What is the purpose of the Louver tool?

a) To create vent-like cuts in sheet metal
b) To apply relief at flange ends
c) To emboss solid shapes
d) To create cross-brakes

Q14. What does the Hem feature create?

a) A bent edge in the sheet metal
b) A cut-out in the sheet metal
c) An embossed shape
d) A solid flange

Q15. What is the purpose of the Close 3-Bend Corner tool?

a) To modify a single bend
b) To modify corners created at the intersection of three bends
c) To create new bends
d) To add a rip corner to a sheet

Q16. In the Cut tool, what does the "Through All Cut" option do?

a) Cuts through a specific plane
b) Cuts through the entire part
c) Cuts between two selected faces
d) Cuts up to a finite distance

Q17. How do you create a normal cutout in sheet metal?

a) By using the Close 3-Bend Corner tool
b) By projecting a sketch onto a solid face
c) By using the Bend tool
d) By applying bulge relief

Q18. What is the primary function of the Bend tool?

a) To add a chamfer to edges
b) To modify bend angles
c) To create bends in sheet metal along specified lines
d) To provide relief at bends

Q19. Which tool is used to apply a round or chamfer to a sheet metal edge?

a Break Corner tool
b) Rip Corner tool
c) Close 3-Bend Corner tool
d) Cut tool

Q20. What is the function of the Treatment drop-down in the Close 3-Bend Corner tool?

a) To add bulge relief
b) To define the corner condition and parameters
c) To modify bend angles
d) To create ripped corners

Q21. In the Cut tool, which option allows defining start and end references for a cut?

a) Through All Cut
b) From/To Extent
c) Finite Extent
d) Through Next

Q22. Which option in the Treatment rollout removes draft or crown treatment?

a) Crown
b) No Treatment
c) Draft
d) Through Next

Q23. What action is performed by the Normal Cutout tool?

a) Creates bends in sheet metal
b) Projects a sketch onto a solid face to remove material
c) Rebends previously unbent features
d) Provides bulge relief

Q24. What happens if the All Bends option is selected in the Unbend tool?

a) All unbend features are rebent
b) All bends in the model are unbent
c) Only the active body is affected
d) A specific bend is modified

Chapter 13

Simulation Study

Topics Covered

The major topics covered in this chapter are:

- *Introduction*
- *Simulation Studies in Solid Edge*
- *Performing Linear Static Analysis*
- *Performing Normal Modes Analysis*
- *Performing Linear Buckling Analysis*
- *Performing Heat Transfer Analysis*
- *Analyzing results*

INTRODUCTION TO SIMULATION

Simulation is the study of effects caused on an object due to real-world loading conditions. Computer Simulation is a type of simulation which uses CAD models to represent real objects and it applies various load conditions on the model to study the real-world effects. Solid Edge Simulation is one of the Computer Simulation programs available in the market. In Solid Edge Simulation, we apply loads on a constrained model under predefined environmental conditions and check the result(visually and/or in the form of tabular data). The types of analyses that can be performed in Solid Edge are given next.

TYPES OF ANALYSES PERFORMED IN SOLID EDGE SIMULATION

Solid Edge Simulation allows to perform various analyses that are generally performed in Industries. These analyses and their uses are given next.

Linear Static Analysis

This is the most common type of analysis we perform. In this analysis, loads are applied to a body due to which the body deforms and the effects of the loads are transmitted throughout the body. To absorb the effect of loads, the body generates internal forces and reactions at the supports to balance the applied external loads. These internal forces and reactions cause stress and strain in the body. Static analysis refers to the calculation of displacements, strains, and stresses under the effect of external loads, based on some assumptions. The assumptions are as follows.

1. All loads are applied slowly and gradually until they reach their full magnitudes. After reaching their full magnitudes, load will remain constant (i.e. load will not vary against time).
2. Linearity assumption: The relationship between loads and resulting responses is linear. For example, if you double the magnitude of loads, the response of the model (displacements, strains and stresses) will also double. You can make linearity assumption if:

• All materials in the model comply with Hooke's Law that is stress is directly proportional to strain.
• The induced displacements are small enough to ignore the change is stiffness caused by loading.
• Boundary conditions do not vary during the application of loads. Loads must be constant in magnitude, direction and distribution. They should not change while the model is deforming.

If the above assumptions are valid for your analysis, then you can perform **Linear Static Analysis**. For example, a cantilever beam fixed at one end and force applied on other end; refer to Figure-1.

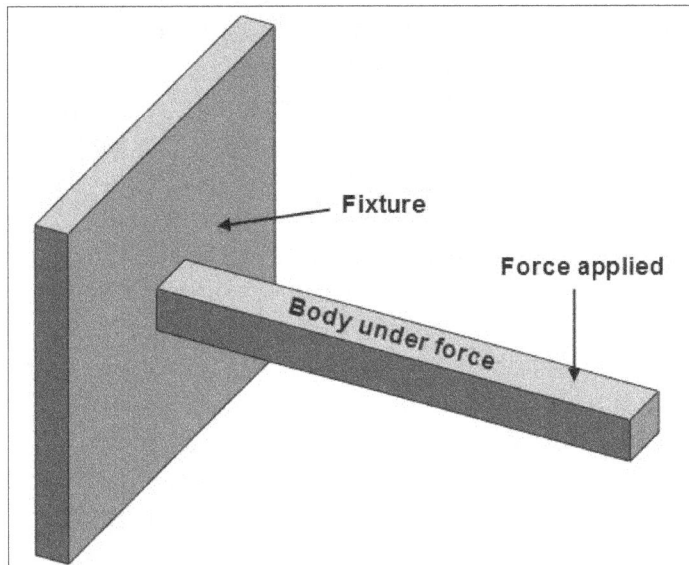

Figure-1. Linear static analysis example

Normal Modes Analysis

The Normal Modes analysis is used to check natural frequency of the model. This study is also called Vibration Analysis. By its very nature, vibration involves repetitive motion. Each occurrence of a complete motion sequence is called a "cycle." Frequency is defined as so many cycles in a given time period. "Cycles per seconds" or "Hertz". Individual parts have what engineers call "natural" frequencies. For example, a violin string at a certain tension will vibrate only at a set number of frequencies, which is why you can produce specific musical tones. There is a base frequency in which the entire string is going back and forth in a simple bow shape.

Harmonics and overtones occur because individual sections of the string can vibrate independently within the larger vibration. These various shapes are called "modes". The base frequency is said to vibrate in the first mode, and so on up the ladder. Each mode shape will have an associated frequency. Higher mode shapes have higher frequencies. The most disastrous kinds of consequences occur when a power-driven device such as a motor for example, produces a frequency at which an attached structure naturally vibrates. This event is called "resonance." If sufficient power is applied, the attached structure will be destroyed. Note that ancient armies, which normally marched "in step," were taken out of step when crossing bridges. Should the beat of the marching feet align with a natural frequency of the bridge, it could fall down. Engineers must design so that resonance does not occur during regular operation of machines. This is a major purpose of Modal Analysis. Ideally, the first mode has a frequency higher than any potential driving frequency. Frequently, resonance cannot be avoided, especially for short periods of time. For example, when a motor comes up to speed it produces a variety of frequencies. So it may pass through a resonant frequency.

Linear Buckling Analysis

If you press down on an empty soft drink can with your hand, not much will seem to happen. If you put the can on the floor and gradually increase the force by stepping down on it with your foot, at some point it will suddenly squash. This sudden scrunching is known as "buckling."

Models with thin parts tend to buckle under axial loading. Buckling can be defined as the sudden deformation, which occurs when the stored membrane(axial) energy is converted into bending energy with no change in the externally applied loads. Mathematically, when buckling occurs, the total stiffness matrix becomes singular. In the normal use of most products, buckling can be catastrophic if it occurs. The failure is not one because of stress but geometric stability. Once the geometry of the part starts to deform, it can no longer support even a fraction of the force initially applied. The worst part about buckling for engineers is that buckling usually occurs at relatively low stress values for what the material can withstand. So they have to make a separate check to see if a product or part thereof is okay with respect to buckling. Slender structures and structures with slender parts loaded in the axial direction buckle under relatively small axial loads. Such structures may fail in buckling while their stresses are far below critical levels. For such structures, the buckling load becomes a critical design factor. Stocky structures, on the other hand, require large loads to buckle, therefore buckling analysis is usually not required.

Buckling almost always involves compression; refer to Figure-2. In mechanical engineering, designs involving thin parts in flexible structures like airplanes and automobiles are susceptible to buckling. Even though stress can be very low, buckling of local areas can cause the whole structure to collapse by a rapid series of 'propagating buckling'. Buckling analysis calculates the smallest (critical) loading required buckling a model. Buckling loads are associated with buckling modes. Designers are usually interested in the lowest mode because it is associated with the lowest critical load. When buckling is the critical design factor, calculating multiple buckling modes helps in locating the weak areas of the model. This may prevent the occurrence of lower buckling modes by simple modifications.

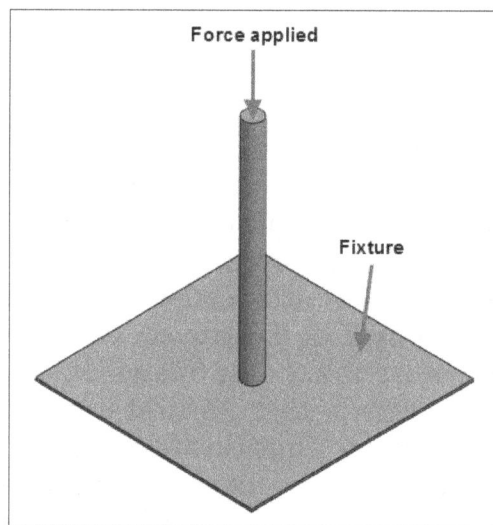

Figure-2. Buckling example

Heat Transfer Analyses

There are three mechanisms of heat transfer. These mechanisms are Conduction, Convection, and Radiation. Thermal analysis calculates the temperature distribution in a body due to some or all of these mechanisms. In all three mechanisms, heat flows from a higher-temperature medium to a lower temperature one. Heat transfer by conduction and convection requires the presence of an intervening medium while heat transfer by radiation does not.

There are two modes of heat transfer analysis.

Steady State Thermal Analysis

In this type of analysis, we are only interested in the thermal conditions of the body when it reaches thermal equilibrium, but we are not interested in the time it takes to reach this status. The temperature of each point in the model will remain unchanged until a change occurs in the system. At equilibrium, the thermal energy entering the system is equal to the thermal energy leaving it. Generally, the only material property that is needed for steady state analysis is the thermal conductivity.

Transient Thermal Analysis

In this type of analysis, we are interested in knowing the thermal status of the model at different instances of time. A thermos designer, for example, knows that the temperature of the fluid inside will eventually be equal to the room temperature(steady state), but he is interested in finding out the temperature of the fluid as a function of time. In addition to the thermal conductivity, we also need to specify density, specific heat, initial temperature profile, and the period of time for which solutions are desired.

Harmonic Response Analysis

The Harmonic response analysis is used to check the effects of response of frequency load on model for specified range of frequencies. Within this specified range of frequencies if natural frequency occurs then it can cause catastrophic failure. So, one of the aim to perform harmonic response analysis is to check whether model is safe from resonance caused by natural frequency loading.

You will learn about performing various studies in Solid Edge later. Various tools in **Simulation** tab are discussed next.

STARTING A STUDY

The **New Study** tool is used to start a new study in the Solid Edge. The procedure to start a new study is given next.

- Click on the **New Study** tool from the **Study** group in the **Simulation** tab of the **Ribbon**. The **Create Study** dialog box will be displayed; refer to Figure-3.

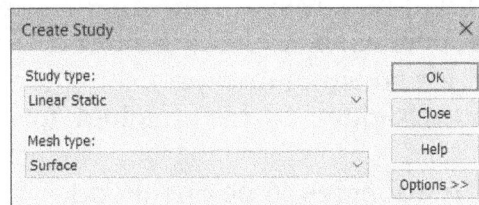

Figure-3. Create Study dialog box

- Select desired option from the **Study Type** drop-down to define which type of study you want to create.
- Select desired option from the **Mesh type** drop-down to define type of elements to be generated in mesh for the study. The **Tetrahedral** element type is used for solid body. Use the **Surface** element type for thin or surface bodies. Select the **General Bodies** option from the drop-down if want to mesh a model which has solid as well as thin bodies.

• Click on the **Options>>** button from the dialog box to display parameters related to study. The options will be displayed as shown in Figure-4.

Figure-4. Options for creating studies

• Select the **Iterative solver** check box to use NX Nastran iterative solver for perform analysis.
• Select the **Large Displacement Solve** check box to perform study on model in which large deformation will be caused (Non-linear analysis).
• Select the **Use multiple processors** check box and then select desired number of processors to be used for running analysis.
• Specify desired number of modes to be checked for Normal Modes or Linear Buckling analyses.
• Specify desired value of frequencies to be checked for finding natural frequencies in the edit boxes of **Frequency range** when performing Normal Modes analyses.
• Select the **On** option from the **Geometry check** drop-down to use default threshold for checking if mesh has desired type and number of element shapes for running the analysis. If there are less number of elements then analysis will fail to run. Select the **Warning Only** option from the drop-down if you want to show warnings in output file when number of elements are less than desired rather than causing the solution to fail.

- Specify desired NX Nastran command in the edit box of **NX Nastran command line options** area. Click on the **NX Nastran Options** button to modify options related to NX Nastran solver. Scan the next QR code to learn more about NX Nastran command lines.

- The **Thermal Load Options** and **Harmonic Response Options** buttons are active when you are performing respective type of analyses. You will learn about these options later in this chapter.

- The options in the **Connector Options** area are active when you are simulating analysis on an assembly. Select the **Create connectors** check box to automatically create connectors when running analysis.

- Select the **Single connection per face pair** check box to create single connection for each face rather than multiple connections for a face.

- Select the **Glue** or **No penetration** option from the **Other connector type** drop-down to apply respective connector in assembly. After selecting desired option, click on the **Properties** button. The **Connector Properties** dialog box will be displayed; refer to Figure-5. Set desired values of friction, search distance, penalty value and so on in the dialog box to define properties of No penetration connector. The coefficient of static friction defines friction between two faces of connector during study. Specify desired value of minimum search distance to define the minimum gap to be used for creating automatic no penetration connector. The value specified in **Penalty value** edit box is used to define stiffness factor for displacement transmission between element pairs. If you are performing thermal study then specify desired value of thermal conductance between elements of connectors. After setting desired parameters, click on the **OK** button.

Figure-5. Connector Properties dialog box

- The **Thermal connector type** drop-down is active when you are performing a thermal analysis. Select the **Thermal Glue** option to make heat transfer between two connected parts based on conduction coefficients of the parts. Select the **Thermal** option to make heat transfer based on specified NX Nastran thermal conductance.

- Select the **Generate only Surface results (faster)** check box to mesh only surfaces of solid and run the analysis as a thin part.

- Select the **Do not process all results after solve (faster)** check box to load only standard results. You can later use **Simulation** pane or tools in **Ribbon** to generate additional results.
- Select the **Check element quality** check box to ignore the elements which are of poor quality due to fine meshing in tight spots of model.
- Select desired check boxes from the **Nodal** section and **Elemental** section to define the results to be generated by the study.
- After setting desired parameters, click on the **OK** button from the dialog box to create a new study.

SWITCHING BETWEEN STUDIES

Select desired option from the **Study List** drop-down in the **Study** group of **Simulation** tab in the **Ribbon** to activate respective simulation study; refer to Figure-6.

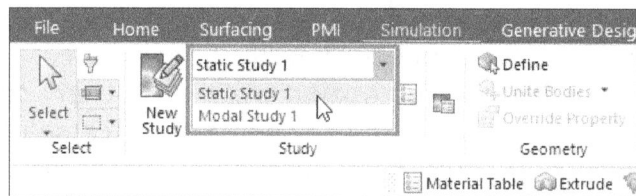

Figure-6. Drop-down for switching study

SELECTING MATERIAL OF MODEL

Click in the **Material List** drop-down and select desired material from list; refer to Figure-7. The selected material will be applied to the part or assembly as default material.

Figure-7. Selecting material for simulation

NASTRAN LOG VIEWER

The **Nastran Log Viewer** tool opens the Nastran Log Viewer pane to view detailed information about your simulation. Click on the **Nastran Log Viewer** tool from **Study** group of **Simulation** tab in the **Ribbon**; refer to Figure-8. The **NX Nastran Log Viewer** dialog box will be displayed; refer to Figure-9.

Figure-8. Nastran Log Viewer tool

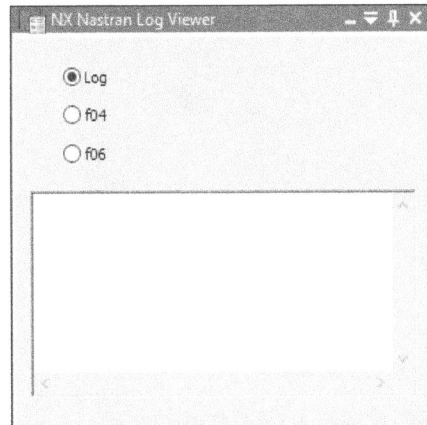

Figure-9. NX Nastran Log Viewer dialog box

SELECTING GEOMETRY FOR ANALYSIS

The **Define** tool is used to select bodies to be used for performing analyses. The procedure to use this tool is given next.

- Click on the **Define** tool from the **Geometry** panel in the **Simulation** tab of the **Ribbon**. The **Define** command box will be displayed; refer to Figure-10.

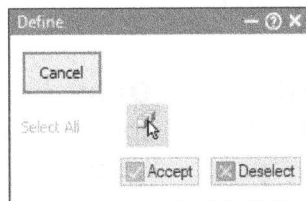

Figure-10. Define command box

- Select desired body to be used for simulation study and right-click in the drawing area.

UNITING BODIES FOR SIMULATION

The **Unite Bodies** tool is used to combine two or more bodies for performing analysis. The procedure to use this tool is given next.

- Click on the **Unite Bodies** tool from the **Unite Bodies** drop-down in the **Geometry** group of the **Ribbon**. You will be asked to select surfaces and solids to be united.
- Select desired surfaces and solids to be united, and right-click in the drawing area. The selected bodies and surfaces will be combined to form single body.

RECOVERING BODIES

The **Recover Bodies** tool is used to extract surfaces and solids from the selected united body. Note that this tool is active only when surface and solid bodies are united in a single body. The procedure to use this tool is given next.

- Click on the **Recover Bodies** tool from the **Unite Bodies** drop-down in the **Geometry** group of **Simulation** tab in the **Ribbon**. You will be asked to select the united body to be split into surface and solids.
- Select desired body and right-click in the drawing area. The bodies will be separated; refer to Figure-11.

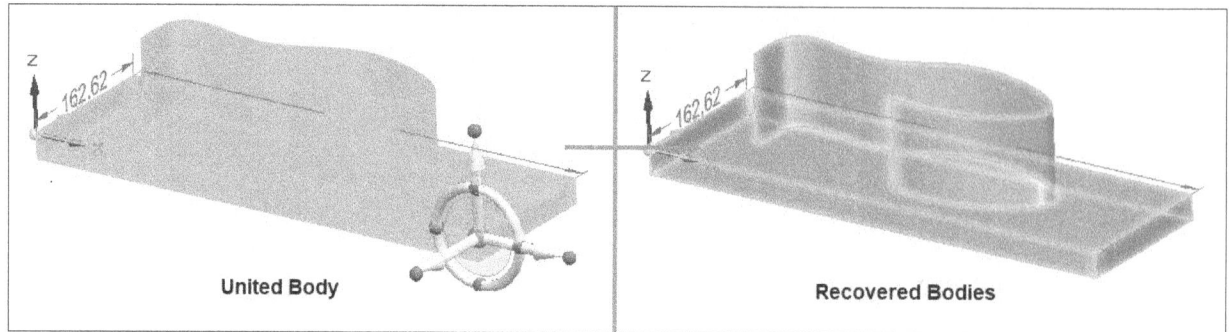

Figure-11. Recovering bodies

• Press **ESC** to exit the tool.

Note that once you have defined bodies for simulation then tools for applying loads will be active in the **Ribbon**. These tools are discussed next.

APPLYING STRUCTURAL LOADS

The tools in the **Structural Loads** group of **Simulation** tab in **Ribbon** are used to apply different type of structural loads like force, pressure, torque, and so on. Various tools in this group are discussed next.

Applying Force Load

The **Force** tool is used to apply force of specified value on selected faces/edges/points/features in the model. The procedure to use this tool is given next.

• Click on the **Force** tool from the **Structural Loads** group in the **Simulation** tab of the **Ribbon**. The **Force** command box will be displayed; refer to Figure-12.

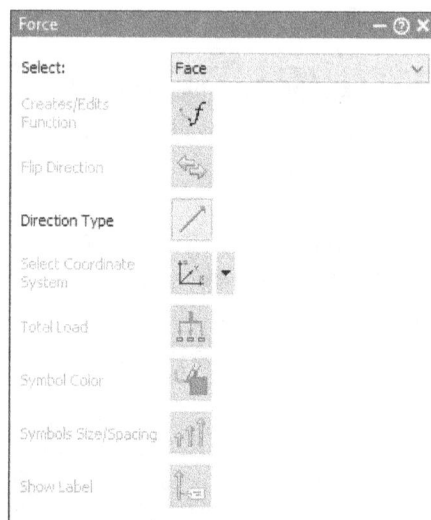

Figure-12. Force command box

• Select desired option from the **Select** drop-down to define which type of objects are to be selected for applying force. (In our case, it is **Face** option.)
• Click on desired faces/edges/points/features on which you want to apply force. Preview of the force will be displayed; refer to Figure-13.

Figure-13. Applying force on face

- Click on the **Flip Direction** button from the command box to reverse the direction of force application.
- Select the **Along a Vector** option from the **Direction Type** drop-down to define direction of load along direction vector of selected face. Select the **Normal to Face** option from the drop-down to apply force perpendicular to selected face. Select the **Components** option from the drop-down to define each component of load separately in respective edit box; refer to Figure-14.

Figure-14. Components of force

- Specify desired load values in the edit box(es).
- Click on the **Select Coordinate System** button to select a different coordinate system for defining direction reference of force.
- Click on the **Total Load** button from the command box if you want to specify combined value of force on all the selected faces. If this button is not selected then specified force value will be individually applied to all the selected faces.
- After setting desired parameters, right-click in the drawing area to apply force.
- Press **ESC** to exit the tool.

Applying Pressure Load

The **Pressure** tool is used to apply the pressure load on selected entity. The procedure to use this tool is given next.

- Click on the **Pressure** tool from the **Structural Loads** group in the **Simulation** tab of the **Ribbon**. The **Pressure** command box will be displayed; refer to Figure-15.

Figure-15. Pressure command box

- Select desired faces on which you want to apply the pressure load and specify desired value in the **Value** edit box.
- The buttons in command box are same as discussed earlier. Right-click in the drawing area to apply the load.

Applying Hydrostatic Pressure Load

The **Hydrostatic Pressure** tool is used to apply pressure load caused by fluid at equilibrium (static) in a container. The procedure to use this tool is given next.

- Click on the **Hydrostatic Pressure** tool from the **Structural Loads** group in the **Simulation** tab of the **Ribbon**. The **Hydrostatic Pressure** command box will be displayed; refer to Figure-16.

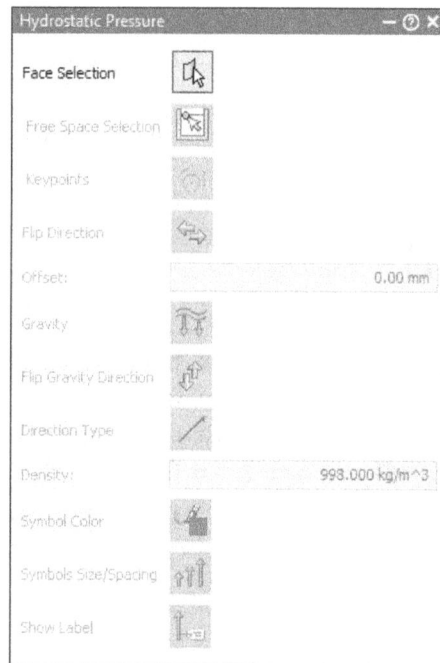

Figure-16. Hydrostatic Pressure command box

- Select the inner face(s) of the model at which you want to apply load and press **ENTER** or right-click in graphics area.
- Specify desired value of density in the **Density** edit box.
- Click at desired key point in the model to specify level up to which fluid will be filled in the container. After specifying point, right-click in the graphics area or press **ENTER**. You will be asked to specify direction and magnitude of gravity.
- Specify desired value and press **ENTER** twice to apply load.

Applying Torque Load

The **Torque** tool is used to apply turning force on selected faces. The procedure to use this tool is given next.

- Click on the **Torque** tool from the **Structural Loads** group in the **Simulation** tab of **Ribbon**. The **Torque** command box will be displayed; refer to Figure-17.

Figure-17. Torque command box

- Select the face on which you want to apply the load and drag the axis of rotation to desired location; refer to Figure-18.

Figure-18. Applying torque on selected face

- Specify desired value of torque in the **Value** edit box and press **ENTER** to apply torque.

Applying Displacement

The **Displacement** tool is used to apply specified value of displacement to selected faces/edge/point/features. Note that there must be matching constraint where selected faces/edge/point/features can move by specified value. The procedure to use this tool is given next.

- Click on the **Displacement** tool from the **Structural Loads** group in the **Simulation** tab of **Ribbon**. The **Displacement** command box will be displayed; refer to Figure-19.

Figure-19. Displacement command box

- Select desired entity of model to be displaced and specify desired value of displacement in the **Value** edit box.
- Set the other parameters as discussed earlier and right-click in the drawing area to apply displacement.

Applying Bearing Load

The **Bearing** tool is used to apply physical loading of bearing on cylindrical and non cylindrical parts. The procedure to use this tool is given next.

- Click on the **Bearing** tool from the **Structural Loads** group in the **Simulation** tab of **Ribbon**. The **Bearing** command box will be displayed; refer to Figure-20.

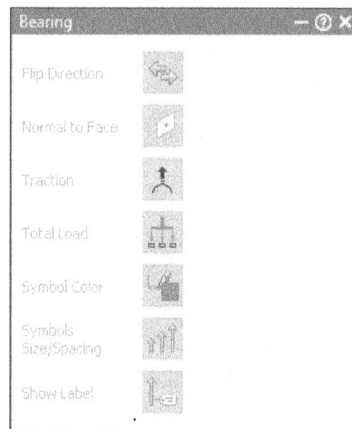

Figure-20. Bearing command box

- Select desired face of model on which you want to apply bearing load and drag the axis center at desired location.
- Select the **Traction** button from the command box if you want to pull the faces opposite to default push motion of the force.
- Specify desired values in the dynamic edit boxes. Right-click in the drawing area to apply the load.

APPLYING BODY LOADS

The tools in **Body Loads** group of **Ribbon** are used to apply loads on whole part/body during analysis. Various tools of **Body Loads** group are discussed next.

Applying Body Temperature

The **Body Temperature** tool is used to apply initial temperature to the model. Note that this tool is useful for transient thermal analysis and radiation analysis. The procedure to use this tool is given next.

* Click on the **Body Temperature** tool from the **Body Loads** group in the **Simulation** tab of **Ribbon**. The **Body Temperature** command box will be displayed with **Value** edit box; refer to Figure-21.

Figure-21. Body Temperature command box

* Enter desired value of temperature in the **Value** edit box to specify initial temperature of model. The temperature will be applied to system.

Applying Centrifugal Load

The **Centrifugal** tool is used to apply centrifugal load on whole model. Centrifugal force is generated when a part rotates about its axis. The procedure to apply this load is given next.

* Click on the **Centrifugal** tool from the **Body Loads** group in the **Simulation** tab of **Ribbon**. The preview of load will be displayed with **Centrifugal** command box and edit boxes; refer to Figure-22.

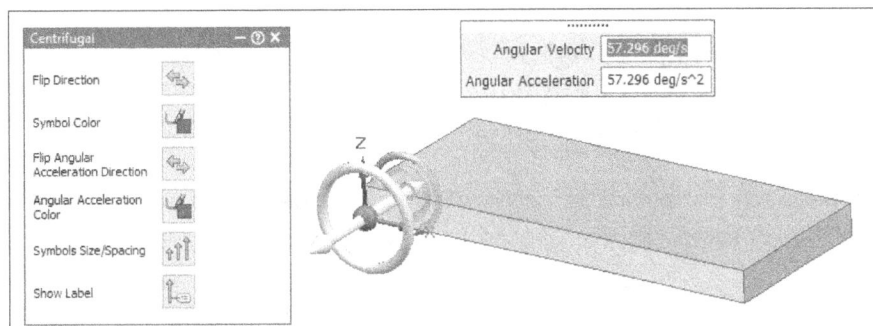

Figure-22. Centrifugal command box

* Drag the center point of axis of rotation to desired location in model. The direction of centrifugal load will be modified accordingly.
* Specify desired values in **Angular Velocity** and **Angular Acceleration** edit boxes.
* Using the **Flip** buttons in command box, you can flip direction of angular velocity and angular acceleration.
* Right-click in the drawing area to apply the loads.

Applying Gravity Load

The **Gravity** tool is used to represent effect of gravity on the model. The procedure to use this tool is given next.

*   Click on the **Gravity** tool from the **Body Loads** group in the **Simulation** tab of **Ribbon**. The **Gravity** command box will be displayed; refer to Figure-23.

Figure-23. Gravity command box

*   Drag the center point of direction vector on desired edge or face to define direction of gravity.
*   Specify desired value in the **Value** edit box to define value of gravity.
*   Right-click in the drawing area to apply the load.

APPLYING THERMAL LOADS

The tools in **Thermal Loads** group of **Ribbon** are active when you are performing thermal analysis on the model; refer to Figure-24. The tools of this group are discussed next.

Figure-24. Thermal Loads group

Applying Temperature Load

The **Temperature** tool is used to apply specified temperature value load on selected face/edge/point/feature. The procedure to use this tool is given next.

*   Click on the **Temperature** tool from the **Thermal Loads** group in the **Simulation** tab of the **Ribbon**. The **Temperature** command box will be displayed; refer to Figure-25.

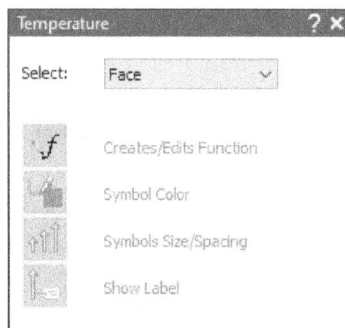

Figure-25. Temperature command box

- Select desired option from the **Select** drop-down in command box to define reference type to be selected for applying thermal load.
- Select the reference face/edge/point/feature from the model. An input box will be displayed for specifying temperature value; refer to Figure-26.

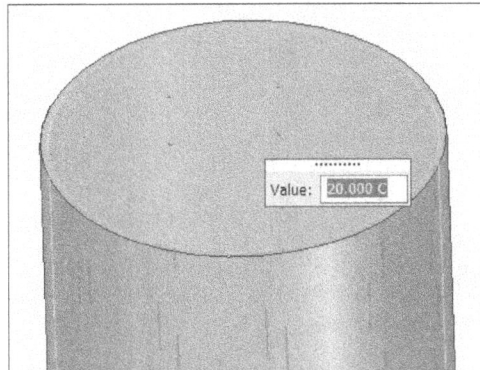

Figure-26. Input box for specifying temperature

- Specify desired value of temperature in the input box and press **ENTER**. The load will be assigned to selected reference.

Applying Heat Flux Load

The **Heat Flux** tool is used to apply heat energy of specified watt value. The energy is applied at the beginning of analysis and it is not a continuous flow of energy. The procedure to use this tool is given next.

- Click on the **Heat Flux** tool from the **Thermal Loads** group in the **Simulation** tab of **Ribbon**. The **Heat Flux** command box will be displayed; refer to Figure-27.

Figure-27. Heat Flux command box

- Select desired reference option from the **Select** drop-down and select the geometry on which load is to be applied. The input box for specifying load will be displayed.
- Specify desired value in input box and press **ENTER**. The load will be applied.

Applying Convection Load

The **Convection** tool is used to apply thermal convection coefficient and ambient temperature. The procedure to use this tool is given next.

- Click on the **Convection** tool from the **Thermal Loads** group in the **Simulation** tab of **Ribbon**. The **Convection** command box will be displayed.

- Select the reference geometry on which load will be applied. Input boxes will be displayed for specifying load; refer to Figure-28.

Figure-28. Input boxes for convection

- Specify desired values of convection coefficient and ambient temperature in the input boxes. Press **ENTER** to apply the load.

Applying Radiation Thermal Load

The **Radiation** tool is used to define radiation coefficient for selected geometry. The procedure to use this tool is given next.

- Click on the **Radiation** tool from the **Thermal Loads** group in the **Simulation** tab of **Ribbon**. The **Radiation** command box will be displayed; refer to Figure-29.
- Select the reference geometry on which radiation parameters will be applied.
- Click on the **Radiation Options** button from the command box. The **Radiation Load Options** dialog box will be displayed; refer to Figure-30.

Figure-29. Radiation command box

Figure-30. Radiation Load Options dialog box

- Select the **Radiation to Space** option from the **Radiation type** drop-down if you want to define parameters of radiation from selected reference to atmosphere. Select the **Enclosed Radiation** option from the drop-down to define radiation parameters between selected reference geometries.
- Specify other parameters as desired in the edit boxes of the dialog box and click on the **OK** button.
- Press **ENTER** to apply the load and exit the tool.

APPLYING CONSTRAINTS

The tools in the **Constraints** group are used to restrict the free motion of part(s) during analysis; refer to Figure-31. Various tools in this group are discussed next.

Figure-31. Constraints group

Applying Fixed Constraint

The **Fixed** constraint is used to stop movement of selected entities in all 6 directions (3 translation and 3 rotation). The procedure to apply this constraint is given next.

• Click on the **Fixed** tool from the **Constraints** group in the **Simulation** tab of **Ribbon**. The **Fixed** command box will be displayed; refer to Figure-32.

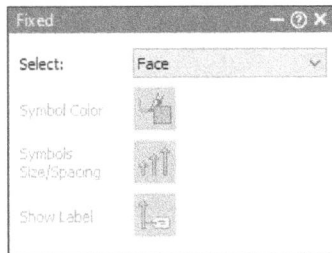

Figure-32. Fixed command box

• Select desired option from the **Select** drop-down in command box to define what type of entities are to be selected and then select desired faces/edges/points/ features to be fixed.
• Right-click in the drawing area to apply constraint.

Applying Pinned Constraint

The **Pinned** constraint is used to allow free rotation of selected entity while restricting translation in all three directions. The procedure to apply this constraint is given next.

• Click on the **Pinned** tool from the **Constraints** group in the **Simulation** tab of **Ribbon**. The **Pinned** command box will be displayed; refer to Figure-33.

Figure-33. Pinned command box

• Select desired option from the **Select** drop-down and then select the respective entities from model to be constrained as pinned.

Applying No Rotation Constraint

The **No Rotation** constraint is used to stop rotation of model while restricting the translation movement of model. The procedure to use this tool is given next.

• Click on the **No Rotation** tool from the **Constraints** group in the **Simulation** tab of **Ribbon**. The **No Rotation** command box will be displayed.
• Select desired entities and right-click in the drawing area to apply the constraints.

Applying Slide Along Surface Constraint

The **Sliding Along Surface** tool is used to allow sliding of selected faces along the selected surfaces. The procedure to use this tool is given next.

• Click on the **Sliding Along Surface** tool from the **Constraints** group in the **Simulation** tab of the **Ribbon**. The **Sliding Along Surface** command box will be displayed; refer to Figure-34.

Figure-34. Sliding Along Surface command box

• Select the planar faces of model that you want to use for sliding and right-click in the drawing area. The constraint will be applied; refer to Figure-35.

Figure-35. Sliding along surface constraint applied

Applying Cylindrical Constraint

The **Cylindrical** constraint is applied to the model for allowing rotation about an axis and translation along that same axis. The procedure to apply this constraint is given next.

• Click on the **Cylindrical** tool from the **Constraints** group in the **Simulation** tab of **Ribbon**. The **Cylindrical** command box will be displayed; refer to Figure-36 and you will be asked to select cylindrical faces for apply constraint.

Figure-36. Cylindrical command box

- Select the cylindrical faces of model to which you want to apply the constraint. The buttons in the command box will become active; refer to Figure-37.

Figure-37. Face selected for cylindrical constraint

- Select the **Constrain Radial Growth** button to restrict radial changes in the model.
- Select the **Constrain Rotation** button to stop rotation of cylindrical part about its axis.
- Select the **Constrain Sliding Along Axis** button to restrict translation movement of part along its center axis.
- After setting desired parameters, right-click in the drawing area to apply constraint.

Applying User-Defined Constraint

The **User-Defined** tool is used to restrict selected degrees of freedom of part. The procedure to use this tool is given next.

- Click on the **User-Defined** tool from the **Constraints** group in the **Simulation** tab of **Ribbon**. The **User-Defined** command box will be displayed; refer to Figure-38.

Figure-38. User-Defined command box

- Select desired option from the **Select** drop-down and then select desired entity of model to which you want to apply the constraint. Preview of constraint will be displayed; refer to Figure-39.

Figure-39. Preview of constraint

- Click on desired axis of triad to restrict movement in respective direction.
- After setting desired parameters, right-click in the drawing area. The constraint will be applied.

MESHING

The tools to create mesh are available in the **Mesh** group of the **Ribbon**; refer to Figure-40. Various tools of this group are discussed next.

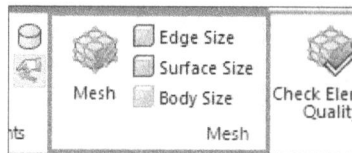

Figure-40. Mesh group

Creating Mesh

The **Mesh** tool is used to create mesh of specified element size. The procedure to use this tool is given next.

- Click on the **Mesh** tool from the **Mesh** group in the **Simulation** tab of **Ribbon**. The **Mesh** dialog box will be displayed related to selected element type in the **Create Study** dialog box; refer to Figure-41.

Figure-41. Tetrahedral Mesh dialog box

- Move the slider to desired side (Coarser or Finer) for creating coarse or fine mesh.
- Select the **Body Mesh** check box to smoothen mesh on faces of bodies. Note that if this check box is selected then options to specify parameters for solid mesh and surface mesh will not be displayed.

- Click on the **Options** button from the dialog box to specify meshing options for different elements. The **Mesh Options** dialog box will be displayed; refer to Figure-42.

Figure-42. Mesh Options dialog box

- Set desired parameters in the **Body Mesh**, **Solid Mesh**, and **Surface Mesh** pages of the dialog box and click on the **OK** button.
- Click on the **Mesh** button from the **Tetrahedral Mesh** dialog box to create the mesh; refer to Figure-43.

Figure-43. Preview of mesh

- Click on the **Close** button from the dialog box to exit the dialog box.

Setting Mesh Size for Edges

The **Edge Size** tool is used to set size of mesh elements along selected edges. The procedure to set mesh size of edges is given next.

- Click on the **Edge Size** tool from the **Mesh** group in the **Simulation** tab of **Ribbon**. The **Edge Size** dialog box will be displayed; refer to Figure-44.

Figure-44. Edge Size dialog box

- Select desired radio button from the dialog box and specify the parameters for mesh elements.
- Select the edges for which you want to specify edge mesh elements' sizes. Preview of the mesh will be displayed; refer to Figure-45.

Figure-45. Preview of edge mesh

- Click on the **Accept** button from the dialog box to apply changes. Press **ESC** to exit the tool.

Specifying Surface Mesh Sizes

The **Surface Size** tool is used to modify size of mesh at surface of model. The procedure to use this tool is given next.

- Click on the **Surface Size** tool from the **Mesh** group in the **Simulation** tab of **Ribbon**. The **Surface Size** dialog box will be displayed; refer to Figure-46.

Figure-46. Surface Size dialog box

- Specify desired element size in the edit box and select desired option from the **Mapped mesh** drop-down to define which type of mesh elements will be created.
- Select the surfaces whose meshing is to be modified. Preview of mesh will be displayed; refer to Figure-47.

Figure-47. Preview of surface mesh

- Click on the **Accept** button from the dialog box to apply changes. Press **ESC** to exit the tool.

Specifying Body Mesh Size

The **Body Size** tool is used set size of body mesh in the model. The procedure to use this tool is given next.

- Click on the **Body Size** tool from the **Mesh** group in the **Simulation** tab of **Ribbon**. The **Body Size** dialog box will be displayed; refer to Figure-48.

Figure-48. Body Size dialog box

- Set desired parameters in the dialog box and click on the **Accept** button.
- Press **ESC** to exit the tool.

SOLVING ANALYSIS

The **Solve** tool in **Ribbon** is active when material, load, and constraints have been applied to the model. After setting the parameters, click on the **Solve** tool from the **Solve** group in the **Simulation** tab of **Ribbon**. The results of analysis will be displayed; refer to Figure-49.

Figure-49. Results of structural analysis

SIMULATION RESULTS

Once you have performed the analysis, the results are displayed in **Simulation Results** environment. Various tools in this environment are discussed next.

Probing Results

The **Probe** tool is used to check analysis results at selected nodes. The procedure to use this tool is given next.

- Click on the **Probe** tool from the **Probe** group in the **Home** tab of **Ribbon**. The **Probe Table** dialog box will be displayed; refer to Figure-50.

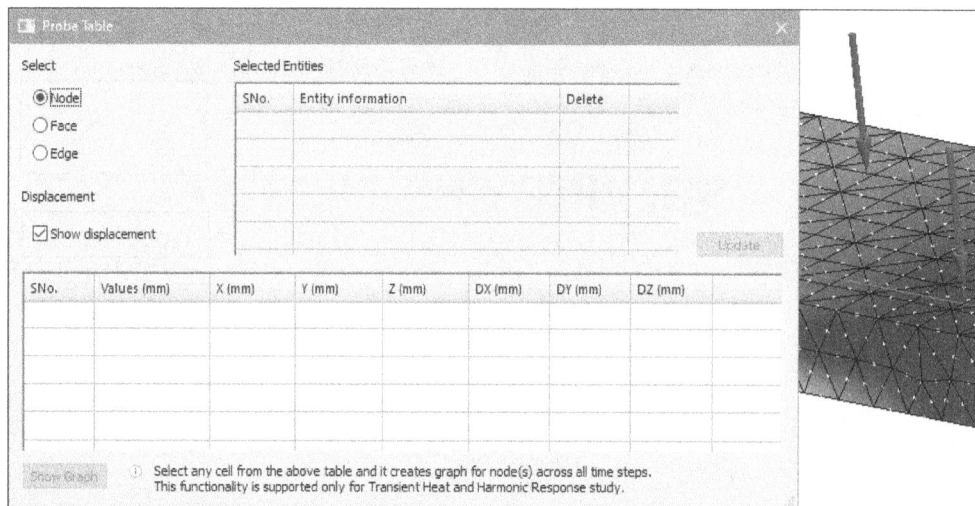

Figure-50. Probe Table dialog box

- Select the **Node** radio button from the **Select** area of the dialog box to check the analysis results of selected node. Select the **Face** radio button from the area to check results at selected face. If you have selected **Face** radio button then all the nodes of selected face will get selected for checking results on clicking **Update** button; refer to Figure-51.

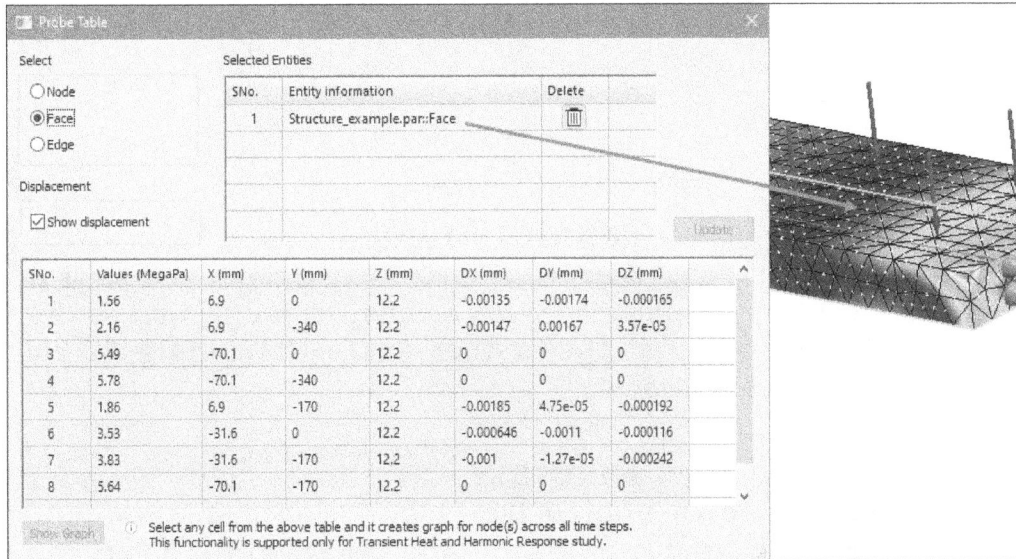

Figure-51. Checking results on nodes

- Select the **Show displacement** check box to also check displacement results of nodes.
- Press **ESC** to exit the tool.

Display Options

The **Display Options** tool is used to show/hide various elements of result. On clicking this tool, the box to define which elements to be shown/hidden in analysis; refer to Figure-52. Select desired check boxes from the box and click on the **Close** button.

Figure-52. Display Options

Equilibrium Check

The **Equilibrium Check** tool is used to check whether forces are balanced on all nodes in the model. On clicking this tool, the **Equilibrium Check** dialog box will be displayed; refer to Figure-53. Click on the **Export** button to export all the results of equilibrium check in the form of a csv file. After checking the results, click on the **OK** button.

Figure-53. Equilibrium Check dialog box

Contour Style

The buttons in the **Contour Style** group are used to display the results in contour styles like smooth contour, banded contour, element contour, and so on. Click on the **Dynamic Iso Contour** tool from the **Contour Style** group in the **Ribbon** to move iso contour lines in the result model.

Deformation

The options in the **Deformation** group of **Home** tab in **Ribbon** are used to define how deformation results will be displayed in the drawing. Select the **Percentage** radio button from the **Deformation** group to display deformation in model as percentage of model size. Select the **Normalize** radio button from the **Deformation** group to set a maximum deformation value for result and adjust rest of the deformation in model accordingly. Select the **Actual** radio button from the **Deformation** group to display real deformation of model.

Animating Results

The **Animate** tool is used to animate the deformation of model. Click on the **Animate** tool from the **Animate** group in the **Home** tab of **Ribbon**. The **Animate** command box will be displayed; refer to Figure-54.

Figure-54. Animate command box

You can use the **Play/Pause**, **Stop**, **Previous Frame**, **Next Frame**, and **Current Frame** options in the **Command box** to surf through the animation. Select the **Animate Full Cycle**, **Animate Sinusoidal Cycle**, **Animate Across Modes**, and **Animate Contours** buttons to set different parameters of animation. Click on the **Save as Movie** button from the command box to save animation as video clip in AVI format. Click on the **Close** button to exit the command box.

CREATING REPORT

The **Create Report** tool from the **Output Results** group is used to generate report of analysis as html, doc, or docx file. The procedure to use this tool is given next.

- Click on the **Create Report** tool from the **Output Results** group in the **Home** tab of **Ribbon**. The **Create Report** dialog box will be displayed; refer to Figure-55.

Figure-55. Create Report dialog box

- Specify desired parameters in various edit boxes of the dialog box.
- Select desired radio button from the **Document** area of dialog box to define format of document.
- After setting all parameters, click on the **Create Report** button. The report will be generated; refer to Figure-56.

Figure-56. Report created

You can use the **Save As Movie** and **Save As Image** tools in the **Output Results** group of **Ribbon** as discussed earlier.

DATA SELECTION FOR RESULTS

The options in the **Data Selection** group of **Ribbon** are used to set the parameters to be displayed in the results; refer to Figure-57.

Figure-57. Data selection for results

Similarly, you can use the other tools of **Simulation Results** environment. Click on the **Close Simulation Results** tool from the **Close** group to exit the environment.

Optimization Study

The Optimization study is used to change the geometry of model based on results of linear static analysis. Using this study, you can get the model with minimum mass which passes the linear static analysis. The procedure to create optimization study is given next.

• After running the linear static analysis on model, click on the **New Optimization** tool from the **Study** group in the **Simulation** tab of **Ribbon**. The **Optimization** dialog box will be displayed; refer to Figure-58.

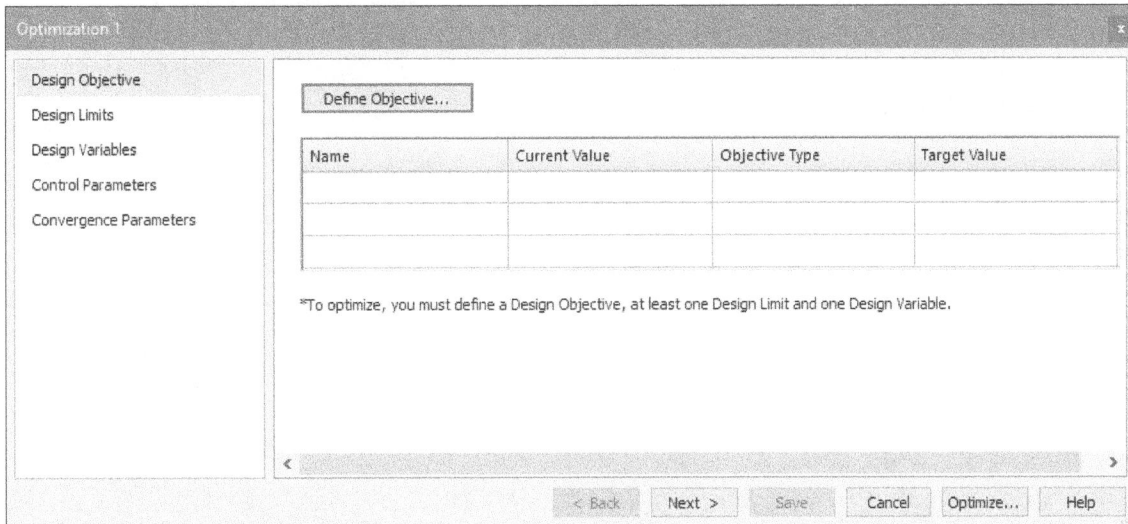

Figure-58. Optimization dialog box

- Click on the **Define Objective** button from the dialog box. The **Define Objective** dialog box will be displayed; refer to Figure-59.

Figure-59. Define Objective dialog box

- Select desired option from the dialog box to define which physical property or simulation result will be used for optimization. For example, select the **Mass** option from the dialog box to get an optimized model with minimum mass which passes the linear static analysis.
- After selecting desired property, click on the **OK** button from the dialog box. The property will be added in the **Design Objective** list; refer to Figure-60.

Figure-60. Objective added in the list

- Select desired option from the **Objective Type** column to specify whether selected property will be minimized, maximized, or need to achieve specified target value.
- Click on the **Design Limits** option from the dialog box to define limit within which the selected property should be when optimizing the model based on analysis.
- Click on the **Design Variables** option from the dialog box to select the design variables to be used for optimizing the model. After selecting this option, click on the click on the **Add Model Variable** button. The **Variable Table** dialog box will be displayed; refer to Figure-61.

Figure-61. Variable Table dialog box

- Select desired properties (while holding the **CTRL** key) you want to add in the list and click on the **Add** button from the dialog box. The list of variables will be added. Click on the button under **Rule** column for desired variable to define rules for properties during analysis. The **Design Variable Rule Editor** dialog box will be displayed; refer to Figure-62. Select desired options from the **Minimum limit** and **Maximum limit** drop-downs, and specify desired values for the parameters.

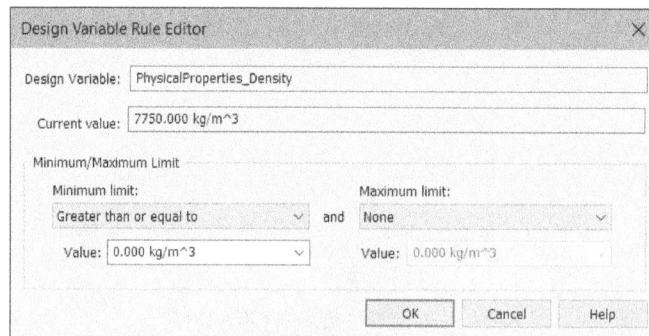

Figure-62. Design Variable Rule Editor dialog box

- Select the **Control Parameters** option from the dialog box to specify maximum number of iterations to be performed for optimization.
- Select the **Convergence Parameters** option from the dialog box to specify the convergence limit at which analysis will be complete; refer to Figure-63.

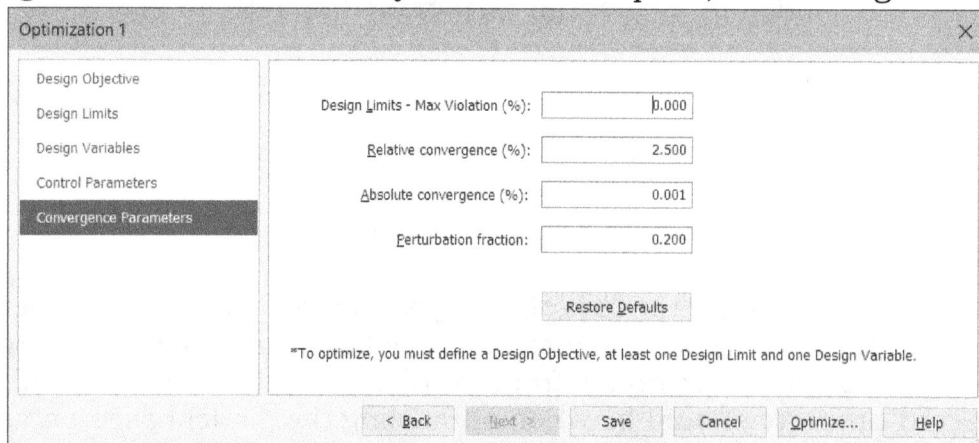

Figure-63. Convergence Parameters options

- After setting the parameters, click on the **Optimize** button to run optimization study. Once the study is complete, an **Optimization** dialog box will be displayed; refer to Figure-64.

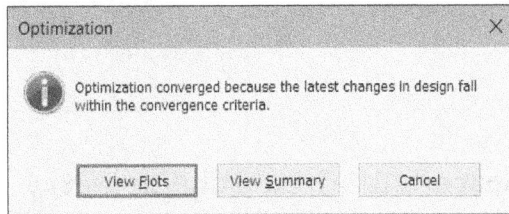

Figure-64. Optimization dialog box

- Click on the **View Plots** button to check plots. Click on the **View Summary** button to check analysis result summary; refer to Figure-65.

	A	B	C	D	E	
10						
11	Design Variables		Type	Name	Value	Range
12		mN	Sim	Force_1_Static_Study_1	1.000E+06	[1,00,000.00 m
13						
14						
15	Optimization Results	Units				
16	Iteration		1	2	3	
17	Design Objective					
18	Mass	kg	3.559	3.559	3.559	
19						
20	Design Variable					
21	Force_1_Static_Study_1	mN	1.000E+06	7.800E+05	7.816E+05	
22						
23	Design Limit					
24	Total Translation	mm	0.016	0.012	0.012	
25						
26						
27	Processed Results	Units				
28	Iteration		1	2	3	
29	Total Translation-Minimum	mm	0	0	0	
30	Total Translation-Maximum	mm	0.016	0.012	0.012	

Figure-65. Optimization summary

Right-click on the iteration of optimization study whose geometry you want to be checked and select the **Show Geometry** option; refer to Figure-66.

Figure-66. Show Geometry option in Simulation Tree

PERFORMING GENERATIVE STUDY

The Generative Study is performed to generate the shape and size of model based on analysis results. The major goal of performing generative study is to reduce mass of model. The procedure to perform the study is given next.

- Click on the **Create Generative Study** tool from the **Generative Study** group in the **Generative Design** tab of **Ribbon**. The tools to perform generative study will be active in the **Ribbon**; refer to Figure-67.

Figure-67. Generative Design options

- Click on the **Design Space** button from the **Geometry** group in the **Generative Design** tab of **Ribbon** to select the bodies to be used for generative study. The **Design Space** command box will be displayed. Select desired bodies and click on the **OK** button.

- Click on the **Preserve Region** button from the **Geometry** group in the **Generative Design** tab of **Ribbon**. The **Preserve Region** command box will be displayed; refer to Figure-68. Select the faces of regions you want to preserve. You will be asked to specify offset value up to which the region will be preserved from selected faces; refer to Figure-69. Right-click in the drawing area to create preserve region.

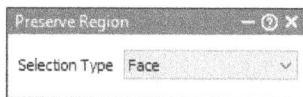

Figure-68. Preserve Region command box

Figure-69. Offset for preserve region

- Apply desired loads like force, pressure, torque, and gravity on the model as discussed earlier.

- Apply desired constraints to the model like fixed, pinned, and displacement.

- After setting all desired parameters, click on the **Generate** button from the **Generate** group in the **Generative Design** tab of the **Ribbon**. The **Generate Study** dialog box will be displayed; refer to Figure-70.

Figure-70. Generate Study dialog box

- Use the **Study Quality** slider to define the quality of analysis. The higher you set quality of analysis, the more time it takes to perform.
- Use the slider in **Mass Reduction** area to define maximum reduction in mass of model to be applied.
- Click in the **Factor of safety** edit box to specify factor of safety to be achieved by model generated.
- After setting desired parameters, click on the **Generate** button. The generated model will be displayed; refer to Figure-71.

Figure-71. Model generated by generative design study

PERFORMING NORMAL MODES ANALYSIS

The Normal Modes analysis is performed to find out natural frequency of model. The procedure to perform this analysis is given next.

Starting Normal Modes Analysis

- Click on the **New Study** tool from the **Study** group in the **Simulation** tab of the **Ribbon**. The **Create Study** dialog box will be displayed.
- Select the **Normal Modes** option from the **Study type** drop-down.
- Select **Tetrahedral** option from **Mesh type** drop-down because we are working on a solid model.
- Set desired number of natural frequencies to be found by analysis in the **Number of modes** drop-down.
- Specify the range of frequencies within which natural frequencies are to be found in the **Frequency range** edit boxes.
- Set the other parameters as desired and click on the **OK** button to start analysis.

Performing Normal Modes Analysis

- Select desired option from the **Material** drop-down in **Study** group of **Simulation** tab in the **Ribbon**. (In our case, it is Steel).
- Click on the **Fixed** tool from the **Constraints** group in **Simulation** tab of the **Ribbon** to fix top face of the model. The **Fixed** command box will be displayed.
- Select the top face of model; refer to Figure-72 and press **ENTER**. The constraint will be applied

Figure-72. Face selected for fixing

- Click on the **Mesh** tool from **Mesh** group in the **Simulation** tab of **Ribbon**. The **Tetrahedral Mesh** dialog box will be displayed.
- Set desired mesh size using slider or edit box in the dialog box and click on the **Mesh** button. The mesh will be created; refer to Figure-73.

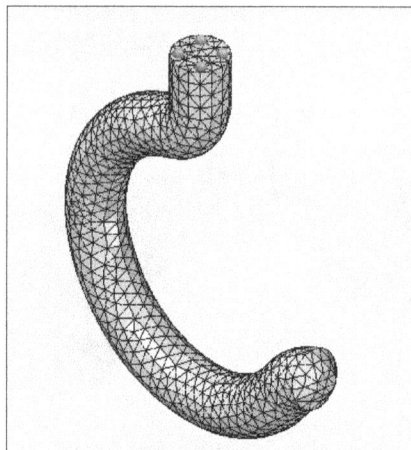

Figure-73. Mesh of model

- Click on the **Close** button to exit the dialog box.
- Click on the **Solve** tool from **Solve** group in **Simulation** tab of **Ribbon** to perform analysis.

- Select the result mode from **Result Mode** drop-down in the **Data Selection** group of **Home** tab in the **Ribbon** to check frequencies and mode shapes of the model; refer to Figure-74.

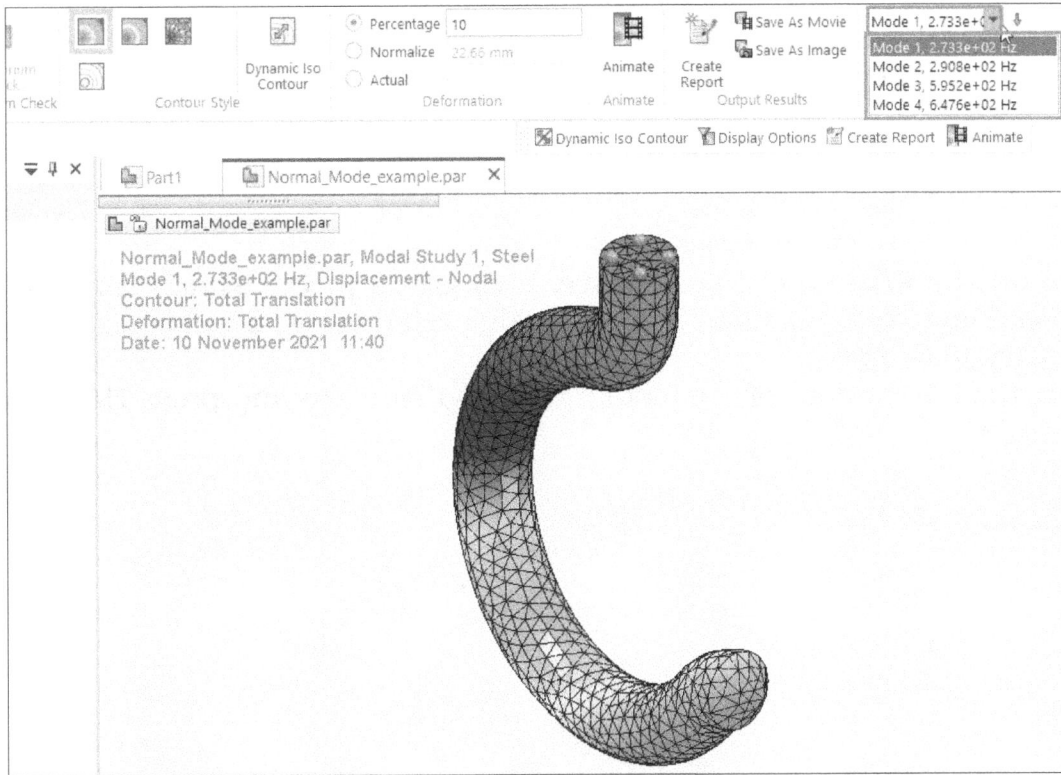

Figure-74. Mode shapes

- Click on the **Close Simulation Results** tool in **Close** group of **Ribbon** to exit results mode.

PERFORMING LINEAR BUCKLING ANALYSIS

Buckling analysis is performed to check sudden buckling failure of the model due to load. In case of linear buckling analysis, non-linearity of material, load, and constraints is not considered. The procedure to perform buckling analysis is given next.

- After opening model for analysis, click on the **New Study** tool from **Study** group in **Simulation** tab of **Ribbon**. The **Create Study** dialog box will be displayed.
- Select the **Linear Buckling** option from the **Study type** drop-down, **Tetrahedral** option from the **Mesh type** drop-down, and **5** in the **Number of Modes** edit box to find out five buckling modes.
- Set the other options as discussed earlier and click on the **OK** button.
- Click in the **Material** drop-down of **Study** group in the **Simulation** tab of **Ribbon** and select desired material (Steel in our case).
- Click on the **Force** tool from the **Structural Loads** group in the **Ribbon**. The **Force** command box will be displayed.
- Select the top face of model and apply a load of 5000 N downward; refer to Figure-75.

Figure-75. Applying load

- Click on the **Fixed** tool from **Constraints** group in the **Simulation** tab of the **Ribbon**. The **Fixed** command box will be displayed and you will be asked to select geometry to be fixed.
- Select the bottom face of the model; refer to Figure-76 and press **ENTER** to apply constraint.

Figure-76. Face selected for constraint

- Click on the **Solve** tool from the **Solve** group in the **Simulation** tab of **Ribbon**. The solution of buckling analysis will be displayed; refer to Figure-77. To learn more about buckling analysis, scan the QR code given next.

Figure-77. Result of buckling analysis

- Click on the **Close Simulation Results** tool from the **Ribbon** to exit result mode.

PERFORMING THERMAL ANALYSIS

The thermal analysis is used to check distribution of heat energy in model under thermal loads. The procedure to perform both static and transient thermal analyses is given next.

- Open the model on which you want to perform thermal analysis and click on the **New Study** tool from the **Study** group in the **Simulation** tab of **Ribbon**. The **Create Study** dialog box will be displayed.
- Select the **Steady State Heat Transfer** option from the **Study type** drop-down and **Tetrahedral** option from the **Mesh type** drop-down in the dialog box.
- Click on the **OK** button from the dialog box to start the study.
- Select desired material (Steel) from the **Material List** drop-down in the **Study** group of **Simulation** tab in the **Ribbon**.
- Click on the **Heat Flux** tool from the **Thermal Loads** group in the **Ribbon**. The **Heat Flux** command box will be displayed and you will be asked to select the geometry on which load is to be applied.
- Select the face of model for applying load; refer to Figure-78 and enter the load as 1500W.

Figure-78. Applying heat flux

- Click on the **Convection** tool from the **Thermal Loads** group in the **Ribbon**. The **Convection** command box will be displayed.
- Select the face through which heat will escape via convection; refer to Figure-79.

Figure-79. Face selected for convection

- Specify the value of film coefficient as **5** W/m² -K and ambient temperature as **25** °C in the input boxes. Press **ENTER** to apply the load.
- After setting parameters, click on the **Solve** tool from **Solve** group in the **Simulation** tab of the **Ribbon**. Result of analysis will be displayed; refer to Figure-80.

Figure-80. Static thermal analysis result

- Click on the **Close Simulation Results** tool from **Close** group in the **Ribbon**.
- To start a Transient Heat Transfer analysis, click on the **New Study** tool from the **Ribbon** and select the **Transient Heat Transfer** option from the **Study type** drop-down in **Create Study** dialog box. The **Transient Heat Options** dialog box will be displayed; refer to Figure-81.

Figure-81. Transient Heat Options dialog box

- Set desired time parameters for analysis step control like analysis end time 10 seconds, calculation time increment 2 seconds, output time increments 2 seconds, and initial temperature 25 °C.
- Click on the **OK** button from the **Transient Heat Options** dialog box and **Create Study** dialog box. Set the loads and constraints as discussed for static heat transfer analysis. Note that when you are performing transient analysis, you can use **Creates/Edits Function** button from the load related dialog box to vary load with time refer to Figure-82. On selecting this button, the **New Function** dialog box will be displayed if you have not created a function earlier; refer to Figure-83.

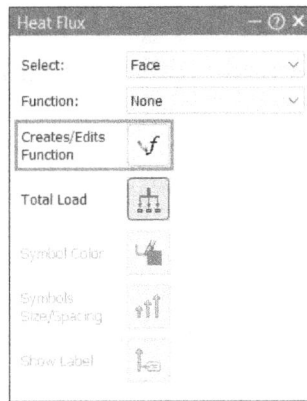

*Figure-82. Heat Flux command box
for transient analysis*

- Set desired parameters in the table to define time and multiply factor for the parameter in the table. The load will vary as per specified values. For example, if you have set time as 1 sec and factor as 0.5 then at the end of 1 second during analysis, the load value will be half (0.5 x load).
- After setting desired parameters, click on the **OK** button.

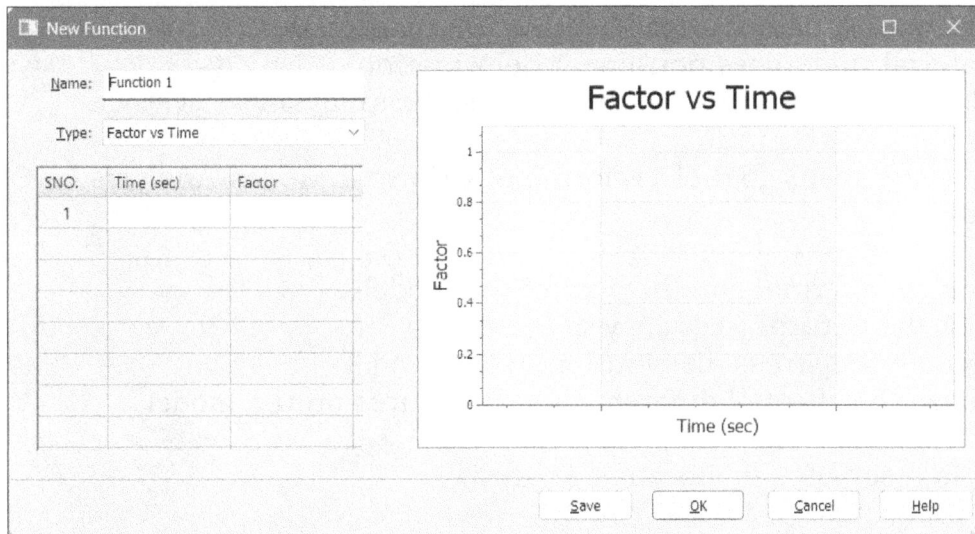

Figure-83. New Function dialog box

- After solving the transient heat transfer analysis, you can check results at different time steps; refer to Figure-84.

Figure-84. Results of transient heat transfer analysis

SELF ASSESSMENT

Q1. Which of the following is a type of analysis available in Solid Edge Simulation?

A) Linear Static Analysis
B) Dynamic Thermal Analysis
C) Aerodynamic Analysis
D) Chemical Stress Analysis

Q2. What does the Linear Static Analysis study in Solid Edge Simulation?

A) The temperature distribution in a model
B) The natural frequencies of the model
C) The displacements, strains, and stresses under applied loads
D) The effect of fluid on the model

Q3. Which assumption is made in Linear Static Analysis regarding load application?

A) Loads are applied suddenly and rapidly
B) Loads remain constant after reaching full magnitude
C) The material undergoes non-linear deformation
D) Boundary conditions change throughout the analysis

Q4. What is the main goal of performing a Normal Modes Analysis in Solid Edge Simulation?

A) To check the thermal conductivity of the model
B) To check the natural frequency of the model
C) To calculate the stress distribution in the model
D) To analyze the effect of different fluid pressures on the model

Q5. What is resonance in vibration analysis?

A) A condition where the model's natural frequency matches an applied frequency, causing failure
B) A condition where the frequency is reduced to zero
C) The frequency at which the entire model moves in a uniform direction
D) The frequency at which the material begins to deform

Q6. Which of the following is NOT one of the mechanisms of heat transfer involved in Thermal Analysis?

A) Conduction
B) Radiation
C) Vibration
D) Convection

Q7. In Solid Edge, what does the Harmonic Response Analysis check for?

A) The temperature distribution across the model
B) The response of the model to varying load frequencies
C) The displacements caused by static loads
D) The forces induced due to vibration in the material

Q8. What happens when the 'Recover Bodies' tool is used in Solid Edge Simulation?

A) It unites two bodies into one for analysis
B) It extracts surfaces and solids from a united body
C) It applies forces to the body
D) It applies thermal loads to the body

Q9. Which type of load is applied using the 'Hydrostatic Pressure' tool in Solid Edge?

A) Pressure due to gravitational forces on a body
B) Pressure load due to fluid at equilibrium inside a container
C) Force load applied to a body
D) Rotational load applied to a body

Q10. What assumption is made in Linear Static Analysis regarding the relationship between loads and responses?

A) Non-linear relationship between loads and displacements
B) The material behavior is ignored
C) The relationship is linear, meaning doubling the load will double the displacement
D) The load will vary with time during analysis

Q11. What happens during the process of buckling in a structural model?

A) The model fails due to excessive stress beyond material limits
B) The structure bends in response to a sudden axial load
C) The deformation of the model occurs at very high temperatures
D) The model becomes unstable due to lateral loads

Q12. Which of the following is NOT a type of heat transfer mechanism involved in Thermal Analysis?

A) Conduction
B) Convection
C) Compression
D) Radiation

Q13. In Steady State Thermal Analysis, which material property is primarily required?

A) Specific heat
B) Thermal conductivity
C) Density
D) Tensile strength

Q14. In Normal Modes Analysis, what are the frequencies associated with each mode called?

A) Harmonics
B) Resonance frequencies
C) Natural frequencies
D) Critical frequencies

Q15. Which element type is used for solid bodies in Solid Edge Simulation when creating a mesh?

A) Surface Element
B) Tetrahedral Element
C) Beam Element
D) Hexahedral Element

Q16. What does the term "critical load" refer to in Linear Buckling Analysis?

A) The load that causes the highest stress in the model
B) The load at which the model starts to buckle
C) The load required to cause a thermal breakdown of the material
D) The maximum allowable load in normal conditions

Q17. What should you do if a model in Solid Edge Simulation has a poor-quality mesh in tight spots?

A) Ignore the poor-quality elements to save time
B) Modify the mesh settings to avoid poor-quality elements
C) Run the analysis without worrying about the mesh quality
D) Apply additional loads to improve the mesh quality

Q18. What is the purpose of the "Nastran Log Viewer" tool in Solid Edge Simulation?

A) To visualize the results of the simulation in 3D
B) To view detailed information and logs about the simulation process
C) To modify the geometry of the model
D) To perform material analysis

Q19. In Solid Edge Simulation, which of the following conditions must be met for a structure to be considered for Linear Buckling Analysis?

A) The structure must have a large displacement
B) The material properties must be nonlinear
C) The structure must be subject to a compressive or axial load
D) The structure must be subjected to only torsional forces

Q20. What does the "Contact Tool" in Solid Edge Simulation help define?

A) The temperature at the contact points between bodies
B) The mesh quality at the contact faces
C) The behavior of bodies when they come into contact with each other during analysis
D) The speed at which contact occurs during simulation

Q21. In Solid Edge Simulation, which mesh refinement technique is used to improve accuracy in areas of high stress concentration?

A) Global refinement
B) Local refinement
C) Default refinement
D) Auto-mesh refinement

Q22. What is the significance of the "Safety Factor" in Solid Edge Simulation results?

A) It indicates the exact load a part can withstand before failure
B) It provides a measure of the reliability of the simulation model
C) It shows how much more load the part can handle before yielding or failure
D) It shows the thermal resistance of the material

Q23. Which of the following methods is used to assess material failure under a given load in Solid Edge Simulation?

A) Modal Analysis
B) Stress-Strain Analysis
C) Von Mises Stress Criterion
D) Buckling Analysis

Q24. In Solid Edge Simulation, when using the "Symmetry" boundary condition, which aspect of the model is being simulated?

A) The exact shape and size of the part
B) Only one side of the model as if it were mirrored
C) The thermal distribution over the entire model
D) The force applied symmetrically to the entire part

Q25. Which analysis type should be used to understand the stress distribution and deformation in a structure subject to varying loads over time?

A) Linear Static Analysis
B) Harmonic Response Analysis
C) Dynamic Response Analysis
D) Transient Dynamic Analysis

Q26. What is the role of the "Solver" in Solid Edge Simulation?

A) To create a mesh for the model
B) To compute the solution of the analysis based on the applied loads and boundary conditions
C) To display results after the analysis
D) To model complex geometries

Q27. When using the Radiation tool, which option allows defining radiation from selected reference to atmosphere?

A) Radiation to Space
B) Enclosed Radiation
C) Surface Radiation
D) None of the above

Q28. Which constraint allows rotation but restricts translation in all directions?

A) Fixed constraint
B) Pinned constraint
C) No Rotation constraint
D) User-Defined constraint

Q29. What does the Cylindrical constraint do in thermal simulations?

A) Allows rotation about an axis and translation along that same axis
B) Restricts rotation of a cylindrical part about its axis
C) Allows only axial translation of the part
D) All of the above

Q30. Which tool is used to set the size of mesh elements along selected edges?

A) Surface Size
B) Body Size
C) Edge Size
D) Mesh Sizing

Q31. When using the Body Size tool, what can you adjust?

A) The mesh size of bodies in the model
B) The element size along model edges
C) The surface mesh size
D) The thickness of boundary layers

Q32. What does the Probe tool do in the simulation results environment?

A) Displays deformation results
B) Checks equilibrium of forces
C) Allows checking of results at selected nodes or faces
D) Animates the results

Q33. What is the purpose of the Equilibrium Check tool?

A) To check if the analysis results are displayed correctly
B) To check whether forces are balanced on all nodes in the model
C) To check the displacement results at specific nodes
D) To export results to different formats

Q34. In the Radiation Load Options dialog box, which option defines radiation between selected reference geometries?

A) Radiation to Space
B) Enclosed Radiation
C) Surface Radiation
D) Radiative Heat Transfer

Q35. What does the Control Parameters option in the Optimization Study allow you to define?

A) The number of iterations for optimization
B) The material properties
C) The convergence limit
D) The design objective

Q36. In the Generative Design Study, what is the purpose of setting a "Preserve Region"?

A) To apply constraints to the model
B) To prevent specific regions of the model from being modified during the optimization
C) To define the number of iterations
D) To set the material properties of the model

Q37. In the Thermal Analysis, what is the role of the "Film Coefficient" parameter in the Convection tool?

A) To define the material properties
B) To set the time parameters for analysis
C) To specify the heat transfer rate from the model's surface to the surrounding air
D) To define the fixed constraints

FOR STUDENT NOTES

Index

Ethics of an Engineer

- Engineers shall hold paramount the safety, health, and welfare of the public and shall strive to comply with the principles of sustainable development in the performance of their professional duties.

- Engineers shall perform services only in areas of their competence.

- Engineers shall issue public statements only in an objective and truthful manner.

- Engineers shall act in professional manners for each employer or client as faithful agents or trustees, and shall avoid conflicts of interest.

- Engineers shall build their professional reputation on the merit of their services and shall not compete unfairly with others.

- Engineers shall act in such a manner as to uphold and enhance the honor, integrity, and dignity of the engineering profession and shall act with zero-tolerance for bribery, fraud, and corruption.

- Engineers shall continue their professional development throughout their careers, and shall provide opportunities for the professional development of those engineers under their supervision.